BEHAVIOR IN NEW ENVIRONMENTS

BEHAVIOR IN NEW ENVIRONMENTS

Adaptation of Migrant Populations

Edited by
EUGENE B. BRODY
Director, Psychiatric Institute,
University of Maryland, School of Medicine

SAGE PUBLICATIONS • Beverly Hills, California

KLINCK MEMORIAL LIBRARY
Concordia Teachers College
River Forest, Illinois 60305

Copyright © 1969,1970 by Sage Publications, Inc.

All rights reserved. No part of
this book may be reproduced or utilized
in any form or by any means, electronic
or mechanical, including photocopying,
recording, or by any information
storage and retrieval system, without
permission in writing from the
Publisher.

For information address:

SAGE PUBLICATIONS, INC.
275 South Beverly Drive
Beverly Hills, California 90212

First Printing

Printed in the United States of America

International Standard Book Number 0-8039-0043-0

Library of Congress Catalog Card No. 70-92359

Acknowledgement

The papers in this volume are adapted from working documents prepared as resource material for a conference on "Migration and Behavioral Deviance" supported by the section on the Center for Studies of Metropolitan and Regional Mental Health Problems of the National Institute of Mental Health (Grant No. MH14632, with the Editor as Principal Investigator). The conference was held November 4-8, 1968, in Dorado Beach, Puerto Rico. Grateful acknowledgement is made to the NIMH and to the participants for the support and effort which made it possible to bring together these contributions to an understanding of behavior in relation to residential change. Several of the papers previously appeared in a special issue of *The American Behavioral Scientist* (September-October 1969) devoted to the subject of "Migration and Adaptation."

CONTENTS

I. DIMENSIONS

1. MIGRATION AND ADAPTATION: The Nature of the Problem   13
   Eugene B. Brody

2. DEPRIVATION AND MIGRATION: Dilemmas of Causal Interpretation   23
   Marc Fried

3. INVOLUNTARY INTERNATIONAL MIGRATION: Adaptation of Refugees   73
   Henry P. David

II. FROM COUNTRY TO CITY

4. ADAPTATION OF APPALACHIAN MIGRANTS TO THE INDUSTRIAL WORK SITUATION: A Case Study   99
   Harry K. Schwarzweller and Martin J. Crowe

5. SOCIAL CLASS ORIGINS AND THE ECONOMIC, SOCIAL AND PSYCHOLOGICAL ADJUSTMENT OF KENTUCKY MOUNTAIN MIGRANTS: A Case Study   117
   Harry K. Schwarzweller and James S. Brown

6. DIFFERENTIAL EXPERIENCE PATHS OF RURAL MIGRANTS TO THE CITY   145
   Robert C. Hanson and Ozzie G. Simmons

7. THE ECONOMIC ABSORPTION AND CULTURAL INTEGRATION OF INMIGRANT WORKERS: Characteristics of the Individual vs. the Nature of the System   167
Lyle W. Shannon

8. COPING WITH URBANITY: The Case of the Recent Migrant to Santiago de Chile   189
Fred B. Waisanen

III. THE SOCIOCULTURE AND INDIVIDUAL BEHAVIOR

9. CONTEXT AND BEHAVIOR: A Social Area Study of New York City   203
Elmer L. Struening, Stanley Lehmann and Judith G. Rabkin

10. MIGRATION AND ETHNIC MEMBERSHIP IN RELATION TO SOCIAL PROBLEMS   217
Elmer L. Struening, Judith G. Rabkin and Harris B. Peck

11. MEXICAN AMERICANS OF TEXAS: Some Social Psychiatric Features   249
Horacio Fabrega, Jr.

12. ADAPTATION OF ADOLESCENT MEXICAN AMERICANS TO UNITED STATES SOCIETY   275
Robert L. Derbyshire

13. IMMIGRATION, MIGRATION AND MENTAL ILLNESS: A Review of the Literature with Special Emphasis on Schizophrenia   291
Victor D. Sanua

14. SOCIAL-PSYCHOLOGICAL ASPECTS OF MIGRATION AND MENTAL DISORDER IN A NEGRO POPULATION   353
Robert J. Kleiner and Seymour Parker

IV. PROGRAM PLANNING AND RESEARCH

15. A SIMULATION MODEL OF URBANIZATION PROCESSES   377
    Robert C. Hanson, William N. McPhee, Robert J. Potter, Ozzie G. Simmons and Jules J. Wanderer

16. POLICIES FOR PLANNING RURAL-URBAN MIGRATION: Urban Villages Reconsidered   395
    Richard L. Meier

17. ALASKAN NATIVES IN A TRANSITIONAL SETTING   405
    Robert L. Leon and Harry W. Martin

18. TO BE OR NOT TO BE POLITICAL: A Dilemma of Puerto Rican Migrant Associations   425
    Lloyd H. Rogler

19. PREVENTIVE PLANNING AND STRATEGIES OF INTERVENTION: An Overview   437
    Eugene B. Brody

APPENDIX

20. RESOURCES FOR REFUGEES   447
    Compiled and Annotated by Henry P. David

ABOUT THE AUTHORS   473

PART I
DIMENSIONS

Chapter **1**

Migration and Adaptation: The Nature of the Problem

EUGENE B. BRODY

This volume is primarily concerned with the human consequences, the socio-behavioral corollaries of migration. The present introductory paper offers the elements of a framework in which these consequences, whether privately experienced or publicly observable may be coherently viewed. It is concerned with how to study or understand the factors important in determining how a migrant makes new friends, finds a job, cares for his family, participates in the informational network and takes advantage of the opportunity structure of his new milieu. It is also concerned with the elements which protect a migrant or make him more vulnerable to the stresses he encounters—so that he may become a winner or a loser, a casualty or a success, in terms of economic absorption, cultural integration and psychological adaptation.

Defense and Adaptation: The Intrapsychiatric and Interpersonal

An approach to these issues begins with the individual and extends to his existence and functioning as part of a sociocultural system. The

---

Editor's Note: *These papers are adapted from working documents prepared as resource material for the conference on "Migration and Behavioral Deviance" supported by the section on the Center for Studies of Metropolitan and Regional Mental Health Problems of the National Institute of Mental Health (Grant No. MH14632, with the Editor as Principal Investigator). The Conference was held November 4-8, 1968, in Dorado Beach, Puerto Rico. Much of the material of this introductory paper is taken from the daily conference discussions. Thus, in many respects, it reflects the thinking and experience of all conference participants.*

individual or intrapsychic viewpoint is epitomized in Freud's concept of ego defense. Defense mechanisms, of which repression is the cornerstone, are activated by signal anxiety, itself a consequence of unresolved conflict between drives or wishes pressing for conscious expression on one hand and forces inhibiting such expression on the other. These mechanisms operate automatically and unconsciously to ward off the awareness or breakthrough into consciousness of unacceptable impulses, and to reduce anxiety, guilt or other psychological tension. They keep the tension at a manageable level.

Intrapsychic and interpersonal equilibrium, however, may not coexist. A paranoid individual, for example, whose self-certainty reflects the anxiety-reducing operations of repression, projection, and other defense mechanisms, may play havoc among his associates and in time evoke such retaliatory anger that his whole life structure lies in ruins. While his ego-defensive processes are functioning well, he is not behaving adaptively.

Adaptation in the psychological sense refers to the process of establishing and maintaining a relatively stable reciprocal relationship with the environment. For human beings this means the human, social, or interpersonal environment. Maladaptive behavior is apparent to outside observers, in contrast to subjective discomfort which may be effectively masked by the various strengths and talents of the individual as well as his more or less advantageous position in a social context. The person with anxiety attacks, phobic fears, depressive episodes or obsessive thoughts will not be identified as a casualty unless these privately experienced states influence his publicly observable behavior, including help-seeking. Even if his private behavior involves action, as for example of a sexual or aggressive nature, it will not contribute material to the investigator so long as it is concealed. In these last instances, help-seeking may be required by some agency of social control following social detection; or if the behavior is not, as it usually is ego-syntonic, it may be experienced as alien, a symptom or foreign body and the sufferer will look for assistance as does any other person who regards himself as sick.

The interplay between defensive and adaptive processes is a function of past history and present environmental circumstances. A person's repertory of defensive and adaptive devices stems from his early socialization experiences, and their use is reinforced or extinguished by current societal sanctions and prohibitions. As he shifts via migration from one socioculture to another, behavioral modes useful in the old setting may prove maladaptive in the new. Acute sensitivity which permits empathic understanding in one group may be perceived as discomfort-provoking vigilance or paranoia in another. The culturally supported tendency to attribute sources of danger to external factors, reinforced by magical belief systems in rural areas, may interfere seriously with the inward-looking or learning about oneself sometimes necessary for survival in the city. A persisting tendency to blame other persons, groups or forces for lack of success impairs both motivation and the acquisition of new, more useful, responses. Under these circumstances adaptive failure may occur without the person's awareness. One index of such failure in a migrant may be incompatibility between his self-image on one hand and the status, of which he is unaware, given him by the social system on the other.

Migration: A Process of Social Change

Migration provides a set of concrete operations for the study of adaptation and defense in relation to social change. A shift in residence involves not only new places, but new faces and new norms. Movement over distance implies the crossing of social system boundaries, whether the systems are defined in terms of national entities, regional subcultures, or immediate friendship and kinship networks. The migrant leaves behind the supports and stresses of the donor system from which he departs, including the push factors which contributed to his decision to move. He loses the support of social and geographic familiarity, of long-term relationships and values which were built into him while growing up. At the same time he is freed of some of the threats of disease and hunger, of the obligation to perform in expected ways and of certain stressful relationships. He is welcomed by the receptor networks or must deal with resistances in the host system to which, lured in part by pull factors, he comes. He is excited by new stimuli and opportunities and fearful of new threats and the unknown. Between the systems, en route, he must cope with a series of transitional factors which color his perceptions, attitudes and capacity to deal with the host environment. And his adaptation throughout is shaped by internal motives for moving which may have little to do with environmental push or pull factors.

Some of these motives are best described as idiosyncratic-psychological. They are the distillate of ungratified wishes and needs, undischarged tensions, and unresolved conflicts. Others are more easily related to the person's place in his social change trajectory. Voluntary migrants anticipate their moves, and as decisions are made and preparations begun they are caught up in a process of change. They are differentiated from their fellows and launched on a social change career with the first stirring of dissatisfaction with the status quo and longing for something new as it leads to the crystalization of resolve. Once in the new environment, the change process may continue indefinitely with the greatest acceleration at the steepest part of an S curve not reached for several years. Continuing change in the new environment is usually adaptive. Inkeles, for example, studying migration into industrial Buenos Aires, found that the first year was devoted mainly to coping with "noise" in the system. After coping skills were acquired there was a dramatic increase in other learning. Change and adaptation, however, are not automatic. They are determined by talent, social context, and the degree to which the immediate consequences of moving fit the premigratory fantasies. The initial encounters in the new environment may be especially important in this regard.

Some migrants are risk-takers, people willing to go a step beyond the ordinary or expected; their vulnerability to adaptive breakdown in the face of unusual stress remains moot. Some are geographical escapers, people who deal with personal or environmental disaster by physical flight; these often carry their problems with them wherever they go. In this last category are those for whom the move can be related to preexisting psychiatric illness. Even here, however, the adaptive consequences are not clear since many move on "rational" grounds in order to be near a source of professional assistance.

The foregoing constitute a mix of social systems and individual determinants of adaptation following a migratory move: the individual's past history and his repertory of defensive and adaptive techniques; his private motivations for the move; the public push and pull factors in the donor and host environments; other elements in the two environments including the consonance of their norms and the resistance or receptor networks encountered in the host system; and transitional factors. The multiplicity of factors and their continuation over time suggest that migration may most usefully be considered not as an act, but as a process. The migrant is an actor involved in this process. He is a mutually interacting member of a subunit of a social system (the system provides output to him and he provides inputs to it).

The process of migration from the system of departure (donor) to the system of arrival (host) begins, as indicated above with the development of anticipations. The process may, hypothetically however, be placed outside of the actor into the donor system itself. Thus with more subsystems in the donor environment the probability of outmigration on the basis of heterogeneity increases. Subsystems by virtue of their differences in density, isolation, and physical or sociocultural barriers to intercommunication, may provide means permitting the actor to move away from involvement with core aspects of the system. Identification with the original system is also weakened by agencies such as schools which hasten upward social mobility. As he moves away from involvement and the core he has more dissociating (mobility) experiences of either a physical or a psychological nature. As the number and frequency of these experiences increase, that is, as he becomes progressively withdrawn from the donor environment, it seems more probable that he will eventually migrate to another social system. An alternative to this last, depending upon a variety of personal factors, might be a reduced probability of geographical movement as his needs are more adequately satisfied within the donor environment and there is, consequently, less strain on his adaptive and defensive techniques.

Once migration has occurred, diminishing commitment, rank and self-esteem in the host's cosmopolitan urban environment may result in a move to enclave (ghetto) residence or a return to the system of departure.

Some indicators reflecting the nature of the process include the following: time in system, such as age, recency of migration and years in the area; participation, such as marital status, family size, contact with neighbors, and office holding; rank, such as age, sex, race and office holding; esteem, such as number of friends, number of godfatherships, perceived sociability; physical mobility, such as visits to an urban center and time lived outside the donor system; and indicators of psychic mobility, such as radio, TV, newspaper, magazine or book reading and contact with friends who have been outside or have read.

Family Ties and Circular Migration

Depending upon donor system economics, families may be organized around work. This is especially true among sharecroppers and tenant farmers, and is also important for homecraft and cottage-industry economies. Migration into urban areas disrupts the familial patterns as individual members work in factories; parents, especially fathers, are separated from children and an important area of shared concern and activity is lost. Systematic changes in family dynamics associated wth migration have not yet been described.

Individual adaptation is also determined by the severance or maintenance of ties to family and friends left behind. Increasingly effective communication and transportation increase the likelihood both of circular movement and of continuing support and information exchange through visits, letters and telephone contact. Young adult family members will often go ahead to the new place to break trail for those who come later. This facilitates job finding for some. For others, without a family connection, it makes it difficult to obtain work in certain factories.

There are also difficulties in the pattern of family movement. Sometimes the young adult males will be left behind to hold onto homesteads, while parents, daughters and younger brothers will go ahead to the new place. The relative preponderance of female migrants into the host system depopulating the donor area is a feature with as yet unidentified consequences. Those left behind may sooner or later become restless and resentful, and for them the migratory process has different psychological meaning than for those who came first.

There is always some circular migration back to the donor environment from the ghetto. Some of those who go back are winners. They have made money and are pulled back by the attraction of becoming landowners. Some of these regard the return as a move to retirement.

Others are pulled back by obligations. They may already own land, and the person left to hold it can do so no longer. A parent may have died, or someone has become ill and they go back to take charge.

A few are unlucky. Sometimes they are casualties of business cycles. They are laid off from work. They are involved in fights started by others. They get into law suits. Among the unlucky ones are women whose husbands have died or deserted them.

The losers are usually of the lowest status among the migrants. (As a general rule the very lowest within the donor system are unable to pull together the money, the energy or the ambition to make the long move.) However, not all losers get back home. Some remain in the lowest rungs of the ghetto. Others are moved into mental hospitals. Some become physically ill and die.

Circular migration influences the nature of the donor system. More or less accurate information gets sent back. Money is brought and sent implementing a redistribution of resources. An example of attitudinal change consequent to circular migration is an acceptance in certain areas of birth-control measures which were not successfully advocated through direct propagandizing from officials. The migrant who returns does so with his former role and value orientation modified.

The Host System in Relation to the Donor

It is difficult to separate the impact of a host environment from the nature of the experience, talent and equipment the migrant brings to it. His initial attributes include those which may be subsumed under personality and health, socioeconomic status, demographic features, and such items as the length of time since his last move. He has certain goals and he searches the opportunity structure of the new environment in order to attain them. He encounters a proximity structure, an institutional structure and a personal network structure. This last may include relatives or friends who have gone before. The range of alternatives open to him depends in part on the perceived support or cohesiveness of others like him. These structures and other aspects of the city, its people and agencies, communicate the opportunities to the migrant and the nature of communications, itself, is a factor. He also encounters resistance, rejection, prejudice and discrimination. These may reflect the limits of the host's absorptive capacity as well as aspects of its culture. Some rejections are experienced and some opportunities are accepted by the migrant. Each accepted opportunity changes his life situation and influences the nature of his continued searching behavior from that point on.

Prejudice and lack of economic opportunity constitute barriers to both acculturation and economic integration. On the other hand, consonance of norms from the donor and host systems increases activity and integration in the new system. Lack of social acceptance may lead to ghetto formation. But previous history determines the response to such resistance and lack. Black migrants, for example, who have had an early environment of constantly reinforced perceptions of themselves as worse off, less capable and less desirable than others may assume that they will be losers and are therefore reluctant to do anything. Thus, the likelihood that they will move into the ghetto is high and that they will eventually move out of it is low.

The ghetto, or "urban ethnic community" is the terminal point not only of those who are unaccepted in the larger host (or cosmopolitan urban) environment, but of most of the upwardly mobile or would-be mobile migrants who come from the original pool—which might be rural Puerto Rico, the United States southern "black belt," or the less-developed areas of any country. The ghetto, whether black or Puerto Rican or white Appalachian, includes temporary sojourners who may come to look after a sick relative and without committing themselves may remain for several years. It is maintained by institutions such as clubs and churches, which may remain even after a spatial dispersion of the ghetto due to invasion by other ethnic elements or even an increasing volume of similar migrants.

The ghetto is one type of mediating organization, although it may not always have that effect. As a geographically and socially defined unit in which the new arrival finds others of his own kind, sharing common norms and language, it can act as a buffer mechanism permitting him a pause for personal and social reorganization before making his way into the larger cosmopolitan urban environment. Even though for many the ghetto is the terminal geographical point, it is not static. It changes in accommodative response to the pressures of the surrounding society. Furthermore, the ghetto home may be the only part-time retreat of those who work outside it during the day. As they are occupationally successful, and acquire new friendships, the process of psychological if not physical movement from the ghetto begins. This aspect of the migratory process may then accurately be said to have been facilitated by the ghetto as a mediating mechanism.

In general the subsequent career of migratory populations is a function of the cohesion developed within the ethnic group and the resources which members of the group make available to one another. They, especially as they may be institutionalized in stable ghettos, provide normative and to some degree comparative reference groups for incoming migrants, as well as an interlocking series of absorbing networks. In some instances ghetto life may promote the evolution of migrants into a collectivity, a group of people committed to action on the basis of a shared value system. When this happens they may cope with and eventually modify the larger or host society.

Within the ghetto the clubs, churches and other groups allow the migrant to achieve his aspirations. Some made up mainly of migrants who have gone before offer nostalgic gratifications and association with compatriots. Others offer opportunities in the larger host society. They all promote a degree of social competence in the new system. The ghetto as a receiving community, however, depends for its own strength and continued vitality upon inflows of new migrants and connections with the donor society. If the inflow stops and there is no marked social visibility or value difference from the host society, the ethnic enclave or absorbing network of the ghetto will gradually lose its reason for existing except as support for unassimilated members. With marked visibility as in the case of American blacks, it may develop a custodial, or even separate identity-promoting role, rather than an integrating buffer or receiving role. These roles are important because as a rule the dominant system is not prepared to meet the needs of the subordinate, made up in large part of immigrants. Thus, the ghetto or ethnic enclave as a receptive system has many functions including the expressive. They are goal oriented and usually include political patronage. In Latin America the rural patron-peon relationship is often recreated in the new community, and this may be seen in North American relationships between the migrant ethnically defined poor and the precinct boss. These considerations are relevant to the problem of developing leadership in urban central cities.

Finally, the host system as well as the migrant can define adaptation. A functionally illiterate Mexican-American in Denver, for example, could not learn furniture repairing because he could not read directions. He lost his motivation, received welfare income and contributions from his children and was able to spend his time drinking beer with friends. This represented adaptation without cultural assimilation. He did achieve a stable, reciprocal relationship with his environment. From the standpoint of the host society, however, he represented a continuing drain. He achieved a kind of social competence in the new system, but it was not one which resulted in personal development. It was not known if his behavior was ego-syntonic or ego-alien. The latter is possible, even though it might have been congruent with the masculine role in the village from which he came. In either case his drinking could lead to the point where he would become a burden to the health-services system of the host society and more of a burden to his family.

Summary

The adaptation of the migrant, his way of establishing and maintaining a stable, reciprocal relationship with his new environment, is viewed in terms of intrapsychic and interpersonal elements. These are constantly in play during the process of migration as it involves changes in attitudes, relationships and behavior and a move across physical space and social system boundaries. Special attention is paid to the interrelationship between donor system and host system factors, as they mutually influence each other, the migrant and his family.

Chapter 2

Deprivation and Migration:
Dilemmas of Causal Interpretation

MARC FRIED

It has often been noted that there is a relationship between deprivation and, particularly, between poverty and migration. Relatively few studies have sought to examine or to clarify this relationship nor has the theme attracted great interest and attention. Historical studies of the vast migrations that peopled the American continent have pointed up, in a fairly general way, the great impetus to migration created by famine and disaster or, conversely, by available land and economic expansion. Many studies have also observed the unfortunate economic and social position of the newcomer to urban, industrial societies. And a few economic and demographic studies have attempted to trace a bit more closely the relationship between expulsive forces in countries of origin and attractive forces in countries of destination. Beyond these studies, which themselves offer a sad commentary on our meager knowledge and understanding, there are only scattered and sparse references to the relationship between deprivation and migration.

The problem arises anew in the large migrations of Negroes from the South over the past five decades. As a social problem, the migration of rural Negroes, largely trained to do only routine agricultural field jobs and

---

Author's Note: *This paper is based on a report of the Office of Economic Opportunity in partial fulfillment of Contract Number B89-4279.*

with virtually insuperable impediments to retraining as an industrial labor force by virtue of their former condition of servitude, arose even before the Emancipation Proclamation. No sooner had the Northern armies entered the South than hordes of Negro slaves flocked to the army camps (DuBois, 1965:222):

> They came at night, when the flickering campfires shone like vast unsteady stars along the black horizon; old men and thin, with gray and tufted hair; women, with frightened eyes, dragging whimpering hungry children; men and girls, stalwart and gaunt, — a horde of starving vagabonds, homeless, helpless, and pitiable, in their dark distress.

As the Northern armies overran vast territories in the South during the Civil War and, even more clearly with the abolition of slavery, a great many former slaves became themselves an army of uprooted. However, in actual numbers, only relatively few Negroes went North or West to settle in the growing urban areas. It was not until the beginning of the twentieth century that a generation of Negroes, born free, who had therefore to learn about the futility of either freedom or opportunity in the South, began to migrate in increasing numbers first to Southern cities and then into the North and Midwest. And it was not until the twentieth century, the first world war and the drastic limitation of foreign immigration that large numbers of Negro migrants from the South began to reach toward the newly developing, albeit limited, chances for jobs and for a dream of dignity in urban society.

Whether we concern ourselves with the European immigration of 1830-1920 or with the Negro migration of 1900 to the present, we must deal with and be guided by fragmentary facts and partial theories. The psychological and social history of migration is even more deeply hidden behind these few statistical facts, incomplete records, selected observations, and bits of theory. In an effort to put these into perspective, we shall examine some of the data on the European migration and then turn to a comparative consideration of the Negro migration from the South. In noting some of the similarities and differences, we may better understand some of the ways in which deprivation functions as both condition and consequence, subjective and objective, of massive population redistributions.

Historical Background

Autobiographical reports of the experience of migration usually submerge the difficulties and tribulations of entering a new society within a sense of overall success and achievement. But the host of studies which investigate a broader range of in-migrant populations documents the enormous set of problems, pitfalls, and often tragedies associated with movement from one society to another.[1] The scattered evidence suggests that these difficulties and problems beset the migrant to any urban, industrial society and not merely those who came to the United States; and, further, that the process varies relatively little whether it involves internal migration from rural to urban areas or, as in the great European migrations of the nineteenth century, emigration from the rural areas of one country to the urban areas of another.[2]

Immigration, the process of geographical and social transition from one society to another is, at best, a drastic experience of cultural change. It requires a shift from embeddedness in the familiar to a constant confrontation with newness and unfamiliarity. More often than not it involves a global experience of being a stranger, an alien at the mercy of an inhospitable, incomprehensible, and uncomprehending foreign population. Even under the best of circumstances when the migrant from one city to another has a relatively clear anticipation of a job or of friends or of housing conditions, migration is a highly disruptive process. Nonetheless, it seems quite clear that the degree of change required in cultural orientations, and in social relationships and patterns is one of the more critical dimensions distinguishing the potential ease or difficulty of adjustment to circumstances of migration. And almost certainly inter-correlated with this and of great relevance is the possibility of anticipating, either because of prior experience or prior arrangements or economic resources, some of the main situations and roles one will encounter in work, in social relationships, and in housing and residence.

Although many lower status immigrants from rural areas in foreign countries to cities in the United States had some prior contact with family or friends who had already migrated, they generally had little accurate information, little basis in past experience for anticipating, and few economic or social resources for coping with these changes. Similarly, among low status Negroes in the United States, the initial and often many subsequent contacts with the city are major transitions for which there can only be minimal preparation. But in addition to the sense of alienation and estrangement, the lack of preparation or anticipation, the absence of fundamental resources for coping with a new environment, the reaction of host societies is almost invariably one of antagonism based on class and cultural differences.

American nativism and its corollary antiforeignisms waxed and waned over the course of several centuries but there were many rumblings of discontent about the freedom of opportunity available to foreigners quite early in American history. By the middle of the nineteenth century, many

Americans became fearful of the ruinous effect of newcomers on our society (Commons, 1920; Handlin, 1959; Hingham, 1955; Stepenson, 1926; Thernstrom, 1964). After the Civil War, the great territorial and industrial expansion led to a further diminution in nativist sentiment (Hingham, 1955). The material need for an expanding population was great; the same national groups, predominantly German, English, and Irish, continued to be the dominant immigrant forces. By the late 1870s and 1880s, however, new patterns of nativism began their opposition to the "new immigration" only to reach a peak during and immediately following World War I.

Viewed as a large, historical phenomenon, nativist sentiment with its strong source in opposition to foreign competition, foreign ideology, and foreign culture may have undergone a slow and variable progression partly modified by the pervasive sense of the United States as both haven and melting pot. Concretely, however, most foreign groups who entered the United States in large numbers and who differed in striking ways from Americans were subjected to extreme difficulties. For the English this was mitigated by the fact that, unlike most other in-migrants, they came to fill relatively skilled jobs in specific industries and the cultural similarities facilitated a rapid transition (Berthoff, 1953). For the Germans and Scandinavians, the movement toward specific regions, generally of low population density, permitted them to follow a path that had already been prepared and despite low rates of assimilation, to experience little long term antagonism (Hingham, 1955; Walker, 1964). But whether the extremely low social position or the cultural and religious distinctiveness, or the competitive economic situation or some larger change in the society as a whole was responsible for the difference, during the latter part of the nineteenth century and the early part of the twentieth century, the Irish, Italians, Slavs, Jews and others who emigrated to the United States were subjected to brutal experiences of isolation, exploitation, and exclusion (Ernst, 1949; Handlin, 1941; Hansen, 1940; Hingham, 1955; Joseph, 1914; Levine, 1966; Potter, 1960; Smith, 1939; Stephenson, 1926; Thomas and Znaniecki, 1918). The more recent experiences of Negroes, Puerto Ricans, and Mexican-Americans follow a well-trodden path of discrimination and segregation in the urban, industrial areas of the United States.

Conditions of Out-Migration

Massive migrations most often occur during intolerable conditions of economic or social crisis. They are always associated with absolute or relative dissatisfaction with the conditions of life or with the opportunities for adaptation which are available in the country of origin (Antin, 1912; Arensberg and Kimball, 1948; Eisenstadt, 1954; Ernst, 1949; Handlin, 1959; Handlin, 1941; Hansen, 1940; Joseph, 1914; Park and Miller 1921; Smith, 1939; Stephenson, 1926; Williams, 1938). The situations of deprivation, oppression, and famine which formed a background for the great migrations of Irish, Jews, Poles, and Italians and, indeed, of English and Germans and, more recently, of southern Negroes, Puerto Ricans, and Mexicans need hardly be elaborated. Only situations as severe as these can begin to account for widespread uprooting by peasants and tenant farmers whose horizons were customarily limited by the spatial environment of a small region and by the social environment of kin and neighbors.

But while these are necessary conditions for an explanation, they are not sufficient to the task. The social and economic deprivations often associated with large migratory movements were severe but they were largely exacerbations of endemic situations rather than wholly unfamiliar catastrophes. Indeed, the currents of migration from various European countries from the middle of the nineteenth century into the beginning of the twentieth century show relatively smooth annual trends of slow increase or slow decrease only occasionally interrupted by large, temporary swings (Table 1, from Jerome, 1926). A similar overall continuity of migration from individual countries makes it evident that cycles of prosperity or depression, at home or abroad, can only account for relatively small proportions of the total volume of migration from any given country. More general longterm trends and policies which affect a country, a region, or a population group over many decades are necessarily implicated in these patterns that characterize the emigration from a country and distinguish it from other countries.

Ordinarily, people are reluctant to move and most reluctant to leave one society, culture, or area for another. This is ever so much more the case among preindustrial people who are bound to long traditions and surrounded by kin and kind. Scattered reports from all over the world indicate that this holds even when the move involves departure from inadequate housing to new developments with many more conveniences only a short distance away in the same country. The data from bombed-out cities in World War II, from the reluctant departure of German Jews and Germans who were potential targets of Nazi aggression, from the reaction of residents of planned relocation in this country and abroad further document the intense and widespread resistance to leaving home (Allport, Bruner and Jahndorf, 1941; Fried, 1963; Hartman, 1964; Watts, et al., 1964; Young and Willmott, 1957). Even the vast migrations

TABLE 1

RECORDED NUMBER OF IMMIGRANTS TO THE UNITED STATES
FROM SELECTED COUNTRIES: 1870-1914, IN THOUSANDS

Year Ending June 30	Germany	England	Ireland	Sweden	Italy	Austria Hungary	Russia	Greece
1870	118	61	57	13	3	4	1	a
1871	83	57	57	11	3	5	1	a
1872	141	70	69	13	4	4	1	a
1873	150	75	77	14	9	7	2	a
1874	87	51	54	6	8	9	4	a
1875	48	40	38	6	4	8	8	a
1876	32	24	20	6	3	6	5	a
1877	29	19	15	5	3	5	7	a
1878	29	18	16	5	4	5	3	a
1879	35	24	20	11	6	6	4	a
1880	85	59	72	39	12	17	5	a
1881	210	65	72	50	15	28	5	a
1882	251	82	76	65	32	29	17	a
1883	195	63	81	38	32	28	10	a
1884	180	56	63	27	17	37	13	a
1885	124	47	52	22	14	27	17	a
1886	84	50	50	28	21	29	18	a
1887	107	73	68	43	48	40	31	a
1888	110	83	74	55	52	46	33	1
1889	100	69	66	35	25	34	34	a
1890	92	57	53	30	52	56	36	1
1891	114	54	56	37	76	71	47	1
1892	119	34	51	42	62	77	82	1
1893	79	28	44	36	72	57	42	1
1894	54	18	30	18	43	39	39	1
1895	32	23	46	15	35	33	36	1
1896	32	19	40	21	68	65	51	2
1897	23	10	28	13	59	33	26	1
1898	17	10	25	12	59	40	30	2
1899	17	10	32	13	77	62	61	2
1900	19	10	36	19	100	115	91	4
1901	22	12	31	23	136	113	85	6
1902	28	14	29	31	178	172	107	8
1903	40	26	35	46	231	206	136	14
1904	46	39	36	28	193	177	145	11
1905	41	65	53	27	221	276	185	11
1906	38	49	35	23	273	265	216	19
1907	38	57	35	21	286	338	259	37
1908	32	47	31	13	129	169	157	21
1909	26	33	25	14	183	170	120	14
1910	31	47	30	24	216	259	187	26
1911	32	52	29	21	183	159	159	26
1912	28	40	26	13	157	179	162	21
1913	34	43	28	17	266	255	291	23
1914	36	36	25	15	284	278	256	36

SOURCE: Jerome (1926). From the reports of the U. S. Immigration Commission, *Statistical Review of Immigration: 1820-1910*; and the *Annual Report of the Commissioner General of Immigration*, 1924, pp. 115-117, U. S. Bureau of Immigration. Prior to 1906, persons entering the United States were recorded by country whence they came thereafter by country of last permanent residence.

[a] Less than 500 recorded immigrants.

from Europe in the nineteenth and twentieth centuries appear often to have been conceived, initially, as temporary or seasonal migrations and, with a few notable exceptions, return emigration was extremely high (Foerster, 1919; Jerome, 1926; Thomas and Znaniecki, 1918; U.S. Immigration Commission, Vol. 1, 1911). Indeed, these vast migrations seem greater by far when the diverse streams of migration are totalled and considered as a contribution to the labor force of the United States than they do when seen as a proportion of the population from the country of origin. Taken as a proportion of population in the country of origin, the largest migration of all, that from Ireland during and immediately following the Great Famine, never rose above three percent during any one year and only rarely reached this level.[3] And no other migration movement ever approached the proportions of the Irish migration during these years.

Thus, whatever the long-run effects of large-scale migration on the population in the country of origin, and no matter how severe the expulsive forces or how seductive the attractive forces, only an extremely small minority of the population in any country is willing or able to leave during any period of time. Any effort to explain large-scale migration must keep this in perspective. On the other hand, even such relatively minor rates of departure, extended over time and selected from particular regions and particular occupational and class groups can have a considerable impact on the residual population. This is of great significance in view of population pressure as a determinant of migration. Brinley Thomas (1954) presents evidence that the single most consistent factor behind the major upswings of European migration during the nineteenth century was the cyclical increase in birthrates. This appears to have been an important factor even in Ireland which sustained an increase in population from 4,389,000 in 1788 to 8,175,000 in 1841, an increase that greatly exacerbated the severity of endemic and epidemic famines. The European population explosion led to an increasingly dangerous subdivision of small holdings which, for the peasantry of Europe, were initially barely sufficient for families of moderate size. Virtually every study of the peasantries of Europe or of emigration from European countries attributes primary responsibility for large-scale migration to these interrelated factors (Balch, 1910; Ernst, 1949; Foerster, 1919; Handlin, 1941; Handlin, 1951; Handlin, 1959; Hansen, 1940; Potter, 1960; Stephenson, 1926; Thomas, 1954; Thomas and Znaniecki, 1918; Walker, 1964; Woodham-Smith, 1962). A similar set of phenomena appears important in explaining internal labor migrations of Italians and Poles, the large migrations of

Puerto Ricans to the United States during the period 1947-1960; the migration of Mexicans to the Southwest; the internal migration of Negroes from the South; and the internal migrations from the Appalachian regions to the Midwest (Balch, 1910; Leyburn, 1937; Myers, 1967; Redford, 1926).

Population pressure and the incapacity of small landholdings to sustain a larger population on the land, thus, appear to be highly general in explaining the powerful expulsive forces that encourage migration. And most of the large migrations consisted mainly of agricultural populations or of laborers from rural regions. At the same time, technical innovations and developments that make possible expanding employment may also displace skilled craftsmen who become potential sources of another stream of international migration. This appears to have been the case during the latter part of the nineteenth century in England when an increasing proportion of skilled laborers migrated from England to the less developed industries of the United States where their knowledge was eagerly sought (Berthoff, 1953; Redford, 1926; Thomas, 1954). Moreover, a host of changes in agriculture itself has often led to the displacement of agricultural labor as in Germany and England. Day's analysis (1967) of United States data reveals that the vast rural to urban migration of the last few decades is directly associated with changes in agricultural technology, the use of fertilizer, and changes in crops farmed, all of which reduced drastically the labor requirements on farms in this country.

These bare but complex facts provide only a rudimentary record of the source of large migrations in serious deprivation. The historical accounts capture more vividly the abject misery created by these events which culminated in the great migrations. Handlin (1951) creates an important image in his conception of the uprooted departing from their homelands all over Europe. The unbelievable tragedy of the Irish famines has been reported often enough to be pathetically believable. Similar, if not quite as severe, situations obtained both earlier and later in much of Europe. Hansen (1940) reports a relatively early event in the migration from Germany in which the demand for berths was so great that fares were boosted beyond the reasonable hope of the bewildered families who had come to Amsterdam and Rotterdam with barely enough money for minimal transportation. And Balch (1910), describing the situation in Eastern Europe points to the strains as the outcome of a long history with its most immediate antecedents in the demise of feudalism without adequate resources to sustain an effective peasant economy.

Although poverty, misery, debt, disappearing land and expanding population stalked the peasantry of England, Ireland, Germany, Poland,

Italy, and Sweden, and must be held to account for the gross trend of out-migration, it is noteworthy that those selective factors that clearly operated gave precedence to the ablebodied, the effective, those who could afford the costs of passage. While the evidence for the superiority of the migrants in the European migrations leaves much to be desired, it is widely accepted by the historians of the European migrations. Balch (1910) reports that those districts in direst and most settled poverty were not the major sources of emigration. Foerster (1919) reports a similar factor in the distribution of emigration from southern Italy. Walker (1964) describes the German emigration as predominantly one of the lower middle classes and of rural craftsmen. And the data from Britain (Berthoff, 1953; Redford, 1926) suggests that, despite strong encouragement to migrate, the paupers were highly resistant to such moves, but the skilled laborer, displaced by machinery, eagerly went to the United States. These views are given greater credence by the more systematic, although fragmentary, data from the United States that indicate the educational superiority of those individuals who migrate compared to the persons who remain in their home communities (Fein, 1965b; Hamilton, 1965; Mauldin, 1940).

Thus, deprivation provides a background but not a specific explanation for the great migrations from Europe. Waves of migration from different European countries rose with increases in population and with technical innovations or changes in crop patterns, that made men redundant or superfluous.[4] Superimposed upon these long term patterns, severe crises, famines, and intensified oppression created great, if short-lived peaks of emigration representing the direct consequences of deprivation. Withal, the strains, difficulties, and demands of migration were sufficiently great, the process sufficiently precarious, that mainly those who could pay for passage, who could willingly look forward to back-breaking labor for the sake of better economic circumstances, who could anticipate the strain without excessive fear appear to have predominated in the migrations. It is certainly reasonable to conjecture that these were also the people most alienated from a rigidly unchanging society or most responsive to the rumors and reports of job opportunities abroad. In this respect, the selective factors involved in migration may provide a fundamental link between the conception of mass migration as a product of expulsive forces and the alternate conception of mass migration as a result of attractive forces.

Whether such selection is the essential link or not, it is important both to distinguish and to relate the several factors associated with mass migrations. The causes of migration have too often been distorted by a simplistic effort to distinguish the relative effects of "push" and "pull." But a model which views "pushes" and "pulls" as an opposition between two distinct and competing forces or a procedure which uses only data from a society or area of origin or only data from a society or area of destination is inadequate for clarifying the operation of a complex, interacting system of forces. One set of these variables may account for large-scale out-migration from a particular country or region and a different set of factors may account for the choice of destination or the timing of migration. At the very least, it is essential to consider, for any instance of mass migration, both the expulsive and attractive forces that affect large-scale population redistribution.

Conditions of In-Migration

Looked at from the vantage point of the country of origin, the pressures of deprivation appear to be the most potent expulsive forces accounting for mass population movements. Looked at from the vantage point of the host country, the pressures of opportunity appear to be potent attractive forces that account for mass population movements.

The classic investigation of the relationship between these forces in the European immigration is the study by Harry Jerome (1926). Data for the earlier half of the century do not permit statistical analysis. Some of these earlier as well as later migration streams were markedly influenced by the availability of land and by land policies in the United States (Hanson, 1940; Walker, 1964). More generally, for those periods after the Civil War in which the economic data are more adequate, the peaks and troughs of economic indicators are accompanied by rises and falls in immigration. However, while the cyclical patterns of employment and immigration were similar, the sheer volume of migration remained high through prosperity and depression and corresponded less closely to the level of employment. These patterns held not only for European immigration as a whole but for each of the separate, large migration streams from different European countries.

Supportive evidence for the conclusion that migration is influenced by employment opportunities in the place of destination can be found in studies of internal migration within the United States. In the temporal series of interrelationships between migration and economic opportunity for the period 1880-1940, Dorothy Thomas finds striking differences in migration between prosperous and depressed decades (Kuznets and Thomas, 1964). Analyzing regional displacements within the United States, she finds a trend toward higher rates of net migration among those

and more highly selective migrations as well as by recent periods of economic expansion, indicate an improvement in rates of occupational mobility for immigrants with a continuing decrement for those immigrants from the least favored countries.[10]

The Myth of Social Mobility and Social Assimilation

The many millions of immigrants from European countries from the middle of the nineteenth century until relatively recent decades bore the full brunt of the low status positions accorded the newcomer, the foreign-born, the rural peasant or worker, the uneducated, and the socially ostracized. Although few studies permit us to clarify the components of background or status most clearly implicated in the demeaning occupational conditions of the immigrant, the data are unambiguous in revealing their lowly position (Carpenter, 1927). At the extreme, immigrants from Ireland, Italy, and Poland were at the bottom of the occupational ladder and worked under conditions of unbelievable degradation and exploitation.[11] However, even the migrants from England, Scotland, and Wales started at levels considerably below the native population (Thomas, 1954; Table 40).

A matter of more serious concern than low initial status is the slow and difficult process of mobility and assimilation. For the British, whose occupational distributions were higher than those of the native Americans by the second generation, assimilation presented no special problems (Thomas, 1954: Tables 40, 43; Hutchinson, 1956: Table 41a). For other groups like the Germans and Scandinavians, who often lived in ethnically homogeneous enclaves, the process of social mobility occurred with moderate facility (Jonassen, 1949; Lieberson, 1963). Among the Jews, who have been the proverbial representatives of rapid social mobility, the process has been uneven and reveals some of the special problems of both mobility and assimilation in the face of overwhelming discrimination and restrictions (Glazer and Moynihan, 1963).[12] And although the Chinese and Japanese, among the non-European immigrants, have achieved extremely high levels of education and occupation, their slow and painful accomplishment is hardly testimony to the open mobility pattern or the ease of assimilation in the United States (Schmid and Nobbe, 1965).

The most recent urban migrant groups, the Negro, the Puerto Rican, and the Mexican-American, have suffered severely, but many earlier immigrant groups have also experienced an extremely slow and precarious process of social mobility. This has been particularly notable among the

Irish (Glazer and Moynihan, 1963; Handlin, 1959). In 1870, more than twenty years after their great migration and about a half century after the beginning of a substantial migration stream, 68.6 percent of the first generation were manual workers or servants; by 1900 it had increased to 71.6 percent of the first-generation Irish. By contrast, the entire population of the United States, including many other low-status immigrants, contained only 45.1 percent in these occupational categories in 1900. Moreover, progress was very slow for the second generation; in 1900, 59.7 percent of the second-generation Irish were classified in manual work or domestic service (Thomas, 1954: Table 40). Indeed, by 1950 the foreign-born Irish, including many who had migrated shortly after the turn of the century during the era of dwindling Irish immigration, were markedly under represented in white collar occupations although they had achieved some status as semiskilled and skilled workers, managers, officials and proprietors. The second generation, however, was moving rapidly toward parity with the native white population of native parentage (Lieberson, 1963: Tables 52, 54).

The situation of the Irish was fairly extreme and was compounded by a long history of degradation and restriction, by the rush to depart that often led them to inappropriate destinations, by the severe anti-Catholicism that met them, and by the dominance of parochial education which sheltered the second generation from the impact of American values and orientations (Glazer and Moynihan, 1963; Handlin, 1941, 1959; Potter, 1960; Thomas, 1954; Woodham-Smith, 1962). But the Italians, arriving more recently than the Irish, were also markedly underrepresented in all higher status occupations as late as 1950 and the second generation was moving more slowly than the Irish to an occupational distribution comparable with native whites of native parentage (Lieberson, 1963: Tables 52, 54). The Germans and the Poles had mobility rates higher than the Irish-Italian pattern but considerably lower than the English-Scotch-Welsh. By 1910, the Germans were showing a modest level of mobility but were still underrepresented in high status occupations. By 1950, the first-generation Germans were close to parity with native white Americans but the second generation was not yet equivalent to native whites of native parentage (Thomas, 1954: Table 40; Lieberson, 1963: Tables 52, 54). The Poles found opportunities for mobility from unskilled to skilled ranks in the major manufacturing industries. But few moved on rapidly to positions as skilled workers (Brody, 1960). By 1950, however, the Poles of foreign birth had moved far and were well-represented among skilled workers, managers, and officials (Lieberson, 1963).

areas of the country which have higher income levels. Lowry's analysis (1966) of migration between metropolitan areas in the United States for the period 1955-1960 also provides support for the importance of economic opportunities in the place of destination. Thus, just as the data from places of origin point to the importance of deprivation and expulsive forces, so do the data based on places of destination support the significance of opportunity and attractive forces in encouraging migration.[5] It is only the larger generalizations about this process that take account of the interactive nature of the process. Thus, as Vance (1938) points out, migration proceeds (a) from lower to higher per capita income areas, (b) from extractive to industrialized economies, and (c) from areas of high natural increase of population to those of low natural increase.[6] The process of migration can only be conceived, in this light, as a shift of population from conditions of disadvantage and restriction to those of relatively greater advantage and potentiality.

Migration is, of course, a continuous process through history. Mass migrations represent only one important form of population redistribution. The most intensive efforts to restrict migratory movement, exemplified in the limitation of the serf or slave to a particular plot of land or a particular master, have never served entirely to eliminate geographical and job mobility. Whether the same factors operate in accounting for the endemic process of geographical migration as those which account for mass migration is unclear. As we have indicated, even mass migrations are highly selective and represent the movement of only a minority of the population from any place. As Eisenstadt (1954) points out, the very choice of migration as a potential mass movement is itself bound up with particular social and economic settings where the rise of autonomous economic motivation and achievement motivation, aspirations for liberalism and universalistic orientations, and the demise of group life and community-embeddedness are moderately widespread. But the selective decision to migrate, whether individually or as part of a massive movement, must still be based on similar orientations and motivations impelled by the contrast between existing deprivations, expectable risks, and anticipated opportunities. And the weight of evidence suggests that the more massive migrations are more heavily influenced by deprivations at the point of origin which disrupt the entire fabric of life or of the very means of living of the vast proportion of low status people in a country or area. It is hardly a surprising consequence that there appears to be a close association between the massiveness of emigration, the severity of deprivation and disruption, and the low status composition of the migration stream.

Certainly, the vast majority of the immigrants to the United States during the nineteenth and twentieth centuries were relatively uneducated people from rural areas: farmers, farm laborers, unskilled or semiskilled workers, and rural or semirural craftsmen (Carpenter, 1927). The United Kingdom, of course, provided a larger proportion of skilled workers than any other region of the world; they were also distinguished by their specific occupational interest, by the eagerness with which their services were sought, and by their continuity in occupations they had held before emigrating (Berthoff, 1953). Between 1875 and 1910, more than half the immigrants from Wales and Scotland were listed as skilled laborers and nearly as high a proportion from England. By contrast, throughout this same period, immigrants from Ireland were preponderantly common laborers or servants and rarely included as many as ten percent skilled laborers.[7] The data for Italian immigrants suggest a picture similar to that for the Irish with agricultural labor substituting for the servant category. The German migration seemed to draw more heavily on farmers and rural craftsmen but, nonetheless, to be predominantly the higher categories of low-status workers (Carpenter, 1927; Foerster, 1919; Walker, 1964).

Although there have been no thorough studies, at least in English, of the occupational histories of migrants compared to nonmigrants, there is a general consensus that low-status newcomers to an area suffer serious disadvantages compared to natives of similar status.[8] These disadvantages occur through entering the lowest status jobs and through a high degree of job insecurity. Thus, migrants have higher levels of unemployment and among the lowest incomes in urban areas. While there almost certainly are large differences depending on educational level, urban experience, and social acceptance of different ethnic or racial groups, the phenomenally low status of migrants appears to be quite general. Vance (1938; also see Tolles, 1937) points out that, while migrants earn less than most workers in the areas to which they move, they earn more on the average than they did before relocating. Leyburn's study (1937) of Southern Appalachian migrants to Cincinnati indicates a marked disadvantage for migrants in employment rates.[9] Lipset and Bendix (1959) show evidence for marked differences in occupational status and mobility depending both on migration status and size of community of origin in several American studies as well as in Sweden and in Germany. Thernstrom (1964) provides data showing the marked differences in occupational mobility of immigrants compared to natives in nineteenth century Newburyport. And Blau and Duncan's analysis (1967) of more recent data, more carefully controlled than previous studies although heavily weighted by more recent

What these facts highlight, imperfect though they may be, is the great gap between an image of continuous and rapid mobility and the reality of slow, arduous, intra generational and inter-generational change in status. There is no question that the process of upward mobility among immigrants has been continuous. But if, sixty to one hundred years after an ethnic group has initiated large-scale immigration into this country and much of that immigration necessarily occurring more than forty years ago, there is still so wide a discrepancy in occupational achievement from the host population, we must alter our conceptions of the process. The rungs of the mobility ladder are wide apart for migrants to an urban, industrial society. Just as the melting pot has failed to melt and consolidate its ethnic prey, so has the mobility process failed to amalgamate its poverty-stricken, uneducated, and unskilled immigrants or their children in a vision of success. The deprived migrants of another era remain relatively disadvantaged and their children suffer the consequences of these deprivations while slowly overcoming their effects.

At the very least, we must consider the conventional conception of mobility and assimilation of ethnic minorities in the United States a myth. Some few ethnic groups have, in fact, been highly mobile particularly if they brought scarce skills or moved into a prepared environment. Other ethnic groups, the large majority, have been slowly mobile and have had to overcome gigantic obstacles in their struggles for educational attainment, occupational status and high incomes. And a few ethnic groups have struggled, virtually in vain, until a new generation, bearing fewer of the marks of ethnicity and in a different social environment, were able to confront the problem without the preformed conviction that they were doomed to failure. We must also forego any ready assumptions about the ease of social assimilation. While social mobility is often a stage in the large assimilation of immigrants, there are large gaps in the process, and mobility achievements among immigrants, as with the Negro, Puerto Rican, and Mexican-American, have often proved necessary but hardly sufficient for social assimilation.

After almost a half century during which there have been no mass immigrations from Europe or Asia, the issue of the social mobility and social assimilation of the foreign born and of their native-born children is no longer as trenchant and pressing a problem as it once was in the large cities of the United States. Although some ethnic groups have not yet reached parity with the population as a whole in education, occupation, or income or have not yet achieved total desegregation in housing, the differences are not large. Buoyed by several periods of great prosperity which have facilitated, probably with disproportionate advantage, the mobility opportunities of immigrants and in the context of a high standard of living in an affluent society, the problem seems academic.

However, the problem is far from academic. If these conclusions are correct, we must not only dismiss the image of rapid mobility and assimilation but must place, in its stead, an image of a moderately restrictive and fundamentally segregationist society. Despite the absence of an overly-structured status system on the model of postfeudal societies, issues of ethnicity, race, and culture have been superimposed on economic and occupational differences to provide a basis for discrimination, prejudice, and social inequality. The labor of millions of poverty-stricken immigrants was necessary for the industrial expansion of the United States and only because of this were its doors open to indentured servants, slaves, serfs and, subsequently, to their descendants. But the people were themselves viewed as a vast and impersonal, low-status labor force to whom society owed nothing. Translations of the Elizabethan Poor Laws discouraging vagrants and indigents and pioneering work in the development of an urban police force were our primary control mechanisms. Little attention was given to the social and personal needs of immigrants until the explosion of urban, social problems made their desperate situations unavoidably evident to a few people. Even then, the society offered the immigrant with less evident needs and who was a less evident threat little or no assistance and placed great impediments in the path of establishing a meaningful and integrated life experience in the new world.

The issue is certainly neither academic nor attenuated when we confront, in this light, the situation of more recent migrants to the urban, industrial environment. We shall focus particularly on the Negro experience because it highlights, in the most extreme fashion, the limits of social mobility and the gap between social mobility and social assimilation in a society characterized by severe prejudice, segregationist policies and an underlying disregard for social justice. Nonetheless, neither the misery and constraint imposed on the Negro nor the general affluence that mitigates the visibility and the most ostensible consequences of underprivilege should allow us to ignore the failures of assimilation and mobility that characterize the history of immigration to the United States. It may be here, in some of the underlying similarities rather than in the many striking differences, that the most severe problems and limitations of our society are buried but only partly concealed.

Negro Migration in the United States

This history of settlement and growth in the United States is dominated by several major trends of population movement. The large-scale movement of American Negroes, one of the most prominent streams of migration in this country since 1910, can be seen as a special case of these trends. The significance of the Negro migration is, in large part, revealed both in the similarities to and in the differences from other patterns of migration in the United States.

The consistency of overall rates of geographical mobility in the United States over long periods of time is striking. Since at least 1850, the proportion of the native population who were living in states different from those in which they had been born has varied little—between approximately twenty and twenty-five percent (Shryock, 1964). This high and stable level of population movement was characterized quite early by expanding populations along the moving western frontiers and by the slow contraction of agricultural populations and the growth of urban populations. These two trends coincided to some extent, not merely in time but in the enormous rates of growth of frontier towns and cities, a frontier which gradually moved further westward over more than a century. During several decades these westward frontier towns and cities shifted from Pittsburgh, Cincinnati, and Lexington to St. Louis, Chicago, Denver, and San Francisco (Wade, 1959). The rapid growth of towns and cities was further implemented, often to the bursting point, by the vast tides of immigration which fed the Northeastern section of the country but gradually expanded to the entire country. At the same time, since 1790, there was a gradual net out-migration from rural areas and a gradual net in-migration to urban areas within the United States. In combination, these forces led to a massive change from a rural to an urban society.

One of the major regional sources of this shift from a rural to an urban society was the slow population decline in the South, a section of the country which had previously harbored the largest rural, agrarian population. From 1880 on, there was a striking transfer of population out of the Southern states, a shift that was eventually to become the largest stream of out-migration in the United States. This is vividly depicted in Figure 1 (from Kuznets and Thomas, 1964: Vol. III, p. 66). Since 1870, this movement has been dominated (and, indeed, was initiated) by the migration of Negroes out of the South to the Northeastern states. After 1910, with an increasing rate of migration out of the South, the North Central states began to receive a large number of Negro migrants. This monolithic and accelerating escape from the South by the large Southern Negro population soon took on the appearance of a dramatic exodus through the addition of almost as large a number of white out-migrants from the South.

It is notable that, between 1950 and 1960 and even earlier, a gradual change in the character of out-migration from the South developed. While

NET MIGRATION OF NATIVE WHITES AND NEGROES, BY REGIONS
1870-1880 to 1940-1950

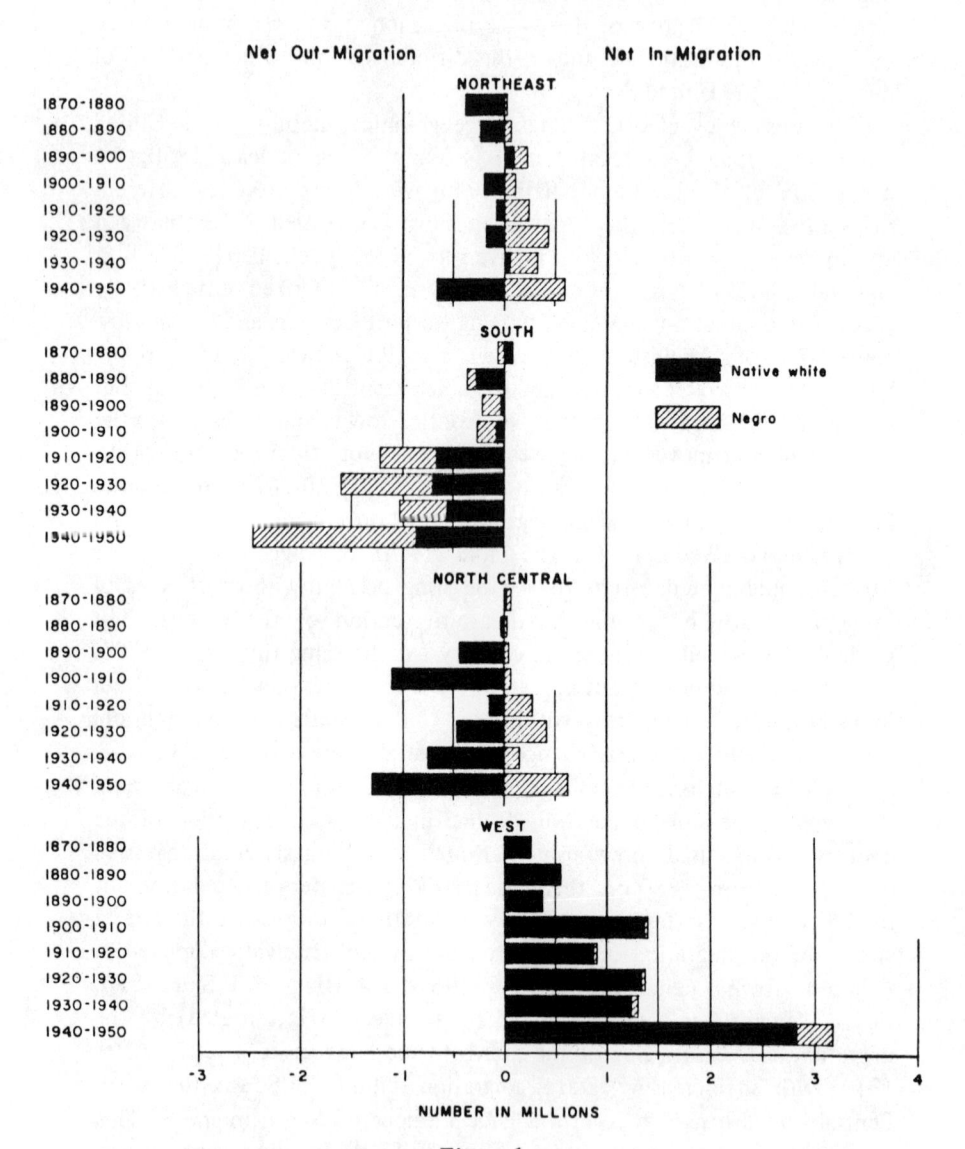

Figure 1

SOURCE: Kuznets and Thomas (1964: Vol. III, p. 91).

the large net out-migration of Negroes continued at only a slightly diminished pace, the net out-migration of whites decreased as a result of a reverse stream of in-migration. In particular, a reverse stream of in-migration developed through the movement of whites from the Northeast to some of the metropolitan and urban areas in the South (Shryock, 1964).

The difference in the destination of the white and Negro migration from the South is also clear in Figure 1. The overwhelming direction of population movement among whites was toward the West throughout the period from 1870 to 1950, a growth that included out-migrant whites from the Northeast and North Central states as well as from the South. By contrast, the Negro out-migration from the South was initially directed almost exclusively toward the Northeast. By 1890 there was a slowly growing movement and, by 1910, a rapidly growing movement toward the North Central states. The West only slowly captured a small part of this migrant stream of Negroes from the South, a direction of movement that did not become notable until the decade 1940-1950.

Viewed only on the basis of the numbers or proportions of Negroes within large sections of the country or moving between them, it is evident that the Negro migration out of the South did not reach striking size until the decade between 1910 and 1920. Nonetheless, this was hardly the beginning of significant Negro migrations. Indeed, conceived only in terms of these regional shifts, the decade between 1900 and 1910 saw a fairly large migration. Between 1870 and 1890, the proportions of Southern-born Negroes among the Northern Negro population remained fairly constant at thirty percent. This itself implies that the rate of in-migration to the North was proportional to the indigenous increase of Northern-born Negroes (Hill, 1924). By 1900, there was a small proportionate increase to thirty-one percent which grew to forty percent in 1910 and fifty percent in 1920. Thus, there was almost as large a percentage increase of Southern-born Negro migrants in the Negro population living in the North between 1900 and 1910 as there was in the frequently mentioned migration of the 1910-1920 decade. Other migratory movements also occurred in the period prior to 1920. While the border states were, at first, most prominent as sources of northward migration of Negroes, there was a strong trend of movement toward the Southwest cotton-growing regions from other parts of the South (Hill, 1924). One of the more striking migrations, although numerically small, was the "Kansas Exodus" of 1879 in which between 25,000 and 50,000 Southern Negroes, disgusted with the failure of reconstruction, moved en masse to Kansas (Woofter, 1920).

A number of important features of the huge migration of Negroes from the Southern states stand out. First and foremost is the fact that by 1960, there had been a remarkable shift of the Negro population out of the South. In 1790, ninety-one percent of the Negroes in the United States lived in the South. By 1910, this was reduced only to eighty-nine percent. But by 1960 (Hauser, 1965), this proportion had dropped to fifty-two percent.[13] While the movement of Southern Negroes followed the gross patterns of population redistribution in the United States since 1910, the proportions of Negroes who left the South were far greater than the proportions of the numerically larger white Southern population. From the vantage point of in-migration in the North, this migration was smaller in numbers than the vast European immigration from 1850 to 1910, annually or in aggregate. But it increased the proportions of the Negro population living in urban areas more dramatically than any equivalent concentration of a single ethnic group during the earlier period. At the same time, it is noteworthy that the Negro population remaining in the South is very large and, indeed, even in 1960 a slight majority of the total Negro population of the United States lived in Southern states.

A detailed analysis of the causes of the great Negro migration of the period 1910-1960 is even more difficult than in the case of the European migrations of 1830-1920. To some extent, the situation of the Southern Negro seems so self evidently miserable that there may be little temptation to investigate it further. Whatever the manifestations of discrimination and unequal opportunities for Negroes in the North and West, objectively discrimination is less pervasive, less extreme, and has less striking consequences for education, jobs, and incomes. This difference alone might account for the mass movement of Southern Negroes even apart from the very slow progress of the South in urbanization and industrialization. Certainly the more recent and rapid growth of industry and of cities in the South, coupled with an expanding economy, has stemmed some of the tide of net out-migration by whites without appreciably affecting the out-migration of Negroes.

Although it is difficult to attribute the geographical displacement of Negroes and the trend of the Negro population out of farms and into rural non farm areas and cities precisely, Dorothy Thomas' data (Kuznets and Thomas, 1964) indicates that economic conditions are almost certainly involved. Rates of migration, according to her analysis, are highest during periods of prosperity and are manifest largely as shifts from areas of relatively low incomes to areas of relatively high incomes. Moreover, at almost all age levels, the migration behavior of the Negro population appears to be affected more severely by cycles of prosperity and depression than are either the native white population or the foreign born. That is, the tide of migration is more markedly diminished during depressions and more markedly augmented during prosperous periods among Negroes than among other sub populations. The implications of these findings are not entirely clear. They may be interpreted as evidence that the Negro population of the United States is less responsive to "expulsive" forces and more responsive to "attractive" forces than native whites or foreign born. Indeed, it is almost certainly the case that, because of discrimination, the Negro's chances of finding a job in the city during depressions are more drastically diminished than for the population as a whole; and, conversely, during periods of prosperity, when the demand for workers is great, job opportunities of Negroes increase disproportionately. Unfortunately, this does not clarify the relationship between the expulsive and attractive forces themselves in generating mass migrations of Negroes.

A similar problem of interpretation arises from the data on the spatial distribution of migration for Negroes and whites. From Thomas' data (Kuznets and Thomas, 1964) on net in-migration, net out-migration, and economic conditions in the areas of in- and out-migration, it is evident that there is a gross movement away from low income areas to high income areas with a slight tendency for the relationship between Negro migration and income differentials to be closer than for whites. Whether this is primarily a function of the expulsive pressures of low income areas or the attractive qualities of high income areas or a combination of both is altogether uncertain. In view of the general principle that migration moves from lower to higher income areas, the relatively weak relationships and the many deviations from the expected pattern are themselves among the most interesting features of the data. Indeed, we might conclude from these results either that purely economic considerations play only a relatively minor role in the migration patterns of both Negroes and whites or that, whatever the ostensible orientation of migration patterns, they are poorly adjusted to the realities of economic circumstances. The looseness of this relationship may be quite unfortunate. In instances of massive migrations the failure of effective correspondence between the timing or direction of movement and economic conditions can be a factor seriously exacerbating the other disruptive potentials of migrant adjustment and discrimination.

Thus, the relationship between migration and economic conditions is less finely attuned than one would hope. For Negro migrants, so much more dependent on flourishing conditions of employment which mitigate the effects of discrimination, even the somewhat greater association between migration and economic circumstances is insufficient to avoid the negative consequences of "misguided" or, more properly, "unguided" migration.

TABLE 2

PERCENTAGE OF NEGRO AND WHITE POPULATION LIVING IN URBAN AREAS BY REGION, 1910-1960

Year	United States Negro	United States White	South Negro	South White	North and West Negro	North and West White
1910	27.4	48.7	21.2	23.2	77.5	57.3
1920	35.4	53.3	27.0	28.5	84.3	61.6
1930	43.7	57.6	31.7	35.0	88.1	65.5
1940	48.6	59.0	36.5	36.8	89.1	67.4
1950	62.4	64.3	47.7	48.9	93.5	70.1
1960	73.2	69.6	58.4	58.6	95.3	73.7

SOURCE: Newman (1965a).

One clear relationship between Negro migration and choice of destination does stand out and appears to be a clear selection of areas with high growth potential: the striking movement of Negroes to cities. This movement toward cities is of equal importance with and as monolithic a trend as the redistribution of Negroes from the South to the North. Table 2 (Newman, 1965a) presents the percentage distribution by urban residence for Negroes and whites for the United States as a whole and by regional division into South and non-South (North, Central, and West) since 1910. The table reveals both the enormous increase in the urbanization of the American Negro since 1910 and the continuous difference between the South and other regions of the country.

In the South, the urbanization of the Negro has paralleled to a remarkable degree the urbanization of the white population. By 1960, as a result of the gradual decrease in agricultural populations, the majority of both Negroes and whites in the South were living in urban areas. Indeed, comparing the data in this table for the South and for the North and West and bearing in mind the enormous migration of Negroes from the South since 1910, it becomes clear that the migration also resulted in a dramatic urbanization of the Negro population in the United States. Even in 1910, before the full swing of the great migration of Negroes from the South, more than three-quarters of the Negro population living outside the South (and predominantly in the Northeast) were living in cities. And with the growing tide of Negro migration from the South since 1910, the proportion living in urban areas has increased continuously. In this respect, the pattern of Negro migration has many similarities to the great wave of foreign immigration particularly during its last fifty years (1870-1920) and has resulted in a predominantly urban ethnic minority (Taeuber and Taeuber, 1958). It is also notable that, while the urban trend of the white population has been quite marked throughout this period, the higher level of urbanization among the Negro population than among the white population in the North and West has been maintained during at least five decades.

The exact characteristics of this urban movement of the Negro population are not altogether clear and the available statistics do not allow more than slightly informed conjectures. Generally speaking, rates of migration have been higher for whites than for nonwhites despite the somewhat higher rates of short-distance moves by Negroes (Shryock, 1964). But the huge loss of farm populations to nonfarm residences has been even greater for nonwhites than for whites and has disproportionately affected the South compared to other regions of the country (Shryock, 1964; Taeuber and Taeuber, 1958). By no means do these short-distance or long-distance moves represent only movement from farm to urban residence either for whites or nonwhites. Indeed, rural nonfarm residence has generally shown the greatest net in-migration rates of any residential group. However, there is some suggestion from these data, to which Figure 2 (from Shryock, 1964: 319) gives visible form, indicating a familiar pattern of migration: that the vast majority of moves are relatively short-distance moves, that these short-distance moves take place over a gradient from sparser to denser population concentrations by stages, and that the large increases in urban populations result from the vast increments due to these smaller and successive migrations.[14] With the gradual depletion of farm populations, with the large shift of the Negro population in the South from rural farm to rural nonfarm residence, to small cities, and to large cities over time, an increasing proportion of the continuing Negro migration becomes closer, both geographically and culturally, to the urban, industrial environment of the Northern, Central, and Western regions of the country.

One final feature of the Negro migration and its effect on the Negro population of urban areas warrants attention: the attributes of migrants and the changing composition of the urban, Negro population. Although there are differences of view concerning the nature of migrant populations, the weight of evidence suggests that the great European migrations were dominated by those people from deprived areas who were most competent, with the highest occupational skills, and with some minimal financial resources to carry them through the earliest phases of migration.[15] Studies of internal migration in the United States give evidence that, indeed, migrants are of higher educational and occupational status than are non-migrants from the same areas (Duncan and Duncan, 1957; Mauldin, 1940).

To some extent, the evaluation of the attributes of migrants has been a function of the vantage point from which observation or analysis was carried out. Seen in the context of an urbanized population, migrants have often appeared to be of lower status, less competent, ill-prepared for dealing with urban complexities and ambiguities. On the other hand, compared to nonmigrants who remained in the areas from which the migrants came, they have often appeared to be those with the greatest opportunity in the area of origin. As Duncan and Duncan's analysis (1957) of Chicago data reveals, both of these observations are probably correct. The educational status of Negro migrants to Chicago in the decade 1940-1950 was higher than that of the states from which they mainly came but lower than that of the resident Negro population in Chicago. However, while both Negro and white migrants tend to be of higher educational level than nonmigrants from the same areas, this is particularly the case for Negro migrants (Hamilton, 1965; Fein, 1965b). There is also evidence of a change in the character of Negro migration which has led both to an increase in inter metropolitan migration and, corresponding to this, to an improvement in the educational and occupational composition of Negro migrants compared to the resident Negro population in the receiving metropolitan areas (Taeuber and Taeuber, 1965).

Taeuber and Taeuber (1964) have given particular consideration to the changing characteristics of Negro migrants and have pointed up the fact that, over the last few decades, the urban attributes of Negro migrants have approximated ever more closely to those that characterize white migrants. However, in this as in many other comparisons which permit a distinction between the South and other regions of the country, there is a marked difference between Southern cities, on the one hand, and Northern, Midwestern, or Western cities on the other. As we might expect on the basis of other information about patterns of Negro migration, a greater proportion of in-migrants to metropolitan areas in Northern and border states come from other metropolitan areas than do in-migrants to

PERCENT OF ALL MOVERS WHO CAME FROM FARMS,
FOR THE WHITE AND NONWHITE POPULATION,
BY SIZE OF PLACE AT DESTINATION: 1949 TO 1950

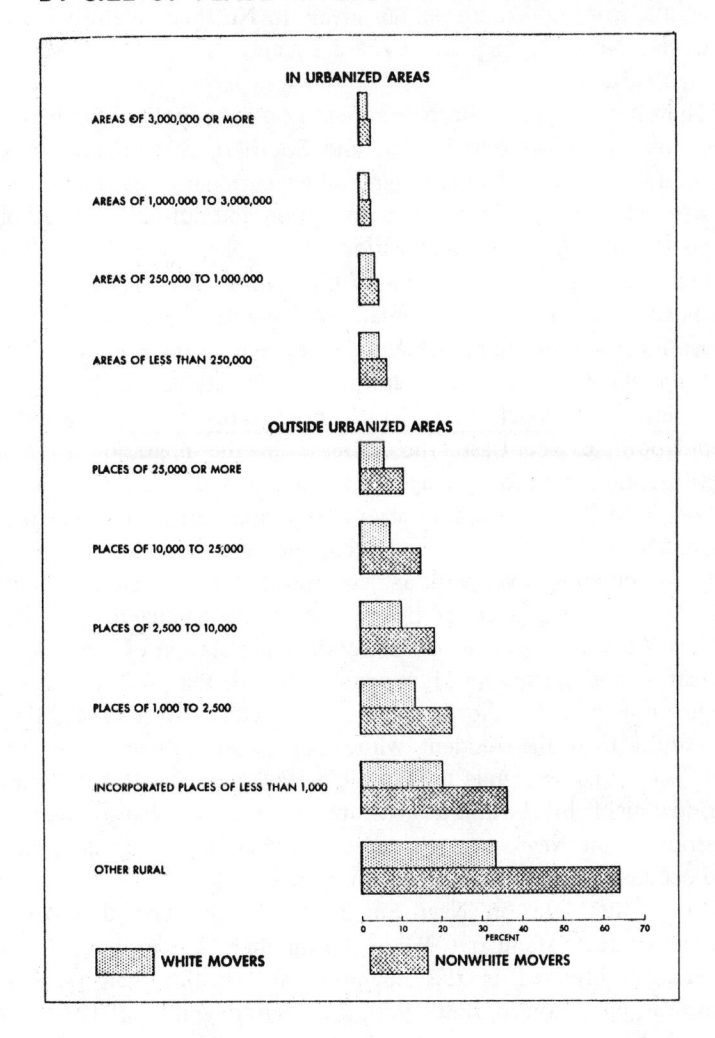

Figure 2

SOURCE: Shryock (1964: 319).

Southern cities. In Southern metropolitan areas, nonmetropolitan in-migrants are considerably more frequent than are in-migrants from other metropolitan areas (Taeuber and Taeuber, 1965).[16] Border metropolitan areas are intermediate and show a slightly greater proportion of in-migrants from non metropolitan areas. In Northern metropolitan areas, however, Negro in-migrants more frequently come from other metropolitan areas.

The differences in metropolitan and non metropolitan origins of Negro migrants into Northern, border, and Southern metropolitan areas, the association of metropolitan origin and educational or occupational status, and the relative significance of in-migration and out-migration account, in part, for the changing composition of the Negro population of metropolitan areas in different regions of the country. In the South, the greater proportion of non metropolitan in-migrants, their relatively lower educational and occupational status compared to the resident population, and the much heavier out-migration of high-status Negroes has led to declining educational and occupational status in the resident Negro population. By contrast, the increase in the migration to Northern Metropolitan areas of young Negro migrants of high educational and occupational status from other metropolitan areas has led to an improvement in the education and occupational composition of the resident Negro population. Even with age controlled, the difference from previous decades is striking since, at the very least, the in-migrant population of 1955-1960 was equivalent to the resident population of the same ages in education and occupation. It is equally notable that, for this same period, Negro in-migrants to Northern cities were equal to or of slightly higher education than the resident white population (Taeuber and Taeuber, 1965). Comparing results for the 1955-1960 period with those for earlier periods highlights both the absolute and relative changes in the composition of the Negro migration resulting in an increase in the educational and occupational characteristics of migrants.

From the data on Negro migration in the United States a few prominent facts stand out. While the numbers and proportions of total population involved in this migration are smaller than those of the European migration of 1830-1920, the Negro migration of 1910-1960 was vast and represented a redistribution of at least six million Negro Americans. However, the loss of Negro population sustained by the South was probably of much greater proportion than the loss of population from any country of origin during the European migrations to the United States. Thus, it is estimated that there was a loss of approximately one-third of the population of Ireland during the half-century after the great famine (Arensberg and Kimball, 1948). The loss of the Negro population from the South during an equivalent period of time was almost certainly greater than one-half. Like most of the streams of European migration, the movement of Negroes out of the South was also largely a rural-to-urban transition. However, the Negro population moving into the North, Midwest and West was even more prominently an urbanizing population with more than ninety-five percent of those Negroes outside the South living in urban areas by 1960.

In discussing the European migration of the nineteenth and early twentieth century, we considered the relative importance of expulsive and attractive forces, concluding that these are inevitably interrelated and that both pushes and pulls were essential components in accounting for the migrations. For the Negro migrations of the past half-century, the data are even less adequate but point in a similar direction. Like the situations in most of the European countries that experienced large-scale migrations, there were endemic forces that operated as continuous expulsive factors: severe poverty, a long process of subdivision of small farms, tenure-farming systems that were often duplicates of slavery or serfdom, increasing population pressure, discrimination and restriction of opportunities, widespread traditionalism opposed to change or deviation. Superimposed on these endemic forces, intensified poverty due to famine or increased discrimination and restriction often led to the marked short-term increases in migration in the European and in the Southern Negro situation alike.

It is clear that, both for the European migrations and for the migrations of Negroes from the South, there were increased rates of movement associated with periods of prosperity in the areas of destination and decreased rates of movement associated with depression or recession in the areas of destination. The endemic expulsive forces are represented, in both instances, by the continuing movement of population throughout except under the most severe economic chaos in potential host areas as exemplified by the depression of the 1930s. On the other hand, through the haze of inadequate data it appears that for most of the streams of European migration and for the Negro migration from the South, there was much misdirection. In general, the trend in both cases was for movement from areas of lower employment and income to areas of higher employment and income. But there is much evidence to indicate that the areas of destination were determined by many factors other than maximal available economic opportunities. For more recent decades, Dorothy Thomas' analysis (Kuznets and Thomas, 1964) suggests that the correspondence of migration and economic opportunity has been somewhat greater for Negroes than for either native-born or foreign-born whites suggesting the greater significance of discrimination in limiting opportunities for Negroes except under conditions of maximal demand for labor.

Despite the disproportionately large number of in-migrants from other metropolitan areas with high levels of education and occupation, a substantial proportion of Negro migrants to Northern cities continues to be of low status and from non metropolitan areas. Thus, while the Negro migration has begun to approximate the character of the white migration in some respects, there remains a substantial problem in the migration of rural, low-status Negroes to metropolitan areas with little preparation either in choice of destination or in the adaptive necessities of life in urban, industrial societies. Moreover, it is not at all clear that the relatively high status inter metropolitan Negro migrants do not suffer some of the consequences of migration, especially under conditions of discrimination, in the form of increased rates of unemployment and diminished occupational status opportunities.

With the extant data, it is not possible to trace these problems further. More detailed data is necessary to examine these questions more fully or with greater analytic precision. However, the patterns of migration are largely a context for inquiring into the fate of the population. While it is not entirely possible to separate issues of geographical migration and social mobility, even to the extent that this could be done with the European immigration, it is important to ask about the rates of social mobility during these decades of migration for the Negro population. We turn, thus, to a consideration of the same set of questions which we addressed about the European migrants: to what extent has there been a pattern of upward social mobility for the rapidly urbanizing Negro American; and to what extent have changes in social status entailed a marked diminution in discrimination?

Social Mobility and Assimilation Among Negroes

In order to provide a meaningful comparison of the situation of the foreign-born immigrant to the United States and that of the Negro American whose migration patterns to urban, industrial areas we have traced in gross fashion, we must now examine rates of social mobility and of social assimilation for Negroes. In examining the social mobility and assimilation of the Negro migrant, whose period of massive movement has occurred since approximately 1910 and shows only slight indications of diminution during the 1960s, we must rely heavily on recent data. Moreover, a consideration of social assimilations must depend almost exclusively on residential segregation. That there have been difficulties in both social mobility and residential segregation of Negroes long before the large-scale migrations of the past half-century, however, is clear. Although several historians point to the fact that, prior to the increased immigration associated with World War II, the resident Negro population in Northern cities had begun to achieve a modicum of occupational and economic advancement, these reports are based on isolated cases rather than systematic population or sample data (Handlin, 1959; Osofsky, 1966). Whatever minimal achievements the settled Negro population in Northern cities experienced, residential segregation was pervasive and more severe than for other in-migrant populations (Drake and Cayton, 1945; Osofsky, 1966; Riis, 1957).

A wealth of data indicate that urban, industrial societies have quite high rates of social mobility and that this is true and has been true for some time in the United States (Lipset and Bendix, 1959; Warner and Srole, 1945; Thernstrom, 1964; Miller, 1960; Blau and Duncan, 1965; Goldstein, 1955). But, apart from the technical difficulties of making even moderately precise estimates, evaluations of mobility rates as "high" or "low" are extremely subjective.[17] Thernstrom's compilation (1964) of ten different studies of occupational changes from fathers to sons covering periods from 1860 to 1956 all evidence a fairly similar pattern. All of these studies vary around estimates that approximately 50-60 percent (ranging from forty-eight percent to seventy-one percent) of the sons of unskilled laborers were themselves either unskilled or semiskilled laborers. That the occupational progress of Negroes has been slow and halting by any criterion, however, is quite evident from the gross estimates of changes in occupational position between 1910 and 1960. Tobin (1965), quoting Hiestand's data shows significant improvement only during the period 1940-1960 although there had been some improvement relative to whites since 1910. Even these changes, which bring the occupational position of Negro males to the level of 82.1 percent of white males by 1960, may overstate the degree of change experienced. On the other hand, most analyses of changes in education, occupation, or income of Negroes over the last few decades fail to distinguish the South from other regions of the country which leads to another serious distortion of the results. Thus, Hare's intracohort analysis (1965) reveals that Negro rates of occupational mobility from 1930 to 1940, 1940 to 1950, and 1950 to 1960 were higher than those for whites in each age group (Hare, 1965). The fact that there was a slight decline in the rate of improvement during the 1950-1960 decade outside the South, coupled with a retrogression in the South during the same decade gave the impression that the improvements of previous decades had not continued since the national figures did not distinguish regional differences.

This pattern of regional differences has been quite persistent. According to Hauser (1965), in 1910, Negro illiteracy was ten times as great as among native whites. However, in the South illiteracy rates for Negroes reached thirty-three percent while in the North they were only ten percent; that is they were more than three times as high in the South as in the North. By 1930 Negro illiteracy rates had been cut in half, but white rates of illiteracy had regressed even more. That the South contributed disproportionately to the slow pace of change for the Negro is revealed by the fact that rates of Negro illiteracy had grown to four times as high in the South in 1930 as they were in the North. Between 1940 and 1960, after the Census Bureau substituted a question on years of schooling for the questions on literacy, it is possible to estimate the change in the form of grades completed. During these two decades, the difference in education between Negro and white males had diminished from 3.3 to 2.6 years and between Negro and white females it had diminished from 2.7 to 2.5 years (Hauser, 1965). Moreover, these gains understate the relative educational achievement in at least one respect. Since the differentials among older age groups are considerably greater than among younger age groups but the figures are based on median years of school completed for all age groups, they do not fully reveal the impact of the gain as a function of changes over time. Throughout, the differences between the South and other regions of the country persisted, further diminishing the overall manifestation of educational gain.

In reviewing studies of educational, occupational, or income mobility of Negroes and whites, one is confronted by contrasting figures and contrasting conclusions which make any simple summary particularly difficult. In part, these differences are based on different sources of data; in part, they result from different statistical analyses. However, some of the most recent figures and reports appear to show greater consensus in finding a consistent improvement in the situation of Negroes relative to whites since 1960. Spady's analysis (1967) of 1964 data, for example, indicates marked educational gains for successively younger age groups among both whites and Negroes. These gains are represented in the differences in educational achievement of men aged 25-34, 35-44, 45-54, and 55-64 compared to their fathers' educational achievements. In virtually all instances of relative achievements of fathers and sons, the improvements among the younger age groups are consistent and relatively greater for Negroes than for whites. On the other hand, looked at from a different vantage point, the youngest age groups among Negroes are approximately at the same educational level as the oldest age groups among the whites suggesting a gap of more than a generation.

Although the general trend of data for earlier periods is fairly consistent in showing relatively greater occupational advances among Negroes than whites, the most recent results show even more striking gains. Much of the occupational gain of earlier years resulted from the shift from agricultural to industrial employment. Thus, between 1910 and 1940, of those gainfully employed, the proportion of Negroes engaged in agriculture dropped from fifty-six percent to forty-one percent; during the same period, the proportion of whites engaged in agriculture diminished from thirty-three percent to twenty-one percent. By 1960, only eleven percent of the Negro labor force and eight percent of the white labor force was engaged in agriculture, a situation of virtual parity (Hauser, 1965). During the latter period from 1940 to 1960, however, other relative gains were achieved by Negro workers and, while these gains were far from what one might have hoped, they did show an increased similarity between the occupational distribution of Negroes and of whites. Moreover, the more detailed analyses available from extensive interview data suggest that, albeit at a lower level of the occupational status scale, Negro advances are considerably greater than those among white workers (Shannon and Morgan, 1966). But the fact remains, as with educational advances, that these gains are occurring at different levels of the occupational hierarchy and, thus, Negro occupational status remains considerably behind that of the white population.

The most recent reports, based on data up to 1965, continue to present a picture of cautious optimism. Several of the higher status level occupations are among the occupational areas which appear to be opening up opportunities for Negroes most rapidly: professional, technical, and kindred occupations, most particularly, but also in other white collar occupations and in the craftsmen and foremen occupational groups (Russell, 1966). To some extent, however, some of the major gains are in occupations which are not rapidly expanding in overall availability of positions and nonwhite gains may have to occur in different occupations to keep pace with changing patterns of industrial development and unemployment. During the expanding economy of 1962-1965, nonwhite workers had an even greater reduction in unemployment than white workers but these marked gains during the period of high employment are not necessarily stable (Hardin, 1967; Johnston and Hamel, 1966; Russell, 1966). At the same time, they highlight a point that Myrdal (1962) has made, that the effects of discrimination can only be markedly diminished through maximizing overall job opportunities. During this same period between 1962 and 1965, the white-nonwhite educational differential for different occupational categories has also diminished. While this repre-

[54] BEHAVIOR IN NEW ENVIRONMENTS

sents, in part, the same phenomenon of increased demand for workers, it does indicate greater parity in the criteria used for employing nonwhite and white workers.

The data comparing incomes of Negro and white families show much the same pattern as do the data for education and occupation and clearly reveal all of the trends previously discussed (U.S. Census Bureau, 1967). Between 1947 and 1966, the proportions of families with incomes of $7,000 per year or more rose considerably among both Negroes and whites. Indeed, they rose more sharply for Negro than for white families. Thus, two and one-half times as many white families had incomes at this level in 1966 compared to 1947, but four times as many Negro families were in this category in 1966 compared to 1947. Despite this remarkable and disproportionately great gain among Negro family incomes, approximately twice as many white families were in this income category in 1966 as were Negro families. As with previous data, the difference between the South and other regions was great. In the South in 1966, three times as many whites had incomes of $7,000 per year as did Negroes; in other regions, the ratio was approximately 3.2 Another feature of the data that warrants attention is that, although the overall gains were proportionately greater for Negroes than for whites, recession years offered minimal interruption to the progress of white families but seriously delayed or retarded the gains of Negro families.

In view of the extremely high levels of discrimination and inequality of opportunity that have plagued the Negro population of the United States and despite the many qualifications that must be made, rates of Negro mobility in education, occupation, and income appear quite high and considerably higher than one might have anticipated. Certainly, the gains have not been great enough to provide any comfortable image of equality. Certainly, there may be serious costs involved in even moderate mobility achievements in the face of immoderate inequality of opportunity. Certainly, marked deficits in the educational, occupational, and economic situation of Negroes remain. And certainly, the evident consequences of current, as well as past, discrimination appear in the discrepancies between achievements and rewards for achievement. At each educational level, Negro occupational status is lower than the corresponding occupational statuses of whites (Fein, 1965a; Johnston and Hamel, 1966). At each educational level, the incomes of Negro families are lower than the corresponding incomes of white families (Lassiter, 1965). And at each occupational level, as well, Negroes receive lower incomes than do corresponding white workers. As Blau and Duncan (1965) point out, "It hardly comes as a surprise that racial discrimination in the United States is reflected in the Negro's inferior chances of occupational success, although the extent to which Negroes with the same amount of education as whites remain behind in the struggle for desirable occupations is striking."

In stating that Negro achievements in social mobility have been surprisingly high in light of marked inequalities of opportunity, we find little cause for optimism. Indeed, in comparing social mobility among foreign-born immigrants and among native Negroes, we have slowly come to the conclusion that the assumed differences in rates of achievement are almost certainly not so great as is usually conceived.[18] But these differences between Negroes and foreign-born whites diminish in the light of these data, not because of the very high levels of manifest achievement but, rather, because both the foreign immigrants of the earlier period and the Negro migrants of the past half century have similarly fought against great obstacles and severe inequalities only to experience slow and meager gains. We have no objective way of measuring or even estimating the differences in opportunities for educational, occupational, or economic mobility for the earlier European immigrant and the more recent Negro migrant from the South. It appears almost indubitable that Negroes have experienced the most devastating forms of prejudice and limitation of opportunity. But this emerges as a serious intensification of continuing patterns of discrimination and inequality of opportunity in our society rather than as a wholly unique phenomenon.

While the earlier literature about the foreign immigrants was invariably hostile and critical and, from our present vantage point, unbelievably insulting, as early as 1890 Riis noted the more severely underprivileged and degrading situation of the Negro. It may, indeed, be the case that the modest degree to which the Negro American population has achieved mobility is all the more remarkable a feat. The scattered evidence of very high levels of motivation and aspiration among Negroes, of widespread and effective community leadership in either collaboration or revolt, of outstanding achievements in numerous fields, even the fact that the most striking occupational gains of the past five years have been in the most highly skilled professional and technical pursuits suggests that we must alter our image of the Negro in the United States. Discrepancies and deficits remain but, in view of the serious impediments in achievement, the level of accomplishment may well be remarkably great. And, in view of the evidence for extremely slow progress among the immigrants of the great European migrations, there may be far less discrepancy in social mobility between white immigrants and Negro migrants than we ordinarily imagine.

Although we do not have any adequate measure of discrimination and the significance of inequality of opportunity as a basis for evaluating mobility achievements, measures of housing segregation do provide some basis for evaluating the role of discrimination in social assimilation. In comparing residential segregation of foreign-born immigrants and their second-generation offspring for earlier periods with the early or more recent patterns of segregation of Negroes, we have only the reports of observers. That such segregation was widespread both along ethnic and social class divisions, however, is quite clear from the literature (Handlin, 1959; Riis, 1957; Woods, 1902). While the Irish, Poles, Italians, Jews, and other groups each tended to form its own little ethnic enclave, ethnic differences often merged on the basis of both social class similarities and time of arrival. Even in the earlier reports (Osofsky, 1966; Riis, 1957), this distinguished the Negro from other ethnic groups. Negro residential areas tended to be ethnically distinctive and to represent a wider range of social class positions as a consequence of pervasive discrimination in housing.

Lieberson's analysis (1963) of residential segregation from 1910 to 1950 highlights the continuities in these patterns. First, he found high levels of residential segregation among all ethnic groups for all the cities studied with little evident difference between the "old" immigration (those who predominated prior to 1880) and the "new" immigration (those who predominated between 1880 and 1920). They were quite uniformly high. Second, over time there was a gradual dispersion but as recently as 1950, patterns of residential segregation by ethnic group were still evident. Finally, depending on time of arrival, there is a gradual dispersion of ethnic groups both in diminished rates of segregation and in movement out of the central city. While there is some evidence in these and in other data of similar patterns among the urban Negro population, rates of residential segregation are consistently higher than for any of the immigrant ethnic groups, rates of change over time are less clear, and associations between residential dispersion and social class achievements are less marked (Lieberson, 1963). Thus, although the conclusions from the analysis of social mobility suggest that Negro rates of achievement may not be so drastically different from those of other earlier ethnic minorities, the patterns of housing segregation point to far more severe discrimination as an index of opportunity for social assimilation.

The more recent analyses by Taeuber and Taeuber (1965:36) both confirm these findings for the Negro and clarify further the patterns of segregation that operate. As they point out,

> In the urban United States, there is a very high degree of segregation of the residences of whites and Negroes. This is true for cities in all regions of the country and for all types of cities large and small, industrial and

commercial, metropolitan and suburban. It is true whether there are hundreds of thousands of Negro residents, or only a few thousand. Residential segregation prevails regardless of the relative economic status of the white and Negro residents. It occurs regardless of the character of local laws and policies, and regardless of the extent of other forms of segregation or discrimination.

It is quite notable that, in addition to initially high levels of segregation between 1940 and 1950 the degree of segregation increased and these increases were fairly evenly spread throughout the country. Indeed, residential segregation of the Negro was generally greater than for other of the most recent urban migrants, the Puerto Rican or Mexican-American populations. However, between 1950-1960, there was a general decrease in levels of housing segregation of Negroes in cities outside the South. As Table 3 (adapted from Taeuber and Taeuber, 1965: Table 4) indicates, the great majority of cities in the Northeast, North Central, and Western regions experienced decreases in residential segregation of Negroes between 1950 and 1960. In the South, on the other hand, the vast majority of cities continued to show increased rates of residential segregation for Negroes.

TABLE 3

CHANGES IN INDICES OF HOUSING SEGREGATION 1940-1950 AND 1950-1960 FOR 109 CITIES, BY REGION

Area	1940-1950	1950-1960
Northeast		
Decreased Segregation	10	23
Increased Segregation	15	2
North Central		
Decreased Segregation	7	21
Increased Segregation	22	8
West		
Decreased Segregation	4	10
Increased Segregation	6	0
South		
Decreased Segregation	5	10
Increased Segregation	40	35

SOURCE: Adapted from Taeuber and Taeuber (1965: Table 4).

What emerges from these data is already clear. Regardless of the high and remarkably persistent forms of residential segregation of ethnic groups, many of which had established residential patterns in these same cities more than one hundred years ago, the Negro has consistently suffered more severe discrimination in housing and has been forced into more pervasive forms of residential segregation. During the decade 1950-1960, there was a minor improvement in spite of the continued out-migration of whites to the suburbs, a factor that tends to increase levels of segregation although it often makes available better housing for Negroes. This improvement, however, did not occur in the South but was limited to cities in the other regions of the country. But even where there was a decrease in segregation, the levels of segregation remained extremely, unconscionably high.

Thus, even if one concludes from the analysis of social mobility that Negro achievements have been comparable to those of the white immigrants from Europe, educational, occupational, and income improvements have not markedly diminished the manifestation of discrimination in housing segregation. And while this is only one of many forms of discrimination, it is one of the better indicators of opportunities for social assimilation. Thus, social mobility for the Negro has not led to commensurate or even reasonably modest changes in equality of opportunity for integrated housing. While high levels of segregation in housing have existed for other ethnic groups, these appear to have been both less severe in general and more responsive to changes in social class position than they have been for the Negro.

Conclusion

It has become conventional to point out the great gap between the achievements of the European immigrants who peopled the United States during more than a century between the end of the Napoleonic wars and the first major restrictions on immigration in 1924 and the failure of achievement, on the other hand, among Negroes whose major entrance into the urban, industrial environment started around 1900 and continues apace. In the course of reviewing the extensive, albeit inadequate, data concerning social mobility and assimilation among both the European immigrants to the cities of this country and the Negro migrants from the South to the industrialized areas of the North and West, we have been forced to challenge this conclusion and to reconsider the implications of the melting pot ideology.

It is difficult to draw unambiguous conclusions about the situation of either the white or the black migrant. But the data appear to provide greater support for a reinterpretation than for the conventional conception of migratory movements and social assimilation in the United States. We have tried to show that the European immigrant most often experienced the transition from rural, preindustrial areas and largely agricultural occupations in Southern, Eastern and Central Europe to the cities and low-status manual occupations in the United States as an extremely painful, difficult, and threatening process. Immigrants left and continued to leave their native homes because they could look forward to nothing but misery and a deteriorating economic and social situation. The United States offered, at the very least, the possiblity of jobs and, no matter how low the wage rates, a higher income than they had previously known. Most immigrants remained in this country despite the long and often bitter struggle to maintain jobs through recurrent depressions, to retain a semblance of self-respect in the face of constant derogation of their abilities, motivation, family relationships, and their assimilability to Western cultural values. Although segregated housing kept them residentially separated, they were constantly accused of separatism and ethnic clannishness.

Although the second generation experienced a less restrictive environment, it was far from the myth of unlimited opportunities for social mobility and assimilation. It is difficult to determine precisely how long it took and under what conditions it was possible to gain a reasonable approximation to equality of status. But for many ethnic groups, full residential assimilation has not yet been realized as late as 1950.[19] The relatively rapid social mobility of the English, Jewish, and Japanese immigrant must, in this view, be treated as one extreme of a continuum although the Jewish and Japanese immigrant suffered a fairly typical history of severe discrimination, restriction, and ostracism. Educational and occupational mobility and cultural parity were not enough to insure social acceptance. And for the largest proportion of the immigrants, educational and occupational mobility were extremely difficult and unduly costly achievements. Thus, the idea of the United States as a melting pot emerges as a mythical elaboration of a fragmentary truth and gives way to an image of widespread inequality, racist attitudes, and ethnic segregation as the dominant reality.

By contrast with contemporary views about former immigrants (but similar to former conceptions of these same European and Asiatic immigrants), the conventional view of the black population emphasizes extremely slow progress, low motivation for achievement, and an unwillingness to share in the responsibilities and concomitant rewards of an urban, industrial society. Even as searching an analysis of the personal and social background of the American Negro as Frazier's (1939) emphasized primarily the historically-determined limitations of the Negro in coping with the urban, industrial environment. Over the last few decades, a number of studies have delineated some of the other factors involved in the situation of the Negro American: the vast migration of Negroes from the pre industrial and racist constraints of the South to the cities of the North and West and the overwhelming impact of discrimination and inequality on occupational and economic achievement among Negroes (e.g., Blau and Duncan, 1967; Fein, 1965a; Newman, 1965b). At the same time, the evidence for quite marked and rapid improvements in educational, occupational, and economic achievement of Negroes during the last few decades has further eroded the conventional view of Negro social immobility and low motivation for achievement (Blau and Duncan, 1967; Hardin, 1967; Hare, 1965; Newman, 1965a; Russell, 1966; Spady, 1967). Indeed, one might well argue that rates of social mobility among Negroes have been remarkably high in view of the inadequate preparation of the rural Southern Negro population who form a very large proportion of contemporary urban Negro Americans, the manifest inequality of opportunity, and the potency of a heritage and current experience of discrimination.

That the Negro is still far from educational or occupational or economic equality with the white population is, of course, clear enough (Blau and Duncan, 1965, 1967; Fein, 1965a; Newman, 1965b). This is revealed both in cross-sectional comparisons with whites and in discrepancies between educational and occupational achievements or between occupational levels and income, discrepancies which are among the more objective stigmata of discrimination. Indeed, similar discrepancies persist for other minorities at higher levels of the status system (Schmid and Nobbe, 1965). It is thus apparent that intense discrimination continues to operate and is not obliterated by social mobility. Not only is each and every advance made slowly and arduously against considerable resistance and almost certainly at considerable cost, but these advances carry with them only a modest part of the rewards that might be expected.

In spite of a real discrepancy in the status achievements associated with these two great rural-to-urban migrations and the more striking discrepancies in other forms of discrimination and inequality, the similarities are considerable and portentous: a modern history of servile status and recent emancipation which created an opportunity to migrate more readily than it provided a basis for economic or social freedom at home; rural origins in preindustrial communities; and the absence of any grounds for hope.[20] The meager opportunities in the industrialized cities of the United States were, thus, a marked contrast with the economic and social vacuum that stretched out before them at home. The struggle for education against overwhelming odds became a most important channel of mobility; and the episodic gains during prosperity were never wholly destroyed by periodic depressions and recessions. But discrimination, segregation, and social rejection were omnipresent for both. Massive migrations like these stem from conditions of dire deprivation. Compared to the resident population of the host society, they eventuate in new forms of deprivation and underprivilege.

But to point up the similarities is far from obliterating the differences. While it is possible to speak of the worst features of the immigrant experience in the past, the realities of the Negro experience are everpresent. Certainly the relative deprivation and the potency of discrimination is far greater for the Negro than that experienced by most of the former immigrant populations. It is, moreover, not at all clear that these differences are due to a change in the economic capacity of our society to absorb newcomers (National Advisory Commission on Civil Disorders, 1968).[21] Nor is there adequate evidence that the majority of the European migrants were any better prepared to deal with the demands of an urban, industrial society than the more recent Negro migrants. While the demands of the economy have undoubtedly changed and a higher degree of skill is required to fill available job opportunities than was the case fifty to one hundred years ago, there is no basis for assuming that rapid upgrading of occupational skills is not entirely feasible. The fact remains, moreover, that there *is* a marked discrepancy between education and occupation and between occupation and income among Negroes which indicates that opportunities are disproportionately *low* relative to preparedness. Thus, we can only attribute the residual problem of Negro achievement to the severity of discrimination.

The differences among European and Asian immigrant groups and the differences between these immigrants and the situation of the Negro provide a basis for subtle analyses of the conditions and processes of rural-to-urban transition. But the similarities point up certain pervasive and underlying characteristics of our society. Superimposed upon striking social class distinctions which continue to function in spite of moderately high rates of social mobility, there is a profound rejection of ethnic, cultural, and color differences in the United States. Despite the importance of social class as a primary dimension, however, and despite the marked differences in immigration experiences and social acceptance associated with social class variations, we cannot wholly subsume other factors within this one basis for social categorization. Rather, if we use the term racism broadly to connote any sharp and pervasive discriminatory behavior toward visible and distinguishable ethnic or cultural groups, a widespread history of racism has marked the trail of the great migrations to the cities of the United States for a century and a half. The Negro is not only among the most recent but the most visible of these minorities and has had the most severely incapacitating history of previous discrimination and caste limitation. These have served further to encourage and rationalize the fundamental inequalities of our society in their manifestation toward the Negro American.

In a different era, when foreign-born immigrants and their children were often viewed as expendable, it was possible to disregard their desperate needs for support in facilitating social mobility and assimilation. Under different economic conditions, when the economy was a captive of the business cycle and its operation less subject to deliberate manipulation, it was more difficult to create jobs, educational opportunities, or other resources for encouraging rapid change. And in an environment in which demands for equality and opportunity were more impetuously violent, more fractionated along ethnic or occupational lines, less broadly goal-directed, and a democratic ideology more limited in conception, it was possible to accept police suppression and the power of the national guard or of the armed forces as an effective means of eliminating a problem by eliminating its manifest expression. All of this has changed. We have an unparalleled potential to create a situation in which rates of achievement among the Negro population can more nearly approximate overt aspirations, needs, and demands.[22] Yet we remain relatively paralyzed in our focus on short-term goals and in our concern with such symptoms as riots rather than with those features of inequality of opportunity that lie beneath these symptoms.

In the deepest sense, our society must undergo radical institutional change in order to eliminate widespread racism and the ready rejection of ethnic and cultural differences. Our society is certainly not unique in its resistance to accepting and integrating great diversity within the province of legitimacy. But one of the cultural consequences of the transition from preindustrial to industrial societies is precisely an increased possibility of achieving an open society. Few other nations are in so ideal a situation for realizing this potentiality. And few other major industrial powers are in so desperate a position of choice between recurrent violence, suppressive blindness, or universal democracy. Unless we appreciate the fundamental importance of these long-term goals that the rhetoric of the great society proposes, more modest goals are likely to be too slow and too meager to offer much hope.

At a more modest level but, nonetheless, one requiring more drastic change than our society appears willing to initiate, far-reaching economic and social legislation could have a major impact. Low status groups have had a long and quite varied history of misery and ostracization (Zaniewski, 1957). While the greatest depths of poverty have been curtailed in our society and a much larger proportion of the population participates in affluence, the economic gap between the lowest and highest income groups remains enormous and has changed little since 1940 and perhaps since 1910 (Kolko, 1962; Miller, 1966). Clearly, in view of the disproportionate number of Negroes in the lowest income groups, a diminution of this discrepancy would most drastically affect the Negro population but, at the same time, would go far in eliminating the marked inequalities of economic status in the country as a whole. Proposals for a guaranteed income move in this direction but, in view of the large gaps in income at the low-intermediate levels, they do not go far enough. A more basic alteration in the entire tax structure appears to be the only solution that is pervasive enough to create a degree of economic equality commensurate with the dream of a great society.

At the same time, in light of the diverse components of deprivation, new forms of social legislation are imperative to provide more adequately for the aged, the ill-housed, the jobless, and the indigent. Each and every form of deprivation more seriously affects the Negro, and, thus, far-reaching economic and social changes most directly benefit the Negro. But it is only by viewing these problems as societal problems that we can hope to achieve the kind of society which is capable of dealing not only with the past but with the present and future. While the Negro migration from the South has begun to diminish, it is far from attrition. Thus, to the extent that the situation of the Negro is sustained by continued migration from the more severe inequalities of the South, policy must be oriented to continued efforts to maximize opportunities and equalities in the industrialized areas of the North and West. At the same time, only major changes at the national level are likely to reduce the severity of Negro underprivilege in the South.

It is paradoxical that, quite often, it is the deprived, the underprivileged, and the alienated who most poignantly demonstrate the major inadequacies of a society. There are many respects in which the Negro social revolution of the last few decades has already pointed up central issues: the pervasiveness of discrimination, the overt and covert forms of inequality of opportunity, the changing political structure which requires greater influence from local communities, the inadequacies of our educational system except in its most privileged sectors, the need for recasting our status system and diminishing its consequences, the desperate necessity for a more searching and egalitarian ethical consciousness in our social behavior. These are problems which affect not only the black person or the individual from other minority groups in our society, but it is the black person and those from other recent minority groups who suffer most severely from these societal failures.

It is an easy escape to define the problem as if its source lay in those who experience the problem most directly and to deal with the "Negro problem" or the "poverty problem" or even the "urban problem." These, however, are simply symptomatic or localized expressions of broader problems of our society. Only by shifting from transitory conceptions to longterm change, from local or focal issues to pervasive difficulties, from an expectation of perpetual progress to an appreciation of the sporadic nature of gains and the return of periodic failures can we hope to achieve a reasonable solution even to immediate and pressing problems. These problems, symptoms though they be, are expressed most strikingly in the poverty and inequality which Negro Americans experience despite a century of individual achievements and social change. But the realistic resolution of these specific and perhaps temporary "problems" requires that our analyses and solutions transcend them and deal with the underlying injustice and restrictions that continue to characterize our society.

NOTES

1. Some of the more thorough reports concerning these problems in the United States can be found in Commons (1920); Davie (1936); Ernst (1949); Fairchild (1927); Handlin (1941); Handlin (1942); Park (1922); Park and Miller (1921); Smith (1939); Stephenson (1926); Warner and Srole (1945); and Wirth (1945).

2. For some of these contrasts, see: Eisenstadt (1954) and Weinberg (1961) concerning Israel; Marris (1961) about Nigeria; Redford (1926) concerning England during the Industrial Revolution; and Taft (1966) dealing with in-migration to Australia.

3. Estimates based on tables in Brinley Thomas, 1954.

4. The most systematic data and the most sophisticated analyses in support of this view are collected by Brinley Thomas (1954). The earlier study of migration and business cycles by Jerome (1926) comes to an opposite view but the data he provides are clearly inadequate to the burden he places on them. In particular, he tries to evaluate the conditions of emigration without adequate emigration statistics (using emigration to the United States rather than total emigration from the country of origin), without an adequate evaluation of the complex factors associated with economic expansion and deterioration other than the depression-prosperity cycle, and with far more adequate indices of these limited variables for the United States than for the countries of origin.

5. Jerome's (1926) analysis does try to take account of both countries of origin and countries of destination but does not consider migration from a country of origin to any place other than the United States. As a consequence, the relative importance of economic changes in the United States is bound to weigh heavily in his statistical analysis. By contrast, Thomas (1954) more systematically considers the process of emigration from the vantage point of the country of origin and immigration from the vantage point of the country of destination and arrives at a more complex conclusion about the conditions under which either set of forces tends to predominate as a determinant of migration.

6. Also see Goodrich et al. (1935); for similar findings in France, see Chevalier (1950).

7. Thomas (1954: Tables 80-84). The amount of distortion in these data is unknown and one may well anticipate some upgrading of occupations generally and the substitution of urban work categories for farmer in anticipation of occupational alterations in the United States.

8. The findings of Taeuber and Taeuber (1965) and of Blau and Duncan (1967) reveal that this does not hold for those migrants who are of relatively high status and predominantly migrate from one city to another. But the disadvantage of migrants does hold, in these more recent data on internal migration within the United States, for migrants from nonmetropolitan areas (Taeuber and Taeuber) and for migrants from farm residence (Blau and Duncan).

9. For some striking differences in unemployment during the recession of 1967 which are associated with rural vs. urban background, see Fried, 1966.

10. While many controls were possible for this study (Blau and Duncan, 1967) which could not be done with earlier data, they do not take account of changes in the occupational structure that have increased the proportion of higher status occupational positions and, thus, necessarily produce a pattern of social mobility which is built into the process of social change. Moreover, even assuming that the occupational distribution approximates an interval scale and, therefore, that occupational changes at both ends of the distribution truly represent similar degrees of mobility, an assumption to be verified, the greater variance in the occupational distributions of immigrants than of natives in their sample suggests that the rates of occupational mobility for immigrants can be more seriously affected than those of natives by differentials in occupational mobility at different levels of the occupational scale.

11. Thomas' (1954) data and some of Lieberson's (1963) findings suggest that many migrants enter the host society at a lower level than that of their former occupations. This tends to confound any simple analysis of social mobility, since,

either for intra-generational or inter-generational mobility, the distance from the lower initial status of the immigrant will exaggerate mobility achievement compared with the distance from the prior, preimmigration status.

12. See Hingham (1955) for a description of some of the current waves of anti-Semitism which led to many of the quota systems, residential restrictions, and recreational exclusions during the antiforeign outbreaks of the 1920's and have begun to diminish only during the past few decades.

13. Taeuber and Taeuber (1958), give somewhat divergent figures suggesting a more marked decline in the Negro population of the South between 1890 and 1910.

14. Redford (1964) demonstrates this phenomenon for Britain in the nineteenth century.

15. The classical study which postulated the greater inadequacy of migrants compared to nonmigrants as the critical selective factor leading to high rates of mental disorder among migrants is that of Odegaard (1932). For more recent reinterpretations of migration and mental illness, see Murphy (1965) and Fried (1964).

16. The cities from different regions for which the more detailed analyses are presented are South: Atlanta, Birmingham, Memphis, New Orleans; Border: Baltimore, St. Louis, Washington; North: Cleveland, Detroit, Philadelphia.

17. Whether these rates are viewed as high or low is dependent on (a) subjective expectations or (b) the application of a lineal model of mobility to a criterion population (e.g., native white Americans). Both of these bases for evaluation are, at best, inadequate for a clear understanding of complex mobility patterns.

18. Blau and Duncan's report (1965) that occupational mobility is similar among native-born whites and foreign-born whites and in contrast with nonwhites does not substantially affect this conclusion. A very large proportion of the foreign-born whites who were still in the labor force in 1962, the year in which this data were collected, represented a wholly different migration from the European immigration of 1830-1920. Not only did it necessarily have a different ethnic composition from the earlier migrations as a result of the immigration quotas, but the countries from which the immigrants came and the conditions under which many of them migrated involved much greater experience with urban, industrial society. In comparing the earlier immigrants with the Negro, we have been trying to assess the differences among groups with similar origins (either by birth or parentage) in rural, agricultural societies.

19. Schmid and Nobbe (1965) present interesting data for several nonwhite ethnic groups. The Chinese situation is of particular significance because Chinese immigration was more forcibly cut off (in 1882) than that of any other nationality. For the Chinese, whose immigration occurred largely between 1830 and 1882, meaningful equivalence to white American statuses in education, occupation, and income was not established until 1930-1940, one hundred years after the beginning and fifty years after the end of large-scale immigration.

20. It is easy to forget the fact that, for many of the countries most markedly affected by massive migrations, serfdom was abolished in the nineteenth century: Italy, Germany, Russia, Poland, Austria. While the Jews and Irish were not, literally, serfs, their servile status and the constraints imposed on their lives were perhaps even more severe.

21. See Thernstrom (1964) for a discussion of the "blocked mobility" hypothesis and the evidence that overall rates of upward social mobility from the lowest ranks have probably increased over time.

22. Lieberson and Fuguitt (1967) point out that, in the absence of discrimination, two generations would bring about a high level of parity in the status of Negroes and whites. In view of the high levels of aspiration and motivation among Negroes, appropriate policy might conceive this as the lowest possible limit.

REFERENCES

ALLPORT, G. W., J. S. BRUNER and E. M. JAHNDORF (1941) "Personality under social catastrophe." Character and Personality 10: 1-22.
ANTIN, M. (1921) The Promised Land. Boston: Houghton Mifflin.
ARENSBERG, C. M. and S. T. KIMBALL (1948) Family and Community in Ireland. Cambridge: Harvard University Press.
BALCH, E. G. (1910) Our Slavic Fellow Citizens. New York: Charities Publications Committee.
BERTHOFF, R. T. (1953) British Immigrants in Industrial America, 1790-1950. Cambridge: Harvard University Press.
BLAU, P. M. and O. D. DUNCAN (1967) The American Occupational Structure. New York: John Wiley.
--- (1965) " Some preliminary findings on social stratification in the United States." Acta Sociologica 9: 2-24.
BRODY, D. (1960) Steelworkers in America: the Nonunion Era. Cambridge: Harvard University Press.
CARPENTER, N. (1927) Immigrants and Their Children, 1920. Washington, D. C.: United States Government Printing Office.
CHEVALIER, L. (1950) La Formation de la Population Parisienne au XIXe Siecle. Paris: Presses Universitaires de France.
COMMONS, J. R. (1920) Races and Immigrants in America. New York: Macmillan.
DAVIE, M. R. (1936) World Immigration. New York: Macmillan.
DAY, R. (1967) "The economics of technological change and the demise of the sharecropper." American Economics Review 57: 427-449.
DRAKE, S. C. and H. R. CAYTON (1945) Black Metropolis: A Study of Negro Life in a Northern City. New York: Harcourt Brace.
DUBOIS, W. E. B. (1965) The Souls of Black Folk: Three Negro Classics. New York: Avon Library.
DUNCAN, O. D. and B. DUNCAN (1957) The Negro Population of Chicago: A Study of Racial Succession. Chicago: University of Chicago Press.
EISENSTADT, S. N. (1954) The Absorption of Immigrants. London: Routledge and Kegan Paul.
ERNST, R. (1949) Immigrant Life in New York City, 1826-1863. New York: King's Crown Press.
FAIRCHILD, H. P. [ed.] (1927) Immigrant Backgrounds. New York: John Wiley.

FEIN, R. (1965a) "An economic and social profile of the American Negro." Daedalus (Fall).
--- (1965b) "Educational patterns in southern migration." Southern Economics Journal 32: 106-124.
FOERSTER, R. F. (1919) The Italian Emigration of Our Times. Cambridge: Harvard University Press.
FRAZIER, E. F. (1939) The Negro Family in the United States. Chicago: University of Chicago Press.
FRIED, M. (1966) "The role of work in a mobile society." In S. B. Warner, Jr. (ed.) Planning for a Nation of Cities. Cambridge: MIT Press.
--- (1964) "Effects of social change on mental health." American Journal of Orthopsychiatry 34: 3-28.
---(1963) "Grieving for a lost home." in L. J. Duhl (ed.) The Urban Condition. New York: Basic Books.
GLAZER, N. and D. P. MOYNIHAN (1963) Beyond the Melting Pot: The Negroes, Puerto Ricans, Jews, Italians and Irish of New York City. Cambridge: Harvard and MIT Press.
GOLDSTEIN, S. (1955) "Migration and occupational mobility in Morristown, Pennsylvania." American Sociological Review 20: 402-448.
GOODRICH, C. A., W. BUSHROD and M. HAYES (1935) Migration and Planes of Living, 1920-1934. Philadelphia: University of Pennsylvania Press.
HAMILTON, C. H. (1965) "Educational selectivity of migration from farm to urban and to other nonfarm communities." In M. Kantor (ed.) Mobility and Mental Health. Springfield, Ill.: Charles C. Thomas.
HANDLIN, O. (l959a) Boston's Immigrants: A Study in Acculturation. Cambridge: Harvard University Press.
--- (1959b) The Newcomers: Negroes and Puerto Ricans in a Changing Metropolis. New York: Doubleday.
--- (1951) The Uprooted. Boston: Little, Brown.
HANSEN, M. L. (1940) The Atlantic Migration, 1607-1860. New York: Harper.
HARDIN, E. (1967) "Full employment and workers' education." Monthly Labor Review 90: 21-25.
HARE, N. (1965) "Recent trends in the occupational mobility of Negroes, 1930-1960: an intracohort analysis." Social Forces 44: 166-173.
HARTMAN, C. (1964) "The housing of relocated families." Journal of the American Institute of Planners 30: 266-286.
HAUSER, P. M. (1965) "Demographic factors in the integration of the Negro." Daedalus (Fall).
HILL, J. A. (1924) "Recent northern migration of the Negro." Monthly Labor Review 18: 1-14.
HINGHAM, J. (1963) Strangers in the Land: Patterns of American Nativism, 1860-1925. New York: Atheneum.
HUTCHINSON, E. P. (1956) Immigrants and Their Children, 1850-1950. New York: John Wiley.
JEROME, H. (1926) Migration and Business Cycles. New York: National Bureau of Economic Research (Publication No. 9).
JOHNSTON, D. F. and H. R. Hamel (1966) "Educational attainment of workers in March, 1965." Monthly Labor Review 89: 250-257.
JONASSEN, C. T. (1949) "Cultural variables in the ecology of an ethnic group." American Sociological Review 14: 32-41.

JOSEPH, S. (1914) Jewish Immigration to the United States: From 1881 to 1910. New York: Columbia University Press.

KOLKO, G. (1962) Wealth and Power in America. New York: Frederick A. Praeger.

KUZNETS, S. and D. S. Thomas, et al. (1964) Population Redistribution and Economic Growth: United States, 1870-1950, Vol. III. Philadelphia: American Philosophical Society.

LASSITER, R. L. (1965) "The association of income and education for males by region, race and age." Southern Economics Journal 32: 15-22.

LEVINE, E. M. (1966) The Irish and Irish Politicians. Notre Dame: University of Notre Dame Press.

LEYBURN, G. (1937) "Urban adjustments of migrants from the southern Appalachian plateaus." Social Forces 16: 238-246.

LIEBERSON, S. (1963) Ethnic Patterns in American Cities. New York: Free Press.

LIEBERSON, S. and G. V. FUGUITT (1967) "Negro-white occupational differences in the absence of discrimination." American Journal of Sociology 73: 188-200.

LIPSET, S. and R. BENDIX (1959) Social Mobility in Industrial Society. Berkeley: University of California Press.

LOWRY, I. S. (1966) Migration and Metropolitan Growth: Two Analytic Models. San Francisco: Chandler Publishing.

MARRIS, P. (1961) Family and Social Change in an African City. London: Routledge & Kegan Paul.

MAULDIN, W. P. (1940) "Selective migration from small towns." American Sociological Review 5: 748-766.

MILLER, H. P. (1966) Income Distribution in the United States. Washington, D.C.: Government Printing Office.

MILLER, S. M. (1960) "Comparative social mobility." Current Sociology 9: Chapter 1.

MURPHY, H. B. M. (1965) "Migration and the major mental disorders." In M. Kantor (ed.) Mobility and Mental Health. Springfield, Ill.: Charles C. Thomas.

MYERS, G. C. (1967) "Migration and modernization: the case of Puerto Rico, 1950-1960." Sociological and Economic Studies 16: 425-431.

MYRDAL, G. (1962) Challenge to Affluence. New York: Pantheon.

National Advisory Commission on Civil Disorders (1968) Report of the National Advisory Commission on Civil Disorders. New York: E. P. Dutton.

NEWMAN, D. K. (1965a) "The Negro's Journey to the city—part I" Monthly Labor Review 88: 502-507.

––– (1965b) "The Negro's journey to the city—part II" Monthly Labor Review 88: 644-649.

ØDEGAARD, Ø. (1932) "Emigration and insanity: a study of mental disease among the Norwegian born population of Minnesota." Acta Psychiatrica et Neurologica, Supplement 4.

OSOFSKY, G. (1966) Harlem: The Making of a Ghetto. New York: Harper & Row.

PARK, R. E. (1922) The Immigrant Press and Its Control. New York: Harper.

PARK, R. E. and H. A. Miller (1921) Old World Traits Transplanted. New York: Harper.

POTTER, G. (1960) To the Golden Door: The Study of the Irish in Ireland and America. Boston: Little, Brown.

REDFORD, A. (1964) Labour Migration in England, 1800-1850. Manchester: Manchester University Press.

RIIS, J. A. (1957) How the Other Half Lives. New York: Sagamore Press.
RUSSELL, J. L. (1966) "Changing patterns of employment of nonwhite workers." Monthly Labor Review 89: 503-509.
SCHMID, C. F. and C. E. NOBBE (1965) "Socioeconomic differentials among nonwhite races." American Sociological Review 30: 909-922.
SHANNON, L. and P. MORGAN (1966) "The prediction of economic absorption and cultural integration among Mexican Americans, Negroes and Anglos in a northern industrial community." Human Organization 25: 154-162.
SHRYOCK, H. S., Jr. (1964) Population Mobility Within the United States. Chicago: Community and Family Study Center.
SMITH, W. C. (1939) Americans in the Making. New York: Appleton-Century.
SPADY, W. G. (1967) "Educational mobility and access: growth and paradoxes." American Journal of Sociology 73: 273-286.
STEPHENSON, G. M. (1926) A History of American Immigration, 1820-1924. Boston: McGinn.
TAEUBER, C. and I. B. TAEUBER (1958) The Changing Population of the United States. New York: John Wiley.
TAEUBER, K. E. and A. F. TAEUBER (1965) Negroes in Cities: Residential Segregation and Neighborhood Change. Chicago: Aldine Publishing.
--- (1964) "The changing character of Negro migration." American Journal of Sociology 70: 429-441.
TAFT, R. (1966) From Stranger to Citizen. Nedlands: University of Western Australia Press.
THERNSTROM, S. (1964) Poverty and Progress: Social Mobility in a Nineteenth Century City. Cambridge: Harvard Unversity Press.
THOMAS, B. (1954) Migration and Economic Growth: A Study of Great Britain and the Atlantic Economy. Cambridge University Press.
THOMAS, W. and F. ZNANIECKI (1918) The Polish Peasant in Europe and America. Chicago: University of Chicago Press.
TOBIN, J. (1965) "On Improving the economic status of the Negro." Daedalus (Fall).
TOLLES, N. A. (1937) "Survey of labor migration between states." Monthly Labor Review 45: 3-16.
U. S. Bureau of the Census (1967) Social and Economic Conditions of Negroes in the United States. Current Population Reports (October).
U. S. Immigration Commission (1911) Reports of the Immigration Commission. Washington, D. C.: Government Printing Office.
VANCE, R. (1938) Research Memorandum on Population Redistribution Within the United States. New York: Social Science Research Council (Bulletin 42).
WADE, R. C. (1959) The Urban Frontier: The Rise of Western Cities 1790-1830. Cambridge: Harvard University Press.
WALKER, M. (1964) Germany and the Emigration 1816-1885. Cambridge: Harvard University Press.
WARNER, W. L. and L. SROLE (1945) The Social Systems of American Ethnic Groups. New Haven: Yale University Press.
WATTS, L. G. et al. (1964) The Middle-Income Negro Family Faces Urban Renewal. Waltham (Mass.): Research Center of the Florence Heller School, Brandeis University.
WEINBERG, A. A. (1961) Migration and Belonging: A Study of Mental Health and Personal Adjustment in Israel. The Hague: Martinus Nijhoff.

WILLIAMS, P. H. (1938) South Italian Folkways in Europe and America. New Haven: Yale University Press.
WIRTH, L. (1945) "The Problem of Minority Groups." In R. Linton (ed.) The Science of Man in the World Crisis. New York: Columbia University Press.
WOODHAM-SMITH, C. (1962) The Great Hunger: Ireland 1845-1849. New York: Harper & Row.
WOOFTER, T. J. (1920) Negro Migration: Changes in Rural Organization and Population of the Cotton Belt. New York: W. D. Gray.
WOODS, R. A. (1902) Americans in Process. Boston: Houghton Mifflin.
YOUNG, M. and P. WILMOTT (1957) Family and Kinship in East London. London: Routledge & Kegan Paul.
ZANIEWSKI, R. (1957) L'Origine du Proletariat Romain et Contemporain. Louvain: Editions Nauwelaerts.

Chapter 3

Involuntary International Migration: *Adaptation of Refugees*

HENRY P. DAVID

The saga of refugees is as old as man's history. From biblical times to modern day, involuntary migration has accompanied the collapse of old societies and the development of new cultures. Refugees have been caught in the turbulent wake of religious and political persecutions, economic convulsions, and demographic upheavals. It is the purpose of this chapter to survey briefly the world refugee situation, note trends in recent immigration to the United States, discuss the dynamics of involuntary migration, summarize reports of adjustment after migration to the United States, Australia, Netherlands, and Israel, and consider implications for intervention. Over 100 selected references are provided. A list of intergovernmental, governmental and nongovernmental voluntary organizations, plus descriptions of their activities, are presented in an appendix on "Resources for Refugees."

World Refugee Situation

At the beginning of 1968 more than seven million people of all races and creeds were considered international refugees. Of these only four

---

Author's Note: *This paper was prepared for the Conference on "Migration and Behavior Deviance," convened by the National Institute of Mental Health at Dorado Beach, Puerto Rico, November 4-9, 1968.*

TABLE 1
TOTAL OF SETTLED AND NON-SETTLED REFUGEES
BY CONTINENTS

Continents	Total Number of Refugees	Total of Settled Refugees	Total of Non-Settled Refugees
Africa	842,550	450,000	392,550
Asia: Far East	2,468,000	767,800	1,700,200
Near East	1,095,000	–	1,095,000
Australia	303,300	303,300	–
North America	1,323,000	1,323,000	–
Latin America	115,000	105,000	10,000
Europe	1,057,050	987,650	69,400
Grand Total	7,200,000	4,000,000	3,200,000
	(7,203,900)	(3,936,750)	(3,267,150)

SOURCE: From the March-April, 1968, issue of *Migration News*, by permission of the publisher, the International Migration Commission (65 Rue de Lausanne, Geneva, Switzerland).

million were deemed more or less permanently resettled. Table 1 shows their numerical distribution in the major geographic regions of the world.

There has been and continues to be much confusion surrounding the term "international refugee." It is generally applied to an individual who has left his homeland under certain pressures that may be political, social, economic, or religious in nature (Wenk, 1968: 62-69). Whether such flight is deemed voluntary or involuntary is often a matter of personal perception.

A refugee may also be defined in terms of the revised 1951 U. N. Convention Relating to the Status of Refugees. It is stated in Article 1 (2) that a refugee is

> an individual who owing to well-founded fear of being persecuted for reasons of race, religion, nationality, membership of a particular social group or political opinion, is outside the country of his nationality and is unable, or owing to such fear, unwilling to avail himself of the protection of that country; or, who, not having a nationality and being outside the country of his former habitual residence as a result of such events, is unable or, owing to such fear, is unwilling to return to it.

As of 1968, more than fifty nations had signed the United Nations Convention. The United States is not among the signatories although it has underwritten the largest share of the costs of intergovernmental relief organizations.

As evident from Table 1, Asia has the greatest number of international refugees and of nonsettled refugees. Of the nearly 3.5 million in Asia, 1.8 million are Chinese and 1 million are Arabs.

More than half the refugees in Africa were officially declared as settled. In contrast with Europe, the African refugee problem is often resolved through voluntary repatriation. Instead of establishing themselves permanently in the host country, many Africans prefer to return to their country of origin when conditions which caused their flight no longer exist.

Of those who chose Europe as a place of asylum after World War II, 70,000 are still benefiting from material, or legal, assistance provided by voluntary agencies. The figure is unlikely to decrease due to the constant influx of new refugees from Eastern European countries who generally require material aid and transportation assistance.

In Australia and North America (Canada and the United States) all refugees are now considered firmly settled. Of the 1,023,000 refugees in the United States, 684,000 are European, 288,000 Cuban, 6,000 Haitian, and 45,000 Asian. The United States has become a country of first asylum, accepting more than 4,000 refugees each month direct from Cuba.

It should be noted that the concept of *international* refugee does not cover those individuals who have been displaced in their own country and are seeking refuge elsewhere within its traditional borders, as for example in such presently divided nations as Korea, Vietnam, and Nigeria. Accepting a less rigorous definition of refugee, the United States Committee for Refugees (1968) estimates that in 1967 more than 15 million persons lived as refugees in eighty counties. The vast majority are in the uprooted areas of Asia, Africa, and the Middle East. This total exceeds the combined population of the Northeast States plus North Dakota, South Dakota, Nebraska, and Kansas. Indeed, there were more refugees in 1968 than during World Refugee Year in 1958.

At year end 1968 the Intergovernmental Committee on European Migration reported that it had assisted 4,500 Czechoslovaks to leave Europe. ICEM forecasts further movement of 10,000 persons in 1969. Another refugee problem concerns the nearly 800 Cubans arriving in Spain each month, hoping eventually to enter the United States. An additional 16,000 have been issued visas and are awaiting transportation to Spain.

TABLE 2
NUMBER AND CHARACTERISTICS OF IMMIGRANT ALIENS ADMITTED TO THE UNITED STATES, 1957 TO 1967

Percent of Total Admissions During Year

Year Ending June 30	Number of Admissions	Male	Ages Under 18	Ages 18-49	Ages 50 and Over	Canada	Mexico	West Indies	Other America	British Isles	Italy	Other Europe	Asia	Other
1957	326,867	47.5	27.5	64.9	7.6	14.2	15.1	5.6	6.2	9.9	6.0	36.0	6.1	0.9
1958	253,265	43.1	26.9	65.0	8.1	17.8	10.6	6.7	9.6	13.3	9.1	23.1	8.2	1.6
1959	260,686	43.9	25.6	64.8	9.6	13.3	8.8	4.6	9.0	9.6	6.5	36.9	9.7	1.6
1960	265,398	44.0	25.7	65.2	9.1	17.6	12.3	5.2	10.0	10.1	5.0	30.2	8.0	1.6
1961	271,344	44.7	26.9	63.9	9.2	17.5	15.3	7.6	11.1	9.0	7.0	23.9	7.2	1.4
1962	283,763	46.4	25.9	65.0	9.1	15.6	19.7	7.4	12.3	8.2	7.1	21.3	7.1	1.3
1963	306,260	45.5	27.0	64.7	8.3	16.5	18.3	7.5	13.2	9.3	5.3	21.0	7.6	1.3
1964	292,248	43.2	27.6	63.7	8.7	17.5	11.8	8.2	16.8	10.9	4.4	21.7	7.3	1.4
1965	296,697	42.9	28.0	63.2	8.8	16.9	13.7	10.5	16.6	9.9	3.7	20.6	6.7	1.4
1966	323,040	43.8	31.7	58.2	10.1	11.5	14.6	11.8	12.4	6.8	8.2	20.9	12.4	1.4
1967	361,972	43.7	30.5	57.9	11.6	9.6	11.9	17.1	8.4	7.1	7.9	20.6	15.9	1.5

SOURCE: Reports of United States Immigration and Naturalization Service, *Statistical Bulletin* (September, 1968).

Recent Immigration to the United States

How many Americans recall that an Irishman, a Jew, and a Negro were members of Christopher Columbus' first crews? Or that the Declaration of Independence protested England's obstruction of the American colonies' naturalization laws? Or that Thomas Jefferson's ringing phrase, "All men are created equal," was a paraphrase of a concept expressed by his friend Philip Mazzei, born in Italy? After 1776, 44 million immigrants came to the United States, perhaps the greatest migration in history. Many were "trash" to the kings and lords and tyrants they fled. "Foreigners" plowed the prairies, built the roads, dug the canals, laid the railroad tracks, and manned the factories (Rosten, 1968).

Millions reached American shores each year from 1900 to 1912. In 1921 Congress established the quota system, allowing only 357,000 annual entries with national restrictions that heavily penalized Eastern and Southern Europe and Asia. In 1924 Congress tightened the law, reducing the number to 164,000.

The influx of intellectual refugees to the United States in the years between 1930 and 1941, much of it the result of Hitler, has been movingly told in *Illustrious Immigrants* by Laura Fermi (1968). America received a whole "culture in exile," including writers, scientists, composers, painters and sculptors, architects, astronomers, conductors, mathematicians, sociologists, psychoanalysts, physicians, and atomic physicists. It was a turning point in American history. After World War II the Displaced Persons Act of 1948 permitted some leeway in admitting political refugees. In 1956, President Eisenhower widened the loopholes to allow entrance to nearly 38,000 Hungarian refugees.

The revised immigration acts of 1965 and 1966 have produced increasing and more diverse immigration into the United States. The number of immigrants entering this country has risen from an average of about 290,000 annually during the years ending June 30 of 1961-65 to 323,040 in 1966 and 361,972 in 1967. Current immigration is at the highest level since 1924. However, the influx was far greater in the first two decades of the twentieth century; between 1905 and 1914 there were six years when more than one million individuals entered the United States.

As indicated in Table 2, reproduced from the September, 1968 *Statistical Bulletin* of the Metropolitan Life Insurance Company, the majority of recent immigrants have been in the prime of life: in 1967 almost three-fifths were men and women at ages 18-49. Children under 18 years of age accounted for less than one-third; one-ninth were age 50 and over. Since 1957 more than two-fifths of the newcomers have been males.

Historically, individuals of European origin constituted the majority of immigrants to the United States. This proportion dropped from one-half a decade ago to little more than one-third in 1967. That year immigration from the United Kingdom made up 7.1 percent of the total immigrants, while East and West Germany combined added 4.6 percent. Immigration from Greece has increased from an average of 1.3 percent annually during 1961-65 to 3.9 percent in 1967; Portugal's contribution rose from 1.0 percent to 3.7 percent.

There have also been significant changes in the pattern of immigration from other countries on the American continent. The proportion from Canada has decreased from about one-sixth prior to 1966 to less than one-tenth in 1967. In that year the growing number of arrivals from the West Indies increased sharply to one-sixth of the total. These 1967 admissions included 26,000 Cuban refugees whose status in the United States was adjusted from parolee to lawful permanent resident. Such adjustments are expected to remain high as 4,000 Cubans arrive in Miami each month.

The less restrictive provisions of the new immigration acts have also resulted in more arrivals from Asia. The proportion of Asian immigrants rose from an annual average of 7.2 percent during 1961-65 to 15.9 percent in 1967. Although Hong Kong and the Philippines were the main points of origin, Taiwan, the People's Republic of China and India also contributed to the total.

In an October 21, 1968 *New York Times* story from Vienna, Tad Szulc reported that the United States is admitting "hundreds of Czechoslovaks, Poles, and other Eastern European intellectuals, artists, and professionals under simplified procedures that allow the immigration of 'involuntary' Communists." A special provision of the 1952 Immigration and Naturalization Act permits members, or former members, of the party to enter the United States provided they can convince the Department of Justice that they were in the party involuntarily. By the end of October 1,500 prospective migrants were expected to be processed for early entry into the United States, with more to follow if the exodus from Eastern Europe continues.

Dynamics of Involuntary Migration

Although migration has played a central role in the history of civilization and in the development of nations and continents, the social-psychological dynamics of the process of migration and resettlement have received relatively limited attention. Traditionally, international migration has been divided into voluntary or planned transplantation and involuntary or forced migration, comprising refugees, displaced persons, and forced laborers. However, voluntary migration is not always so voluntary. It is at times difficult to determine whether an individual emigrates of his own free will or because of anxiety over anticipated persecution which may, or may not, materialize. Psychologically, a voluntary migrant may be as much a refugee as an involuntary migrant (Weinberg, 1955). Migration represents an interruption and frustration of natural life expectations, with all the related anxieties and potential damage to the self-concept. Migration induces cognitive stress, forcing the immigrant to change his familiar images and build a new cognitive map (Bar-Yosef, 1968).

There is much evidence from clinical psychological and psychiatric observations that individuals in severe conflict situations, when facing seemingly insurmountable obstacles, tend to regress to former patterns of behavior. The ego's integrative capacity is under particular strain when an immigrant starts life in a new country, away from the familiar surroundings which had previously provided some protection or nurture. A mentally healthy person will create conditions to maintain his inner security by satisfying his need for belonging and acquiring the esteem of others. Those who latently or overtly are not in good mental health will be in need of special attention.

Studies of the mental health and/or adaptation problems of refugees have usually been concerned with prevalence, incidence, etiology, and symptomatology of mental disorders. Most frequently, researchers have attempted to investigate relationships between mental illness and the tendency to migrate, between displacement and flight and mental ill health, between immigrant's adjustment and the incidence of mental disorders, and between social class and deviant behavior. Others considered general social psychological problems, including selectivity of migrants; their sense of security or insecurity; relationships between the host society and the newcomers; intergroup tensions and resulting problems of anomie, marginality, and delinquency. These studies have been fully reviewed by Murphy (1955, 1961, 1965) and by Weinberg (1961). The diversity of reported observations on major mental disorders and the shifting theoretical positions are noted by Sanua in Chapter 13.

A decade ago, at the 1958 meeting of the World Federation for Mental Health in Vienna, Erikson (1960) discussed identity and uprootedness. He noted the universality of types of transmigration which, like all catastrophes and crises, produce different traumatic world images, and within them, seem to demand the sudden assumption of new and often transitory identities. Erikson considered the situational determinants: what had motivated and moved the transmigrant; how he had been excluded or had excluded himself from his previous home; how he had been transported or had chosen to traverse the distance between home and destination; and how he had been kept or had kept himself separate, had been absorbed or had involved himself in his new setting. These determinants, Erikson noted, do not account for those inner mechanisms which permit man to maintain and regain in this world of contending forces an individual sense of centrality, of wholeness, and of initiative, the attributes of what Erikson defined as identity.

Migration is one of the most obvious instances of complete disorganization of the individual's role system and some disturbance of social identity and self-image is to be expected. In this sense, migration has a desocializing effect (Eisenstadt, 1954). The psychological dangers of uprooting, as observed in refugees during and after World War II, have been well described by Pfister (1949; 1958; 1960; 1967) and by Weinberg (1949, 1961). While conducting a psychotherapeutic service for refugees in Switzerland, Pfister noted that the risk of psychological disorder was especially great when conditions before and after flight were particularly stressful, ranging from sudden disaster to prolonged persecution and danger to life. High rates of transitional emotional reactions were apparent in the initial phases of rescue. These seemed to have a self-protective quality in response to stress and bewilderment. Refugees appeared to sense this danger intuitively, seeking confreres with whom to cluster in groups. However, the danger to psychological well-being increased when the period between immigration and definitive resettlement was protracted and diffused by rumors and uncertainties.

Another factor influencing adaptation or maladaptation is the possibility or impossibility of returning to the country of origin or migrating to a third country. An immigrant who knows, or believes, he can return to his old environment and maintains close contacts abroad, may be hesitant to change. Individuals migrating against their will may sabotage themselves unconsciously, refusing to succeed in a country of asylum they did not choose. For example, many persons who left Hitler Germany to go to Palestine before World War II began serious study of Hebrew only after Israel attained independence and they came to realize that they could not or did not want to leave the country.

The sudden influx of nearly 200,000 refugees from Hungary to Austria in late 1956 offered a unique opportunity to explore the utility of mental health services. Strotzka and his colleagues established a Working Group for Refugees in the Austrian Society for Mental Health. Emergency funds were supplied through the World Federation for Mental Health. In his 1958 report to the Federation, Strotzka (1960) described the numerous difficulties encountered in providing emergency rehabilitation and prevention services, and the problems resulting from hasty selection or faulty training of personnel expected to deal with improvised programs. Most important for prophylactic purposes were arrangements for a regular flow of correct information, the encouragement of camp self-government, and programmatic efforts to maintain and/or strenghten the self-respect of refugees.

At the time of the Hungarian influx, the United Nations High Commissioner for Refugees was also attempting to cope with approximately 18,000 "hard core" refugees who had remained in "temporary" European camps after the massive population displacements of World War II. Most of these persons were believed to be mentally and/or socially handicapped. Following his experiences with Hungarian refugees Strotzka was invited to serve as Mental Health Advisor to the High Commissioner for Refugees. This marked the first time that a mental health specialist was given an opportunity to participate in broad-scale social action planning at an international level.

In presenting a report of his experiences at the 1960 meeting of the World Federation for Mental Health, Strotzka (1961b) described the enormous variations in the social atmosphere of the camps, ranging from very abnormal to nearly normal living conditions. The differences seemed to be related to the number and composition of the population; to the economic, social, and cultural conditions inside and outside the camps; and to the personalities of camp officials and counseling personnel. Only about ten to fifteen percent of the camp residents appeared to have integration difficulties which met the criteria of social or mental handicaps. While the case load included psychotics, mostly schizophrenics, and all kinds of neurotics, the majority of patients were alcoholics and/or individuals who had lost, or believed they had lost, their ability to work.

It became apparent that the well-recognized negative aspects of camp life, e.g., isolation, insecurity, unrealistic attitude, apathy, mistrust, and discrimination, could be partially balanced by positive influences, especially when group cohesion was strong. In some camps Strotzka noted a kind of "pseudo-security" which protected the residents but rendered reentry into normal life situations very difficult. This phenomenon is not unlike that encountered in slum clearance efforts. In subsequent years excellent progress was made in clearing the last camps, as reported by Berner who succeeded Strotzka as Mental Health Advisor to the United Nations High Commissioner for Refugees (David, 1968).

The experience of the postwar era has taught the importance of organizing the flow of refugees in such a way as to avoid the debilitating effects of camp life. Although approximately 34,000 persons continue to escape annually from Eastern European countries (U. S. Committee for Refugees, 1968) they are being processed so rapidly in the countries of first asylum, especially in Austria and Western Germany, that temporary settlement in camps is avoided. At this writing (November, 1968), the Intergovernmental Committee for European Migration is flying about 1,500 Czechoslovak refugees per week from Vienna to Australia, Canada, and the United States. Similarly, the 4,000 persons arriving in Miami from Cuba each month are dispersed to new homes throughout the United States within a few days after arrival.

While much has been learned about the dynamics of involuntary migration, there is no single source book summarizing the experience of sophisticated field workers and/or making recommendations for dealing with future waves of refugees. The literature is widely scattered. Specialists appear too preoccupied with immediate service demands to take the time needed for reflection. And yet, if past is prologue, the utility of a source book on the mental health needs of refugees would appear to be self-evident.

Adjustment after Migration

Meaningful reports on psychosocial adjustment after migration to another country are sparse. This section will summarize highlights from four articles, including a ten year follow-up study of Jewish Hungarian refugees coming to New York City, a report on Southern Europeans to Australia, a longitudinal study of Dutch-Indonesian refugees in the Netherlands, and a depth psychological study of immigrants to Israel.

Of the more than 200,000 individuals who fled Hungary in 1956, about 38,000 came to the United States. Approximately 5,100 of the arrivals in New York City were Jews, and 3,100 of these were served by the New York Association for New Americans (NYANA). A decade later NYANA surveyed 200 families representing 568 individuals selected at random from case files. Responses to sixty-three questions were received from 163 families (Soskis, 1968).

On arrival in New York the adults were relatively young. The majority were under 40 and only two were over 60 years. Unlike earlier waves of refugees, the Hungarians were generally healthy, skilled, and well educated. Approximately eighty percent of the adults had gone beyond elementary school; twelve percent were university graduates. Another important factor was that the majority had relatives in New York City to whom they could and did turn. Most of the relatives were themselves comparative newcomers. While they could not provide extensive financial assistance, they did offer a warm welcome and served as guides through the complexities of city life. Although few of the arrivals spoke English, all were literate in one or more languages; almost all continued some form of education, even if only attending evening English classes.

About sixty-six percent of the families attained economic independence after less than four months' residence; forty-one percent were completely on their own in less than two months. Only seven percent required help for longer than a year, and these included the older people and widows with small children.

A decade after arrival, forty-one percent of the 200 families reported incomes of over $9,000 a year, with six percent exceeding $20,000. The largest clustering (twenty-four percent) is in the $5,000 - $7,000 group with an additional twenty-two percent in the $7,000 - $9,000 bracket. Only five families reported incomes under $3,000 a year. All but two of the 200 families were fully self-supporting. About seventy percent of the women were working. More than forty new businesses were founded, employing over 200 persons. Just over sixty-five percent of the newcomers believed that their current standard of living was higher than it had ever been in Hungary.

High among the reasons given for leaving Hungary was "an opportunity to educate my children." How well this aim was accomplished is apparent from the finding that thirty-eight of the sixty-three college-age young people are now in college; two are attending schools of art and music; nine are college graduates, and twelve are graduate students. Of those working, all were in skilled jobs or in the professions; four were in the Armed Forces.

One of the questions asked was, "How long did it take you to feel at home in the United States?" Of the 154 replies, 58 persons reported they felt at home in less than one year; 71 mentioned periods from one to three years; and 25 indicated that they felt a sense of belonging only after acquiring citizenship. All but eight had become citizens. Only 17 of the 200 stated that their expectations of the United States had not been realized.

Few of the questions dealt with early adaptation or maladaptation. While the overall results are most encouraging there is little reported evidence of the human stories behind the statistics or the assistance rendered by a generally favorable economic climate. More and more intensive follow-up studies of differing groups would be useful in gaining a better understanding of the adaptation process in the United States.[1]

The search for a clearer definition of the concept of assimilation, along with concern about the social implications of immigrants' integration, led Price (1968) to report on the problems faced by Southern Europeans migrating to Australia. He considers a number of different variables affecting and measuring integration or assimilation, including urban concentration, the role of the ethnic community, intermarriages, political involvement, the functions of religious organizations, and intergenerational conflict. A valuable review and critique of immigration research conducted in Australia up to about 1964 has also been published by Price (1966).

In *From Stranger to Citizen*, Taft (1966) summarizes a series of studies of immigrant assimilation in Western Australia. Social acceptance and respect for the vocational and cultural aspirations of immigrants were noted as the key to assimilation. Formal instruction in English and the mores of the host country were found to have only limited effects. As noted elsewhere by Richardson and Taft (1968), there is growing evidence of a liberalization of attitudes towards immigrants among Australians, even if the newcomers do not fit the core of traditional Australian culture.

A longitudinal study of the psychosocial adjustment of refugees to the Netherlands was reported by Ex (1966) in the series issued by the Research Group for European Migration Problems (REMP). As noted in Engel's (1968) review, the same migrants were interviewed at intervals of four months, one year, two years, and three years after their migration from Indonesia to the Netherlands. Each of the forty families consisted of a young husband (age 25-40), a wife, and two to four children, all of whom had been born and raised in the Dutch East Indies and fled to the Netherlands in 1958 because of threats to their material livelihood and/or life.

In general, Ex concluded that the refugees had adjusted to panoramic aspects of their environment (e.g., the scenery) within a year of their migration but that adjustment to other facets of life and meaningful integration within Dutch society required more time. While the migrants became quickly habituated to their Dutch environment, acculturation and assimilation were a slow process. After three years twenty-five percent of the refugee families considered themselves inferior and believed that they lacked contact with Dutch citizens. While the small sample precludes major conclusions, the longitudinal method, based on repeated and systematic interviews, deserves further testing and refinement.

One of the landmarks in the sparse literature on migration and mental health is Weinberg's (1961) classic study of *Migration and Belonging*, reporting a research project conducted in 1954 among students of two courses of the Ulpan "Etzion," Jerusalem, an institution providing intensive Hebrew courses for immigrants to Israel. The methodological procedure consisted of assembling, compiling, computing and discussing data obtained by a focused depth interview and/or a written questionnaire. The ninety-nine interviewees were encouraged to impart free associations. The questionnaire was presented to groups of 313 students at the beginning and end of the course. The procedures used and statistical analyses are reported in exemplary detail.

Among the many observations made, several are particularly worth noting. General persecution in adulthood and severe persecution in camps and prisons by the Nazis before immigration were related to better general adjustment. Severe persecution did not affect subsequent mental health, but discrimination and persecution were related to worse mental health. First experiences after immigration had a distinct impact on general adjustment, mental health, and psychosomatic complaints. The need to defend against uprootedness and to retain, or obtain anew, a feeling of having roots, of belonging, is seen as a key concept of good mental health. An immigrant needs to find a group with a similar outlook on life. Well-planned and properly designed immigration policies should include education of the resident population for receiving immigrants, fostering contacts between old and new inhabitants.

Weinberg conceives of acculturation as the active and passive personal adjustment to the culture of the society of resettlement. Integration occurs when adjustment has been successful and the immigrant is accepted by and feels "belongingness" to the receiving society. Adjustment, acculturation, and integration depend on the immigrant's active adjustment to alien conditions of life. He has to struggle for economic security, social acceptance and the respect of persons in his immediate environment. When the sociocultural distance between the immigrant and the host society is great, adjustment becomes an arduous task. The innerly insecure immigrant, in a difficult and not immediately successful struggle for existence, may escape into mental or psychosomatic disturbance.

Of major importance in preventing disorders is the benevolent but not overprotective reception of the newcomer by the resident population. A sympathetic and understanding attitude is essential, as suggested, for example, in Pfister's (1958) 18 points *Help for Refugees*, prepared in 1945 for Swiss authorities.

Weinberg's pilot investigation of the relationship between the mental health of immigrants and their personal adjustment in Israel opens the door to more sophisticated research of mental health as a state, and personal adjustment as a process, allowing the use and analysis of a broad range of variables. He makes an excellent case for more research into the mental health aspects of migration and the absorption of immigrants, as well as better delineation of both mental health and general adjustment.

The studies cited suggest that the way is open for major collaborative research in several countries of the process of migration and adjustment. If we could devise a more effective procedure for matching the assets and psychological needs of refugees with the requirements and absorptive capacities of host countries, it is conceivable that policies could be implemented to facilitate more economic placement, reduction of mental health hazards and promotion of general adjustment.

Implications for Intervention

In considering current programs of assistance to refugees, the most important characteristic seems to be that they are *planned*. This means that every effort is made to meet as adequately as possible the individual needs of the refugee and the expectations of the receiving community. Families are kept together as a unit; children are not separated from parents. The long range goal is effective adaptation and integration into the host country.

Planned migration begins abroad with premigration counseling and orientation. When the refugee has determined that he wishes to go to the United States, a voluntary agency counselor will consult with him, provide assistance in completing visa application forms, and transmit to the U. S. home office background information designed to elicit appropriate sponsorship in cases where families or friends are not available to plan resettlement. Informational materials about the United States are distributed and English language classes are organized. Beginning with advice on which belongings the immigrants should pack and which to leave behind, premigration orientation covers American life and customs, including the employment situation, licensing requirements, housing, education, recreational facilities, religious institutions, medical care, insurance, legal obligations and selective service. Information is distributed in multilingual format. A medical examination is conducted (Schou, 1968).

In recent years much emphasis has been placed on the absorptive capacity of the receiving community and its socioeconomic and psychological readiness to accept immigrants. If involuntary migration is perceived as a process of desocialization, then adaptation may be seen as a process of resocialization (Bar-Yosef, 1968). The community must be prepared to allow the migrant time to learn—by trial and error—before he can be expected to assume a role and social identity meaningful in terms of the new society. This has been the experience of most immigrant-oriented societies, particularly Australia, Canada, Israel, and in earlier years, the United States.

Receiving communities should provide a warm reception and immediate access to social networks, combating the isolation and loneliness which engender psychological disturbances and psychosomatic disorders. Suitable means of communication should be fostered, including newsletters in the migrants' native tongue. Desocialization tendencies are slowly eliminated while resocialization forces expand. An effort is made to reestablish the role-set, to rebuild the connections between self-image and the role-image, and to achieve a real and acceptable social status (Bar-Yosef, 1968). Adaptation is not a well-ordered temporal sequence of phases of adjustment, but a fluid exchange between the immigrant and society. Inputs are determined by the social situation and also by the changing ability of the immigrant to accept change. Particularly instructive are the reports on the social absorption of immigrants in Israel (Bar-Yosef, 1968; Eisenstadt, 1954; and Weinberg, 1961).

There is general consensus that migrants well briefed on their new social-cultural environment tend to adapt more rapidly than those who are ill informed. Similarly, those for whom life is better than, or in accord with, their expectations tend to feel more at home in their new community and constitute less of a psychosocial risk. Immigrants must be given time to reflect on the novelty of new experiences and to regain the inner security and self-respect so essential to effective continuation of normal life processes.

The vast majority of refugees coming to the United States have adapted remarkedly well (Bernard, 1967; Davie, 1947; Handlin, 1951). They have learned to cope with new symbols and social patterns and have made major educational and economic contributions (Fermi, 1968). However, when unsuccessfully navigated, transcultural migration can lead to emotional stress and disorders. A predictable percentage of refugees do become psychological casualties, engaging in anxiety reducing behavior that may range from antisocial acting out to withdrawal into psychosis.

In recent years there has been growing recognition that the development of behavior disorders in migrants is not so much due to innate qualities or characteristics, but rather the result of interaction between individuals and the receiving community. Stress seems most severe for those whose native culture differs radically from that of the adopting community. Strain is further intensified when there is pressure for rapid assimilation and the migrant is unable, or unwilling, to join a familiar group, membership in which might offer tension reduction and flexibility in coping with cultural change. Awareness of the social components in the psychological adjustment of migrants coincides with the growing discomfort in accepting the traditional medical model as the primary basis for mental disorders.

Governmental and private agencies concerned with successful resettlement and the prevention of uprootedness and mental ill health might well heed Weinberg's (1961) observation of the remarkable similarity between the needs of the new immigrant and those of the newborn human being. The need for belonging, the need to be loved, understood and supported, but not to be dominated, pampered or spoiled; these needs are similar to those enabling the child to develop into a sound, mature person, satisfactorily integrated with his family, community and society. Capably functioning mental health and social services will assist the new immigrant to become a well adjusted and physically healthy citizen, satisfyingly and productively adapted to his new environment.

Discussions with representatives of international, national, and local service and relief organizations suggest that they are well experienced and aware of the problems. There is general agreement that intervention is likely to be effective at an early stage when the migrant is most vulnerable and eager for help. Later attempts to change behavior patterns may be far more difficult to implement. What is lacking are reported experiences with alternative strategies of intervention and systematic approaches to matching the resources of immigrants with the needs and attitudes of receiving communities. Despite the accumulated wisdom of experienced field workers, there is as yet no compendium of lessons learned with what kind of migrants under which conditions. With the severe limitations of manpower and financial resources it seems essential to utilize already available information more effectively.

So far, intervention programs have been general and reactive in the sense of fostering an atmosphere of acceptance in the community and responding to specific needs of individual migrants. Research studies have demonstrated that factors predisposing individual migrants to adjustment or breakdown in given circumstances can be identified and that community attitudes toward newcomers can be reasonably well ascertained. But, rare indeed are applied efforts attempting to match the adjustment potential of selected migrants with the absorptive capacity of a receiving community. The need for coordinated efforts with built-in research components is apparent.

Summary

This study has presented a review of the world refugee situation and a survey of immigration to the United States, followed by a discussion of the dynamics of involuntary migration and ajustment after migration, including summaries of reports from the United States, Australia, Netherlands, and Israel. Implications for intervention conclude the study. Over 100 selected references have been provided. A summary of "Resources for Refugees" is appended following Chapter 19.

Considering the long history of migration, the magnitude of the refugee problem, and the near certainty that periodic crises will continue, more effective utilization of already available information appears paramount for the development of responsible policy. It is time to bring together, in a single source book, what has been learned about the psychological dynamics of migration, develop hypotheses based on the experience of field workers, and plan coordinated empirical research on an international basis. Perhaps we could then determine which social behavioral theoretical concepts are most pertinent to voluntary and involuntary migration, and are also amenable to predictive and empirical evaluation.

Similarly, interventive strategies with differing migrants can be explored on a research basis. Who should intervene with whom, when, and how? What are the gaps in information which, if resolved, would permit more effective consultation with policy makers? Should we train "migrant counselors" to work with migrants? If so, what sort of person should be trained, at which level, with what sort of materials?

To what extent are the mental health professions prepared to cope with calls for assistance in meeting the next wave of involuntary migrants? What experience-based recommendations can we offer to assist policy makers and those working with refugees? It is time to utilize already available information more effectively. In the words of Hillel, "If not now, when?"

NOTE

1. Prof. Joseph W. Eaton has called my attention to a similar study by Mrs. Helen Glassman of the Jewish Family Service of Cleveland, Ohio on "Adjustment in freedom: a follow-up study of 100 Jewish displaced families," published by the United Hias Service in New York in 1956.

REFERENCES

ABRAMSON, J. H. (1966) "Emotional disorder, status inconsistency, and migration, a health questionnaire survey in Jerusalem." The Milbank Memorial Fund Quarterly 44: 23-48.
ADLER, P. L. and R. TAFT (1966) "Some psychological aspects of immigration assimilation." Pp. 75-92 in A. Stoller (ed.) New Faces: Immigration and New Family Life in Australia. Melbourne: F. W. Cheshire.
BARNON, W. V. (1954) "The social structure of a Sindhi refugee community." Social Forces 33: 142-152.
BAR-YOSEF, R. W. (1968a) "Social absorption of immigrants in Israel." Pp. 55-70 in H. P. David (ed.) Migration, Mental Health and Community Services. Geneva: American Joint Distribution Committee.
--- (1968b) "Desocialization and resocialization: the adjustment process of immigrants." International Migration Review 2: 27-45.
BELINE, E. (1967) Social Integration of German Immigrants in Israel (in German). Frankfurt: Europaische Verlagsanstalt.
BERNARD, W. S. (1967) "The integration of immigrants in the United States." International Migration Review 1, No. 2: 23-33.
BERNER, T. (1965) "The social psychopathology of refugees" (in French). Evolution Psychiatrique 30: 633-655.
BETTELHEIM, B. (1943) "Individual and mass behavior in extreme situations." Journal of Abnormal and Social Psychology 38: 417-453.
BORRIE, W. D. (1959) The Cultural Integration of Immigrants. Paris: UNESCO.
BOWER, E. M. (1967) "American children and families in overseas communities." American Journal of Orthopsychiatry 37: 787-796.
BRODY, E. B. (1967) "Transcultural psychiatry, human similarities and socio-economic evolution." American Journal of Psychiatry 124: 616-622.
--- (1966) "Recording cross-culturally useful interview data: experience from Brazil." American Journal of Psychiatry 123: 446-456.
--- [ed.] (1968) Minority Group Adolescents in the United States. Baltimore: Williams & Wilkins.
CANTRIL, H. (1965) The Pattern of Human Concerns. New Brunswick: Rutgers University Press.
CHAMPION, Y. (1968) "Social-psychiatric aspects of migration." Pp. 34-45 in H. P. David (ed.) Migration, Mental Health and Community Services. Geneva: American Joint Distribution Committee.
CNOSSEN, T. (1964) "Integration of refugees: the Hungarians in Canada." International Migration 2: 135-153.

COLLOMB, H. and H. Ayats (1962) "Migration in Senegal: a psychopathological study" (in French). Cahiers d'etudes africains 2, No. 4: 570-597.
DAUMAZON, G., Y. CHAMPION and J. CHAMPION-BASSET (1955) "Incidence of psychopathology in a transplanted North African population" (in French). In H. Duchen (ed.) Etudes de socio-psychiatrie. Paris: Institut National d'Hygiene.
DAVID, H. P. [ed.] (1968) Migration, Mental Health and Community Services. Geneva: American Joint Distribution Committee.
DAVID, H. P. and D. ELKIND (1966) "Family adaptation overseas." Mental Hygiene 50: 92-99.
DAVIE, M. R. (1949) World Immigration. New York: Macmillan. 2nd ed.
——— (1947) Refugees in America. New York: Harper.
——— (1936) World Immigration. New York: Macmillan.
DECARO, D. (1965) "Psychiatric problems of emigration." Minerva Medicolegium 85: 161-165.
DIEGUES, J. M., Jr. (1955) The Education of Immigrants in Brazil. Paris: UNESCO Education Clearing House.
DUNCAN, H. G. (1933) Immigration and Assimilation. Boston: Heath.
EATON, J. W. and R. J. WEIL (1955) Culture and Mental Disorders. New York: Free Press.
EISENSTADT, S. N. (1968) Israel Society. New York: Basic Books.
——— (1954) The Absorption of Immigrants. London: Routledge & Kegan Paul.
EITINGER, L. (1966) "Psychoses among refugees in Norway." Acta Psychiatrica Scandinavica 42: 315-328.
——— (1959) The incidence of mental disease among refugees in Norway. Journal of Mental Science 105: 326-338.
ENGEL, M. H. (1968) Review of J. Ex Adjustment After Migration. International Migration Review 2, No. 6: 73-75.
ERIKSON, E. H. (1960) "Identity and uprootedness in our time." In WFMH Uprooting and Resettlement. Geneva: World Federation for Mental Health.
EX, J. (1966) Adjustment After Immigration. The Hague: Martinus Nijhoff.
FERMI, L. (1968) Illustrious Immigrants: The Intellectual Migration from Europe, 1930-1941. Chicago: University of Chicago Press.
FERRACUTI, F. (1967) European Migration and Crime. Strasbourg: Council of Europe, DPC/CDIR (67) 9.
FITZPATRICK, J. P. (1966) "The importance of 'community' in the process of immigrant assimilation." International Migration Review 1, No. 1.
GAERTNER, M. L. (1955) "A comparison of refugee and nonrefugee immigrants to New York City." In H. B. M. Murphy (ed.) Flight and Resettlement. Paris: UNESCO.
GIBBENS, T. C. and R. H. AHRENFELDT [eds.] (1966) Cultural Factors in Delinquency. London: Tavistock.
GLASSMAN, H. (1956) "Adjustment in freedom: a follow-up study of 100 Jewish displaced families." New York: United Hias Service.
GLAZER, N. and D. P. MOYNIHAN (1963) Beyond the Melting Pot. Cambridge: MIT Press.
GORDON, M. N. (1964) Assimilation in American Life. New York: Oxford.
HANDLIN, O. (1951) The Uprooted. Boston: Little, Brown.
HINST, K. (1968) Relationship Between West Germans and Refugees (in German). Bern: Huber.

HOCHBAUM, J. (1967) "Social planning for immigrant absorption." International Migration 5, No. 3-4: 176.
HOFSTEE, E. W. (1952) "Some remarks on selective migration." Research Group for European Migration Problems, No. 7.
KANTOR, M. B. [ed.] (1965) Mobility and Mental Health. Springfield, Ill.: Charles C. Thomas.
KENNEDY, J. F. (1964) A Nation of Immigrants. New York: Popular Library.
KLEINER, R. J. and S. PARKER (1959) "Migration and mental illness: a new look." American Sociology Review 24: 687-690.
KLINEBERG, O. (1968) "Intercommunity relations and mental health." Pp. 83-89 in H. P. David (ed.) Migration, Mental Health and Community Services. Geneva: American Joint Distribution Committee.
KORANYI, E. K., A. KERENYI and C.J. SARWER-FORNER (1959) "On adaptive difficulties of some Hungarian immigrants: social and psychotherapeutic aspects." Progress of Psychotherapy, Vol. 4.
KRUPINSKI, J. and A. STOLLER (1966) "Family life and mental ill health in migrants." Pp. 136-150 in A. Stoller (ed.) New Faces. Melbourne: Cheshire.
LARARUS, J., B. Z. LOCKE and D. S. THOMAS (1963) "Migration differentials in mental disease." The Milbank Memorial Fund Quarterly 41: 25-42.
LEET, G. (1967) Full Employment for Refugees. New York: Community Development Foundation.
LEIGHTON, A. H. (1959) "Mental illness and acculturation." Pp. 108-128 in I. Galdston (ed.) Medicine and Anthropology. New York: International University Press.
LEIGHTON, A. H. and J. HUGHES (1961) Culture as a causative of mental disorder. In Causes of Mental Disorder: A Review of Epidemiological Knowledge. New York: Milbank Memorial Fund.
LIPSKY, L. (1955) "The children." In H. B. M. Murphy (ed.) Flight and Resettlement. Paris: UNESCO.
MCDONALD, J. S. (1964) "Chain migration, ethnic neighborhood formation, and social networks." The Milbank Memorial Fund Quarterly 42: 82-97.
MALZBERG, B. and E. S. LEE (1956) Migration and Mental Disease: A Study of First Admissions to Hospitals for Mental Disease, New York 1939-1941. New York: Social Research Council.
MARIATEGUI, J. and F. SAMANEZ (1968) "Sociocultural change and mental health in the Peru of today." Social Psychiatry 3, No. 1: 35-40.
MAST, W. VAN DER (1954) "Interlinked emigration." R.E.M.P. Bulletin, No. 20.
MEAD, M. (1953) Cultural Patterns and Technical Change. Paris: UNESCO.
MURPHY, H. B. M. [ed.] (1955) Flight and Resettlement. Paris: UNESCO.
--- (1965) "Migration and the major mental disorders." Pp. 5-29 in M.B. Kantor (ed.) Mobility and Mental Health. Springfield, Ill.: Charles C. Thomas.
--- (1961) "Social change and mental health." The Milbank Memorial Fund Quarterly 39: 385-434.
MURPHY, R. Z. and S. G. BLUMENTHAL (1967) "History of voluntary efforts in the United States." Migration News 16, No. 4: 37-39.
ØDEGAARD, Ø. (1936) "Emigration and mental health." Mental Hygiene 20: 546-553.
--- (1932) "Emigration and insanity." Acta Psychiatrica et Neurologica, Supplement IV.

PACHECO E SILVA, A. C. (1961) "Immigration and mental health in Brazil." In E. Thornton (ed.) Planning and Action in Mental Health. Geneva: World Federation for Mental Health.
PALGI, P. (1963) "Immigrants, psychiatrists, and culture." Israel Annals of Psychiatry 1: 43-58.
--- (1968) "Cultural components of immigrants' adjustment." Pp. 71-82 in H. P. David (ed.) Migration, Mental Health and Community Services. Geneva: American Joint Distribution Committee.
PARKER, S. and R. J. KLEINER (forthcoming) "Current status and new directions in mental health research." American Journal of Orthopsychiatry.
--- (1966) "Migration and mental illness: some reconsiderations and suggestions for further analysis." Paper presented at 1966 World Congress of Sociology, Evian, France.
--- (1966) Mental Illness in the Urban Negro Community. New York: Free Press.
PETERSON, W. (1961) "A general typology of migration." In S. Lipset and N. Smelser (eds.) Sociology: The Progress of a Decade. Englewood Cliffs, N. J.: Prentice-Hall.
PFISTER, M. (1967) "Community mental health work for migrants." Migration News 16, No. 15: 1-5.
--- (1960) "Uprooting and resettlement as a sociological problem." In WFMH Uprooting and Resettlement. Geneva: World Federation for Mental Health.
--- (1958) "Help to refugees." World Mental Health 10: 16-20.
--- (1955) "The symptomatology, treatment and prognosis in mentally ill refugees and repatriates in Switzerland." In H.B.M. Murphy (ed.) Flight and Resettlement. Paris: UNESCO.
--- (1949) "On the psychology of the refugee" (in German). Gesundheit and Wohlfart 29: 552-563.
PRICE, C. A. (1968) "Southern Europeans in Australia: problems of assimilation." International Migration Review 2, No. 3: 3-26.
--- (1966) Australian Immigration: A Bibliography and Digest. Canberra: Department of Demography, Australian National University.
RICHARDSON, A. (1968) "A theory and a method for the psychological study of assimilation." International Migration Review 2, No. 1: 3-30.
RICHARDSON, A. and R. TAFT (1968) "Australian attitudes towards immigration: a review of social survey findings." International Migration Review 2, No. 3: 46-55.
ROSTEN, L. (1968) "Magic island." Look Magazine, December 24, 1968.
RUESCH, J., A. JACOBSON and M. B. LOEB (1948) "Acculturation and illness." Psychological Monographs 62: 1-40.
SANUA, V. D. (1967) "The social adjustment of Sephardic Jews in the United States." Jewish Journal of Sociology (June).
SCHOU, C. (1968) "To cope with a crisis: a medical report on the Hungarian emergency." International Migration 6: 129-150.
SELLIN, T. (1938) Culture Conflict and Crime. New York: Social Science Research Council.
SHUVAL, J. T. (1963) Immigrants on the Threshold. New York: Atherton.
STOLLER, A. (ed.) (1966) New Faces: Immigration and Family Life in Australia. Melbourne: Cheshire.

STROTZKA, H. (1961a) "Migration and mental health." In E. Thornton (ed.) Planning and Action for Mental Health. Geneva: World Federation for Mental Health.
——— (1961b) "Action for mental health in refugee camps." In E. Thornton (ed.) Planning and Action for Mental Health. Geneva: World Federation for Mental Health.
——— (1960) "Observations on the mental health of refugees." In WFMH Uprooting and Resettlement. Geneva: World Federation for Mental Health.
SZULC, T. (1968) New York Times, October 21, 1968.
TAFT, R. (1966) From Stanger to Citizen: A Survey of Studies of Immigrant Assimilation in Western Australia. London: Tavistock.
TAFT, R. and A. G. DOCZY (1962) "The assimilation of intellectual refugees in western Australia." REMP Bulletin, 1961, No. 4; and 1962, No. 42. Perth: University of Western Australia Press.
TAFT, D. and R. ROBBINS (1955) International Migration. New York: Ronald Press.
TYHURST, L. (1955) "Displacement and migration: a study in social psychiatry." In H. B. M. Murphy (ed.) Flight and Resettlement. Paris: UNESCO.
——— (1951) "Displacement and migration." American Journal of Psychiatry 107: 561-568.
UNESCO (1955) The Positive Contribution of Immigrants. Paris: UNESCO.
United Nations (1960) Report of European Seminar on the Social and Economic Aspects of Refugee Integration. Geneva: United Nations Technical Assistance Office.
United States Committee for Refugees, Inc. (1968) World Refugee Report: 1968. New York: U. S. Committee for Refugees.
WEINBERG, A. A. (1961) Migration and Belonging: A Study of Mental Health and Personal Adjustment in Israel. The Hague: Martinus Nijhoff.
——— (1955) "Mental health aspects of voluntary migration." Mental Hygiene 39: 450-464.
——— (1949) Psychosociology of the Immigrant (in Hebrew with English summary). Jerusalem: Heilinger.
WENK, M. G. (1968) "The refugee: a search for clarification." International Migration Review 2: 62-69.
WINNICK, H. Z. (1957) "Psychological problems of immigrants" (in Hebrew). Ofakim 11: 138-144.
WOLFGANG, M. E. and F. FERRACUTI (1967) The Subculture of Violence. London: Tavistock.
WONG, P. (1967) "The social psychology of refugees in an alien social milieu." International Migration 5, No. 3-4: 195.
World Federation for Mental Health (1960) Uprooting and Resettlement. Geneva: World Federation for Mental Health.
ZUBRZYCKI, J. (1956) Polish Immigrants in Britain: A Study of Adjustment. The Hague: Martinus Nijhoff.

JOURNAL RESOURCES

International Migration:—Quarterly review, published by the Intergovernmental Committee on European Migration in Geneva in cooperation with the Research Group for European Migration Problems at The Hague.

International Migration Review:—Published three times a year by the Center for Migration Studies, 209 Flagg Place, Staten Island, New York, 10304; $4.50 per year. Includes research studies, notes and statistics, book reviews, and a review of reviews.

Migration News:—Bimonthly review in English, published by the International Catholic Migration Commission, 65 rue de Lausanne, Geneva, Switzerland; $2.50 per year. Includes research studies, notes, Facts and Figures (a statistical supplement), and a detachable bibliography on migration.

Migration Today:—Published twice yearly by the Secretariat for Migration, World Council of Churches, 150 Route de Ferney, Geneva, Switzerland. Special focus on movement of semi- or unskilled workers. Includes a selected bibliography, arranged by topics and country.

HCR Bulletin:—Published quarterly by the United Nations High Commissioner for Refugees, Palais des Nations, Geneva, Switzerland. No charge. Reports on activities of the UN High Commissioner for Refugees and those of international nongovernmental organizations.

PART II

FROM COUNTRY TO CITY

Chapter 4

Adaptation of Appalachian Migrants to the Industrial Work Situation: *A Case Study*

**HARRY K. SCHWARZWELLER
and MARTIN J. CROWE**

One of the more important sources of potential strain in the transitional adjustment of rural-to-urban migrants is the process of adaptation to the industrial work situation encountered in the area of destination.[1] This may be expecially true for migrants reared within a familistically oriented social organization and accustomed to the self-directed work routine characteristic of economic pursuits in relatively isolated, subsistence farming localities of Appalachia. An individual migrant from Appalachia, for example, has little opportunity prior to migration to acquire industrial-type work experiences in the area of origin. Upon arrival in the area of destination he seeks out and assumes a work role for which he may have very little, if any, preparation and which, moreover, is at once sharply differentiated from family activities. It seems inevitable that some kinds of strain result in the process of adaptation.[2] To the extent that the

---

Author's Note: *This is a revised version of a paper presented at the Conference on Migration and Behavioral Deviance, on which this volume is based. It is one of a series of papers from the Beech Creek Study sponsored by the National Institute of Mental Health in cooperation with the Kentucky Agricultural Experiment Station. The study was designed and directed in collaboration with James S. Brown of the University of Kentucky and Joseph J. Mangalam now of the University of Guelph.*

family-kinship system is responsive to the changing needs of the migrant—and we pursue this theme more specifically in another paper (Brown, Schwarzweller and Mangalam, 1963; see also Schwarzweller and Seggar, 1967)—serious adjustment difficulties are probably avoided. For a proper understanding of the adaptation process, however, it is also necessary to take into account the nature of the industrial work situation in the area of destination. Indeed, the social context in receiving areas, which includes the industrial work situation *and* other aspects of the local community situation, probably determines the effectiveness and, perhaps, the very form of response by the family-kinship system to the changing needs of the migrant.

Our aim in this paper is to explore, by descriptive analysis with some historical depth, the patterns of adaptation and reaction to the industrial work situation among a selected group of male migrants drawn from an isolated rural mountain locality in eastern Kentucky. We are, of course, especially concerned with the functions performed by the kinship structure. Our thesis is that these, and perhaps other migrants from rural Appalachia, have adapted with a minimum amount of strain to existing circumstances in the host communities as the result of a combination of favorable factors: the particular and particularistic nature of the industrial work situation (which is the focal point of this paper), the normative equivalencies which exist in the donor and recipient subsystems both at the place of origin and destination, and the supportive functions performed by the kinship network during the transitional period.

Research Design and Study Population

The research findings reported here are drawn from a separate study (Crowe, 1964) designed to supplement the survey phase of a larger project (The Beech Creek Study).

The Beech Creek Study, in turn, is based upon and is an extension of an earlier study in 1942 by James S. Brown who, in the anthropological tradition, had as his main purpose "a description and analysis of the social organization of an isolated rural neighborhood in the Kentucky mountains." He found that Beech Creek, as he called the locality, was a family-centered social system. Kinship units, in effect, tended to be culturally insular groups, kinship relations the more meaningful interactional patterns, and familistic norms the more important mechanisms for social control. Familism, as a traditionally sanctioned value orientation, dominated the cultural configuration (Brown, 1950; 1952a; 1952b; 1952c).

For the Beech Creek Study, (i.e., the larger project), persons who were residents of Beech Creek in 1942 were followed up and interviewed at their places of residence in 1961. Since Brown's original study, as one familiar with demographic transitions occurring in the southern Appalachians would surmise, there had been a considerable stream of out-migration from the mountain locality. By 1961, of the 319 Beech Creekers still living, 178 (or fifty-six percent) were residentially relocated in areas outside eastern Kentucky; interview data were obtained from 161 (or ninety percent). Most of the migrants (sixty-three percent) live in and around the major metropolitan areas of southern Ohio and almost all of the remainder (thirty-five percent) are located in other industrialized areas of Ohio, Indiana, and central Kentucky. Well over half of the migrants have lived for ten years or longer in areas outside eastern Kentucky. Most of the male migrants (fifty-nine percent) are employed in manufacturing industries and generally at semi-skilled or unskilled jobs.

In order to explore more deeply the problem of the occupational adaptation of rural migrant workers, focused interviews of some length were conducted during the summer of 1962 with a selected group (N=30) of male migrants from Beech Creek who were residing in or near the city of Cincinnati.[3] Almost all of these men were married and had children. The median length of time they had lived outside eastern Kentucky was twelve years with a range of from five to twenty years. The median years of schooling they had completed was eight; their median annual income in 1961 was $5500. Two were unemployed at the time. Most of the others were employed in factories, at skill levels ranging from "packer" to "finish-grinder" to "precision inspector." In general, the selected group was fairly representative of the male migrant population from Beech Creek.

A series of interviews were also obtained during the spring of 1963 from a number of industrial relations personnel, foremen, and union representatives in the various factories where many of the Beech Creekers worked (Crowe, 1964). As informants, they provided additional information about the characteristic and stereotyped "traits" of eastern Kentuckians within the industrial work situation.

Data collection focused upon: (a) the social setting within which the institution of work is located (in terms of both the areas of origin and destination); (b) the period during which a migrant's reaction to and evaluation of his job and work situation occurred (which dealt with three "occupational time periods," namely, prior to migration, immediately after migration, and at the time of the interview); and (c) the specification of factors explaining occupational adaptation. This approach to the general problem was modified somewhat, on the basis of field experience and data interpretation.

The present paper summarizes our observations, emphasizing those more relevant to the stem-family hypothesis, which we have elaborated and discussed at length in an earlier paper (Brown, Schwarzweller and Mangalam, 1963), and to a fuller understanding of the social context within which this form of familistic adjustment to changing environmental circumstances is located. Our implicit objective is to suggest a useful sociological approach (in the holistic tradition) for the study of this multifaceted phenomenon, namely, the occupational adaptation of rural migrants.[4]

From Field to Factory

The work situation in the mountain area of eastern Kentucky was, and to a considerable extent still continues to be, the antithesis of the work situation encountered by migrants in urban Ohio. Prior to migration most male migrants from Beech Creek were engaged in farming, either on a full-time or part-time basis, and this work was very often a family endeavor with responsibilities divided according to age and sex, and clearly articulated with other life activities. Some men, to be sure, had held supplementary jobs in the log-woods, in the mines, or as laborers on county road-construction projects and a few had worked at one time or another in factories in southern Ohio. But the rhythm of work life was, in the main, organized around, and tempered by the seasonal demands of subsistence agriculture. "Public work," as Beech Creekers called almost any kind of off-farm employment, required only a temporary separation of the individual from the family homestead and a man's obligation to do his share of the farming remained foremost.

Male migrants in Ohio generally reflect favorably and with considerable nostalgia upon their early life and work experience in the mountains. "Farming," they feel, "was good to grow up on." They recall the independence and sense of security it accords and the fact that farming is an outdoor activity with a great variety of tasks. Many would agree with the Beech Creeker who said, "If I could take my present job and move it back to the hills, I'd go in a minute."

Of course, as the latter suggests, farming in the Beech Creek area did not and does not offer the possibility of an adequate cash income. The difficulties of "making it on the farm" and the lack of occupational alternatives in the mountain region function as important "push" factors in stimulating out-migration and, likewise, provide a subsequent basis for comparison with the work situation in the area of destination. Most Beech Creekers, consequently, are not unhappy with their new work situations. As one migrant succinctly put it: "Any job here beats hell out of pounding rocks in Kentucky."

Migration to Ohio is an old pattern for the Beech Creek neighborhoods. Contemporary migrants undoubtedly moved with the comforting knowledge that many before them—kinsfolk and neighbors—had been successful in making this transition and in adapting to the industrial work situation. Their predecessors' obvious mastery of the situation (relatively few returned permanently) coupled with the visible spoils of victory (many visited home with new cars and other symbols of affluence) bolstered the confidence and undergirded the fortitude of these "new recruits" to the Ohio labor market.

Professor Slotkin (1960:99-100), in his book dealing with new factory employees, describes one type of migrant as the "permanently uprooted." These migrants perceive the donor culture (our terminology) in the area of origin as substantially and permanently inadequate, hence migration from the area is undertaken with expectations of permanency. We suggest, as a further elaboration of Slotkin's theory, that such "expectations of permanency" are directly tied in with the supportive functions performed by the kinship structure. From significant kin-group members in the area of destination, for example, the potential migrant secures information about the kinds of jobs available as well as some idea about the work expectations connected with these industrial occupation roles. In this way, the potential migrant is significantly aided in formulating an image of work requisites in the factory vis-a-vis those of the farm. His kin are often able to supply details about a specific job "opening." The Beech Creek migrant therefore, who had fairly accurate information about the job situation in Ohio, was able to at least partially anticipate the industrial occupation role prior to migration; the event of migration was, from his point of view, the end-result of a rational decision and the manifestation of a firm resolve to accept the "punishment" which would be entailed in pursuing the "reward." Our data converge upon this conclusion. Indeed in numerous ways the transition from field to factory and the process of adaptation to the industrial work situation were begun long before the migrant left Beech Creek; the kin structure (i.e., the stem-family system) stabilized and managed the process.

Initial Job Situation

Few if any Beech Creekers had kinsfolk in the area of destination who were in a position to actually hire them. Employers in Ohio, however, not only recognized the importance of kin ties among Kentuckians but utilized the migrant kin network in securing an adequate labor supply, especially at the laborer and unskilled job levels. When job vacancies occurred, the word was passed along within the shop and, via the kin communication network, soon became common knowledge in the migrant community, quickly trickling down to families in the coves and hollows of eastern Kentucky. Such personalized appeals were, and continue to be, far more effective than mass media forms of communication for drawing out job applicants from the mountain "labor pool." Moreover, a worker who is hired on the basis of references supplied by kinsfolk in that same factory is bound to be more reliable; family obligations are involved and family honor is at stake.

"We have special appeal for Kentuckians," said one company official, "because of our reputation for hiring many Kentuckians through the years and the fact that we have many family ties in the company over a period of three generations." Indeed, it is common knowledge among migrants that some employers favor job applicants who have family connections within the plant; "unless your brother or your brother-in-law is working for them," said a Beech Creeker bluntly about a high-paying factory in the area, "there is no use in trying to get on." It is not unusual, therefore, to find many members of a family group working for the same firm; three brothers, for example, and some of their cousins from Beech Creek are employed in one of the larger factories.

A few managerial personnel whom we interviewed were of the opinion that hiring along kin lines is less prevalent nowadays than formerly. It had tended to create certain, rather unique problems. For example, as one informant put it: "We used to hire close relatives of our employees and, especially with Kentuckians, if there was some emergency back in the mountains we would have a whole group of workers who took off to visit a sick aunt. This paternalistic attitude can backfire on you." Nevertheless, we found that kin-hiring still appears to be an important technique used to secure employees for lower status jobs. As a result of this practice over the years a type of homogeneity with respect to the workers' backgrounds and normative expectations was fostered in many work situations. Southern Appalachian migrants, for example, predominate in the light and heavy manufacturing industries around Cincinnati; in some plants the proportion of Kentucky-born workers is reported to be as high as fifty to seventy-five percent. It is not surprising, therefore, that few Beech Creekers encounter difficulties in getting along with native Ohioans in the work situation for, as they often exclaim, "there ain't no Buckeyes to get along with."

Beech Creekers generally, like most migrants from rural Appalachia, found their initial jobs in factories which did not require at the time of hiring any previous industrial experience nor a high school diploma. They were often hired to perform simple assembly-line tasks that were quickly learned with a minimal amount of on-the-job training. For instance, one Beech Creeker recalled that, "the boss took me to the place where I'd work and told a guy there to explain what I'd do . . . he did . . . it took about ten minutes." Similarly, a foreman explained that his plant "doesn't require any polish or a lot of education and Kentuckians know this by word of mouth and a lot of them come here." Another foreman reported that, "anyone can get a job here. They give an aptitude test, but hell, the whole thing depends on whether they have an opening or not." The type or work required by these industries seemed to have been designed to make use of the potential labor force in the nearby southern Appalachian region. One company official declared pointedly: "Our strongest appeal to the Kentuckian worker is our proximity to Kentucky."

This proximity to eastern Kentucky, which permits the Beech Creeker to maintain visiting ties with the family homestead, coupled with the supportive kin network in the area of destination and the minimal skill requirements demanded of the Kentucky migrant by Ohio industry, facilitated the intial entry of the Beech Creekers into the industrial labor market. To be sure, the work that newcomers were expected to perform (for example, punch press operator) and the job context (a factory or shop situation) constituted new experiences for most Beech Creekers. It was, nevertheless, a relatively simple transition under the circumstances. Indeed most Beech Creekers seem to have been quite satisfied with their first jobs in Ohio. Although their starting wages were not high (ranging, for example, from $0.60 to $2.60 an hour with a median of $1.25 during 1941-1956 for the thirty men interviewed), what they earned was a great deal more than they could have expected in the mountains; and, more important, they *were* employed. Management, at least in terms of its past policies, tended to be paternalistic in its dealings with Appalachian migrants and, perhaps because of this, Beech Creekers regarded the initial work conditions as quite satisfactory. Getting along with co-workers offered no special difficulties; after all, most of them too were "Briarhoppers" from the mountains. The initial situation, from the migrant's perspective, provided an effective mechanism for allowing him to adapt gradually, and with integrity, to the demands of a machine technology.

Advancement and Stability

While the industrial work situation in southern Ohio was generally in accord with the needs and unskilled talents of beginning workers from rural Appalachia, these same initially favorable conditions made it possible, perhaps even necessary, for ambitious migrants after a year or so "to look around for better jobs." Some, to be sure, were encouraged to rise up through the ranks within the factory where they had started. Seniority rules, however, and other factors tied in with a particular firm's organization of manpower made such movement difficult. For the most part, and especially in the case of those who had begun at unskilled laborer levels, who had managed to acquire the basic industrial training for subsequent advancement, and who were eager to capitalize on that experience, upward mobility toward higher paying, more skilled jobs often meant seeking out new employers. The relatively high rate of job turnover, i.e., interplant mobility, by Appalachian migrants in Ohio and elsewhere (a phenomenon quickly noted by observers) should not be interpreted as a sign of occupational insecurity or instability (a trait often attributed to these newcomers). Rather, it is more likely a consequence of the migrants' desire to get ahead, a behavioral manifestation of the "maturing" workers' realistic appraisal of the situation, and indeed an indication of the newcomers' adaptation to the demands and opportunities of the industrial labor market.

As a matter of fact, the Appalachian migrant is rather reluctant to change jobs because it not only entails moving into an unfamiliar situation but also means that he must give up the security of accrued seniority rights. A foreman explained: "They have a great value for security and once they get to know their work group and boss they don't want to move. Also, they are sensitive about their lack of educational skills, which may be required in another job, so they tend to stay on the same job." To become upwardly mobile, however, the Appalachian migrant often must seek-out a new job.

The general advancement in occupational status (and, of course, level of skill) over the years by Beech Creek migrants is striking. Of the thirty men, for example, whom we interviewed at length during this phase of our research, twenty-one had begun work in Ohio as unskilled laborers, four as semiskilled, two as skilled, and three as farm workers. In 1962, ten were still at an unskilled level (one temporarily unemployed), but nine were semiskilled (one temporarily unemployed), nine skilled, one a salesman, and one a poolroom attendant (service). The proportion who were able to command a more skilled job had tripled and their wages reflect this increased status (ranging, in 1962, from $1.25 to $4.37 an hour with a median of $2.67).

During their relatively short work careers in the urban area (from five to twenty years) these men had found it useful or necessary to make a number of place of work changes. One migrant, in fact, had worked for thirteen different employers during his fifteen years in Ohio; six migrants, on the other hand, were still in the same factory where they had started and were apparently quite satisfied. The median number of employer changes for this representative group of thirty male migrants is three. More significantly, the median length of time they had held their current (1962) jobs is over four years; in fact, the man (mentioned above) who had exhibited the most "unstable" employment pattern had, nevertheless, worked for his current employer for more than two years.

Beech Creekers in Ohio, then, had manifested some degree of occupational "restlessness" but most of this seems to have occurred early in their work careers. Perhaps it was a function of youth, or represented the rural migrant's way of "testing" his abilities on the urban labor market, or maybe it was linked with social-psychological changes that had come about as a result of migration. In any event, although relatively frequent job changes appear to have been the norm during the initial period of transition that followed migration from the mountains, the later period of a Beech Creeker's work career had become markedly stabilized. He had, it seems found his place in the industrial order — a niche that was in reasonable accord with his talents and ambitions.

One additional point is especially relevant here: Most male migrants from Beech Creek, as noted earlier, secured their initial jobs in Ohio through the aid or influence of kinsfolk. Those who subsequently changed jobs — and most of them did — more likely did so "on their own" without help from kinsfolk. After having been exposed to the urban occupational subculture for a period of time and having become familiar with the industrial work situation, Beech Creekers were in a much better position to personally pursue and evaluate job opportunities in light of their own occupational mobility aspirations. Changing jobs at that time was not of the same order of crises as finding the first job; individualism, not familism, was the appropriate orientation called forth in this situation.

Reaction To Lay-Offs

The threat to being "laid-off" (i.e., an involuntary, though temporary, loss of job for a period ranging anywhere from one week to six months or longer) is an ever-present fact of life among manual workers, especially those employed in manufacturing and construction industries. A great many Beech Creekers (over half of those interviewed during this phase of our study) had experienced a "lay-off" at some time during their industrial work careers. The economic recession of 1957-58 was a particularly difficult time. More commonly, however, lay-off periods were normally associated with massive retooling operations or production "change-overs" such as occur, for example, every two or three years in the automobile industries. To be sure, a Beech Creeker now and then "quit" or was "fired" for personal reasons or for reasons of incompetency. But the lay-off pattern, either as an actuality or as a threat, was a prevailing norm in the industrial work situation of Ohio during the 1950s and early 1960s, and we shall confine our brief remarks to this form of unemployment and the Beech Creekers' reactions to it. Their reactions (in retrospect) ranged from a deep sense of frustration on the part of a few to the more typical attitude of regarding a lay-off period as a vacation and a chance to do some work around the house or to visit with the family in the mountains.

In general, Beech Creekers accept the threat of a lay-off as one of those annoying conditions of industrial work, like punching a timeclock and working indoors, that have to be tolerated much as the vagaries of weather must be tolerated in farming. As a worker gains seniority on the job, of course, the threat is reduced; men hired last are the first to be "bumped." But even those with considerable seniority are attitudinally prepared for the eventuality: they too may be included in the next round of lay-offs. Most Beech Creekers feel fairly secure in the knowledge that unemployment compensation will hold them over in good stead; if a lay-off period turns into chronic unemployment, for whatever reason, they can always return to the mountains and wait out the crisis on the family homestead.

During a lay-off period, then, Beech Creekers try to make the best of it. They draw unemployment compensation, attempt to find other jobs as they must under existing regulations, and wait for their old jobs to reopen. In the meantime, it provides an opportunity for them to visit kinsfolk in Kentucky and in the surrounding Ohio communities, do chores around the house, work in the garden, fix up the back porch, go fishing, or simply loaf. There is no question that the Beech Creeker, in his own way, has found it rather easy to adapt to this potentially disturbing feature of industrial work life.

Attitude Toward Unions

The Beech Creeker supports union activities in much the same way as do the majority of rank and file union members in American industry. His general opinion of union activities is on the whole favorable; his participation in union activities is in most cases minimal. The Beech Creeker's attitudinal support tends to focus on the "practical" functions of unionism, i.e., so-called "bread and butter unionism," such as protection of the worker from arbitrary acts of management that can result in loss of job or pay. In many ways he is like the American workingman described by Schneider (1957:305), who "expects his union to secure for him (1) above all, better wages; (2) more favorable hours; (3) job tenure; and (4) congenial work rules and conditions of work."

In spite of this basically favorable attitude, a general behavioral apathy nevertheless prevails. Practical issues are rarely regarded as sufficiently important for personal involvement. There appears to be an undercurrent of fear of managerial reprisal for active union involvement, especially among older migrants; as a matter of fact, one Beech Creeker had indeed lost his job as a result of union organizing activities. Moreover, Beech Creekers just aren't very good joiners; they feel uncomfortable in a formal gathering. Participation in union meetings and activities outside of the immediate job situation tends to interfere with home life and most Beech Creekers are unwilling to allow this to happen unless such union activity involves and serves the needs of the whole family. One man, for example, reported that he used to take his family to all appropriate union events but had ceased to do so because these events often become "beer blasts." Of the thirty men interviewed during this phase of our field study, sixteen were union members but only four were active in the sense of having attended a number of union meetings the previous year. For most Beech Creekers, union membership is a nominal status.

On occasion a Beech Creeker may voice some negative comments about unions: "You don't get anything for the dues you pay;" "You take a gripe to the shop steward and that's the last you hear of it;" "If you do your job right and work hard, you don't need a union;" "I think someone ought to crack down on both the union and management. They spend too much money fighting each other when they could be helping the worker."

Most Beech Creekers, however, do not seem to question the right or place of unions in the industrial work situation. Although generally apathetic about getting involved with union activities, Beech Creekers, like rank and file union members elsewhere, are advocates (passively) of pragmatic unionism.[5] They accept union membership in much the same way as they accept other, more discomforting aspects of factory work life, and they obey union dictums in much the same way as they obey shop regulations or the orders of a foreman. Whether a Beech Creeker's initial motivation to join derived from his employment in a factory that was bound by a union shop contract (in which case new workers must join the union within a stipulated time, usually thirty days after being hired) or from informal pressures by co-workers who insisted that "to be a union member is to be a right guy," further involvement (attending meetings, assuming a leadership role, proselytizing, etc.) demands an emotional or intellectual commitment over and above that for which the Beech Creeker is prepared. In that respect the Beech Creeker is not very different from the majority of American industrial workers. His apathy is mixed with allegiance. Indeed, one might say that he has adapted to the form of industrial work life without having become uncomfortably involved in its complexities.

Job Satisfactions

Among American workers generally, the pattern of responses to such questions as, "taking into consideration all the things about your job, how satisfied or dissatisfied are you with it?," invariably indicated a high degree of satisfaction.[6] Similarly, virtually all employed Beech Creekers in Ohio (survey phase of study) say they are quite satisfied with their current jobs. Of course, the meaning of "satisfaction" is inherently vague;[7] during the focused interviewing phase of our study, therefore, we pursued this aspect of occupational adaptation a bit further.

Most of our informants (male migrants from Beech Creek) emphasize that they like the kind of work they are doing because it is "interesting" or they are "learning something different." They talk a lot about the working conditions; it is "clean work" or they are working with a "nice crew." The amount of take-home pay, of course, and the degree of security accorded (in the form of seniority rights, adequate compensation during lay-off periods, etc.) are important considerations in assessing the job situation. But pay and security factors are fairly standardized in terms of skill levels among the industries in southern Ohio; hence, if dissatisfactions about a particular job exist they usually focus upon specific working conditions and especially the interpersonal relationships among work crew members and with the boss. As one foreman explained: "They are very sensitive to the kidding from other workers. Then too, they seem to have a holy fear of the boss. After about six months they adapt to the kidding but it seems to be a general characteristic that they are more afraid of the boss than other workers." Another foreman put it more strongly: "They don't like to be bossed and they seem to be afraid or shy in front of the boss. Then too, you have to ask them to do the work rather than tell them." To the highly individualistic, personalistically-oriented Beech Creeker, social relationships with fellow workers and immediate supervisors are a major source of potential strain; the fact that most Beech Creekers work with other Appalachian migrants from similar sociocultural origins contributes to the stability and, from the Beech Creeker's point-of-view, satisfactoriness of the work situation.

Advancement opportunities would certainly be a factor in the overall evaluation of any job; here too Beech Creekers are quite satisfied. Few feel "trapped" or "held down;" few feel that their job is a "dead-end." In general they seem aware of existing opportunities. Those who have attained skilled levels feel they might eventually move on to supervisory or "office" position. Those who are at semiskilled levels, although cognizant of opportunities and confident of their abilities to attain higher levels, apparently prefer (so they say) to avoid the "headaches" and responsibilities that inevitably accompany higher rated jobs. Laborers, on the other hand, more often than not simply feel that further advancement is not important, especially if it means (as it often does) giving up the security of the moment for the uncertainties of occupational mobility. Beech Creek migrants, in these respects, are not unlike American industrial workers in general; over the years, undoubtedly, a sorting-out along the lines of relative ambition and talent has occurred.

The basic satisfaction with job and work situation is further reinforced by, on the one hand, the migrant's favorable attitude toward management (a naive-like trust whose roots perhaps are to be found in the patriarchical tendencies of mountain society) and, on the other hand, the migrant's conviction that employers in Ohio are quite satisfied with the work performance of Kentuckians. Indeed most Beech Creekers feel that factory supervisors consider Kentuckians to be "better and harder workers" than native Ohioans. The personnel managers and foremen whom we interviewed tend to validate the Beech Creekers' own favorable self-image vis-a-vis hard work; but they add, often in the same breath, that the Kentucky mountaineer appears to be a bit too docile for his own good in the industrial labor market.

Occupational Adaptation in Context

Beech Creek men who had migrated to Ohio had been able, over the years, to make a satisfactory and, as they see it, satisfying transition "from field to factory." In the process, it seems they did not encounter, and therefore did not find it necessary to cope with, those difficult tension-producing conditions that are so often associated with rural to urban migration and the phenomena of industrialization in other parts of the world. Their record of upward occupational mobility in the urban area, which we regard as impressive under the circumstances, and their relatively long tenure in current (1962) jobs which we regard as a sign that stability has been normalized, attest to their confidence in and acceptance of the industrial work role, and their successful adaptation to the industrial work situation.

Initially, of course, the migrants had encountered some difficulties as beginning workers. The formal schedule and rigid authority system of the factory, for example, was particularly irksome, and working with and around complicated machines was for many quite confusing and sometimes even frightening at first. Yet these men, reared in an isolated mountain locality of eastern Kentucky, few of whom had been fortunate enough to get beyond the eighth grade in school, were able, after a relatively short period of time, to master the technical details of their new jobs, to familiarize themselves with the industrial arts and the formalized procedures of factory work and, indeed, to feel rather comfortable in the midst of industrial complexity. Perhaps, during the transitional period, their frontier-bred fortitude and willingness to work hard had compensated in part for their initial lack of skills on the job. Other factors, such as the labor market situation at that time, must be considered in venturing an explanation of why the process of adaptation in this case was not more difficult and disturbing. We, however, chose to focus our inquiry on the kinship factor which we believe offers a valid, though partial, explanation of the relative "success" of Beech Creekers as industrial workers.

The stem-family form of kinship structure, as pointed out earlier, helped to stimulate out-migration from the mountains, directed and "cushioned" the relocation of Beech Creekers, and facilitated, in various ways, the entry of migrants into the industrial work situation. Through the kin network, information about jobs and working conditions in the area of destination were made known to potential migrants in the mountain neighborhoods. Kinsfolk in the host community assisted newcomers in finding the initial jobs and, thereafter, served as advisors and instructors in the process of urbanizing their "greenhorn" kinsmen. More important, the "branch-family network" in the area of destination, which is linked directly with the family homestead in the mountains, provided the newcomer with a measure of assurance that, in the event of some unforeseen crises, he would not stand alone. The Beech Creek stem-family system, in short, served to stabilize the migrant's social world external to the factory and, consequently, helped to keep "off-the-job" problems and anxieties from entering into and disturbing the migrant's "on-the-job" performance. (If the Beech Creek kin system had been a more nucleated form, the migrant worker, we believe, would have experienced greater difficulty in adapting to the industrial work situation and, as a consequence, factory managers in the area would have had many more labor problems and far greater labor costs. The contribution of Appalachian mountain families to the economy of Ohio, other states, *and* the Federal Government that resulted from extended family normative obligations "to take care of their own," if it could be measured, would undoubtedly stagger the imagination of many government officials.)

In the Beech Creek case, perhaps the most abrupt, immediate change (i.e., system-disturbing change) that occurred and was experienced by the Beech Creeker as a result of migration was the distinct separation of occupational activities from family activities. For many of the sociocultural elements characteristic of the Beech Creek neighborhoods had been transferred to (or recreated within) the area of destination via chain migration of kinsfolk and neighbors over the years. Furthermore, a kind of residential segregation has given rise to a number of "little Kentucky" neighborhoods in and around the major metropolitan centers of southern Ohio. The host neighborhood in the area of destination is therefore, very often structured in the image (sociocultural) of a Kentucky mountain community, and because kinsfolk are near at hand, the newcomer from Beech Creek is, in many respects, "at home."

For most Beech Creekers, then, the abrupt separation of family life from work life was, in the normative sense, the biggest change that had come about as a result of moving to Ohio. Some men, to be sure, had been employed off the farm in "public work" prior to migration. But, as we have explained, this was generally defined as a temporary activity, peripheral to the family work activity configuration, and very often undertaken on a seasonal basis; farming, for most of these men, continued to be the main enterprise and management of the homestead and its lands the primary obligation. That attitude had to be and was modified in confrontation with the industrial situation. After migration, work for wages in a shop or factory became the family's only means of support, and a man's job (about which his wife had little comprehension) became, without question, his primary responsibility.

Adaptation to an industrial occupation role, therefore, undoubtedly had some stress-producing potential because Beech Creekers were not well prepared for this experience and its immediate and obvious consequences. Yet the potential, so far as we could discern, was not manifested to any unusual degree (e.g., through instances of marital discord, criminal behavior, alcoholism, mental illness).[8] Supportive functions performed by the kin network, we believe, had much to do with keeping resultant tensions within manageable bounds. Moreover, because the kin network tended to isolate the newcomer from other segments of the urban community, it tended to perpetuate the Beech Creek value system and to provide the migrant with a means for self-expression and for the satisfaction of culturally-derived needs. Adaptation, then, to the industrial occupation role required merely the acceptance of new standards in an isolated area of behavior, namely work; it had little effect upon other and to them more important areas of life. The tensions aroused by these "minor" changes in the migrant's life were more than adequately compensated for by the obvious rewards which were forthcoming. Over time, of course, these same "minor" changes may build into system-disturbing influences which affect more fundamental changes; at that point the Beech Creek sociocultural system will have been absorbed into the great "melting pot" of American Society.

NOTES

1. For an extensive bibliography used in conjunction with the design of this study, and for a statement of our theoretical guidelines, see Mangalam (1968).

2. Specifically, we are referring to the process of adaptation by rural migrants to occupational roles (i.e., changes in and demands of) within the industrial work situation.

3. In addition to focused interview data reported here and findings from the survey phase of the Beech Creek Study, information and insights were also gained by the research staff during three months of residence and quasi-participant observation of selected migrant families in a migrant community in Ohio.

4. To be sure, there still exists much confusion over the meanings and proper usages of terms such as adaptation, adjustment and accomodation. For the exploratory purposes of this study we have defined "occupational adaptation" as a process by which an individual approaches, evaluates and accepts a new occupational role. Our definition was intended as a research guide, not as a conceptual clarification. See Crow (1964: 23-26) and Mangalam (1962).

5. The best documented exception is members of the International Typographical Union. See Lipset, Trow and Coleman (1962).

6. See, for example, Hoppock (1935); Shister and Reynolds (1949); Morse and Weiss (1955); Palmer (1957). See also Brown (1954: 190-191), who points out that American research supports the generalization that, "Even under the existing conditions, which are far from satisfactory, most workers like their jobs. Every survey of workers' attitudes which has been carried out, no matter in what industry, indicates that this is so."

7. Drucker (1954: 303), for example, argues that "satisfaction as such is a measureless and meaningless word."

8. This is *not* to say that behavioral deviancy is not associated with migration, nor that marital discord, crime, alcoholism, mental illness, and other signs of unmanaged tension are absent in migrant neighborhoods and "ghettos." To the contrary, there is much evidence to suggest that migration fosters the kinds of social conditions and situational circumstances from which deviant behaviors emerge. What we are saying is that where the family-kin network intervenes as a stabilizing instrumentality, as it did in the Beech Creek case, the individual migrant is more likely to remain anchored into a normative system which discourages deviancies. This is not a new idea: see, for example, Thomas and Znaniecki (1958); Zimmerman and Frampton (1935); and LePlay (1878).

REFERENCES

BROWN, J. C. (1954) The Social Psychology of Industry. Baltimore: English Pelican Edition.

BROWN, J. S. (1952a) The Farm Family in a Kentucky Mountain Neighborhood. Lexington: University of Kentucky Agricultural Experimental Station, Bulletin 587 (August).

--- (1952b) The Family Group in a Kentucky Mountain Farming Community. Lexington: University of Kentucky Agricultural Experimental Station, Bulletin 588 (June).

--- (1952c) "The conjugal family and the extended family group." American Sociological Review 17 (June).

--- (1950) "The social organization of an isolated Kentucky mountain neighborhood." Ph.D. dissertation, Harvard University.

BROWN, J. S., H. K. SCHWARZWELLER and J. J. MANGALAM (1963) "Kentucky mountain migration and the stem family: an American variation on a theme by LePlay." Rural Sociology 28 (March): 48-69.

CROWE, M. J. (1964) "The occupational adaption of a selected group of eastern Kentuckians in southern Ohio." Ph.D. dissertation, University of Kentucky.

DRUCKER, P. (1954) THe Practice of Management. New York: Harper.

HOPPOCK, R. (1935) Job Satisfaction. New York: Harper.

LEPLAY, F. (1878) Les ouvriers européens. (Second ed.) Six volumes. Paris: Tours A. Mame.

LIPSET, S. M., M. TROW and J. COLEMAN (1962) Union Democracy. New York: Doubleday.

MANGALAM, J. J. (1968) Human Migration. Lexington: University of Kentucky Press.

––– (1962) "A reconsideration of the notion of adjustment." Proceedings, Southern Agricultural Workers Conference, Jacksonville, Florida.

MORSE, N. C. and S. WEISS (1955) "The function and meaning of work and job." American Sociological Review 20 (April).

PALMER, G. L. (1957) "Attitudes toward work in an industrial community." American Journal of Sociology 63 (July).

SCHNEIDER, E. V. (1957) Industrial Sociology. New York: McGraw-Hill.

SCHWARZWELLER, H. K. and J. F. SEGGAR (1967) "Kinship involvement: a factor in the adjustment of rural migrants." Journal of Marriage and the Family 29 (November): 662-671.

SCHISTER, J. and L. G. REYNOLDS (1949) Job Horizons: A Study of Job Satisfaction and Labor Mobility. New York: Harper.

SLOTKIN, J. S. (1960) From Field to Factory. Glencoe, Ill.: Free Press.

THOMAS, W. I. and F. ZNANIECKI (1958) The Polish Peasant in Europe and America. (Revised ed.) New York: Dover.

ZIMMERMAN, C. C. and M. E. FRAMPTON (1935) Family and Society: A Study of the Sociology of Reconstruction. New York: D. Van Nostrand.

Chapter **5**

Social Class Origins and the Economic, Social and Psychological Adjustment of Kentucky Mountain Migrants: *A Case Study*

HARRY K. SCHWARZWELLER
and JAMES S. BROWN

Most researchers studying the "success" or "failure" of rural migrants in cities have viewed rural migrants as an undifferentiated group or category, and though several excellent studies have compared rural migrants from various ethnic or racial backgrounds or from various geographic locales, rarely have social differences within a group of rural migrants who stem from a particular sociocultural situation been considered.[1] This is perhaps especially true of migrants from the Southern Appalachians, whose people have been falsely assumed to be much more homogeneous culturally and socially than they actually are.

The research reported here attempts to take into account one kind of intragroup social diversity that may affect the adjustment of rural migrants. We are dealing with a specific population of persons who reside in or near major industrial centers of the Ohio Valley and who had migrated from a particular mountain locality in eastern Kentucky. We are exploring the hypothesis that the social class positions of families in the

---

Author's Note: *This is one of a series of papers from the Beech Creek Study sponsored by the National Institute of Mental Health in cooperation with the Kentucky Agricultural Experiment Station. The authors gratefully acknowledge the assistance and suggestions of their colleague, Joseph J. Mangalam, now of the University of Guelph.*

area of origin significantly influenced the patterns of out-migration as well as the economic life chances and various other social and psychological aspects of adjustment of individual migrants and families in the areas of destination.

Although our study focuses upon migration and migrants from three small, contiguous, mountain neighborhoods (Beech Creek) which had a total population in 1942 of less than 400 persons, and hence we refer to our research as a "case study," one should note that the experience of Beech Creek migrants is more or less representative of hundreds of thousands of others who have left eastern Kentucky. From 1940 to 1960, the eastern Kentucky area had a *net* loss through migration estimated at 490,000 persons; in 1940 the area's population was only 750,000 (see Brown and Hillary, 1962). Even though the Southern Appalachian area contains much social and cultural diversity, many of the migrants making up that region's net loss of some two million persons during the forties and fifties came from sociocultural backgrounds similar to that of Beech Creekers. With migration of such magnitude affecting hundreds of towns and cities throughout the nation—the receiving areas for the "Great Exodus" from the mountains—it is obvious that if social class origins do influence the patterns of adjustment of Appalachian migrants such knowledge and understanding can be very helpful to urban agencies and organizations in developing programs based on the special problems and needs of migrants from different class backgrounds in areas where they tend to cluster.

Our data were derived from a larger project (The Beech Creek Study) which was based on, and was an extension of, an earlier study (see Brown, 1950, 1952a, 1952b, 1952c). The families in this relatively isolated, three-neighborhood mountain locality, referred to as Beech Creek, were the subjects of an intensive anthropological type field investigation by the junior author in 1942. Of those residents of Beech Creek at the time of the original study, about ninety percent of all still living were located and interviewed in 1961. The study population, then, constitutes in effect a total "migration universe"; for many reasons we have found it useful to look upon this migration universe as a "migration system" (Brown, Schwarzweller and Mangalam, 1963: 66).

In a recently published article, we have discussed the social class origins of Beech Creek migrants in relation to patterns of out-migration, the social structure of the migration process, and economic life chances (Schwarzweller and Brown, 1967: 5-19). Since the first paper was designed as an introduction to the discussion of economic, social and psychological adjustment given in the present paper and is therefore essential to understanding and interpreting Beech Creek migrants' adjustment, we are summarizing the earlier paper here.

In the earlier study (1942), Parsons' definition (1940: 850) of class was used: "the group of persons who are members of effective kinship units which, as units, are approximately equally valued." The techniques used by Warner and his associates in their earlier works were followed in determining which families were equally valued: People are considered to be "of the same class when they normally (a) eat or drink together as a social ritual, (b) freely visit one another's family, (c) talk together intimately in a social clique, or (d) have cross-sexual access to one another, outside of the kinship group" (David and Dollard, 1940: 261).

Through listening to what Beech Creekers said as well as observation of their activities, the investigator was able to determine much consensus in the evaluation of families, though the final division of families into classes was "not so much...one made by the people themselves as one made by the investigator based on the informants' doings and sayings." It was hard to draw definite class lines between families close in rank; the classes were not sharply defined (after all, vagueness about class differences in such a social setting has a positive functional importance). But because it was necessary to draw lines somewhere in order to present quantitative data on the characteristics and differences of the various classes, Beech Creek families were eventually grouped into three classes designated high, intermediate, and low (Brown, 1950: 272-273).

Without going into a detailed description of economic, familial, educational and other attributes of the three classes delineated in the original study, we may briefly "characterize the high-class families in the Beech Creek neighborhood as being long-resident families of good background, 'moral athletes,' hard workers and 'good livers,' less isolated and more modern than other families in the area and as people who emphasized self-improvement and who participated more widely in neighborhood affairs. The low-class families, on the other hand, tended to be newcomers with 'shady' pasts, morally lax, economically insecure, not ambitious, old-fashioned and 'backward,' and people who participated relatively little in many neighborhood activities. The intermediate-class configuration was not so much a distinct pattern as a combination of the high-class and the low-class configurations" (Brown, 1951: 233).

Social Class Origins and Patterns of Out-Migration

By 1961, of the 271 persons in the study population, only about twenty-five percent were still resident in the original Beech Creek neighborhoods; twenty-five percent had moved a short distance to a small town or neighborhoods adjacent to Beech Creek; but sixty percent had moved to areas outside of eastern Kentucky (Table 1).

TABLE 1
RESIDENCE LOCATION IN 1961 OF PERSONS
IN BEECH CREEK STUDY POPULATION,
BY SOCIAL CLASS ORIGIN IN 1942

Social Class Origin in Beech Creek, 1942
(percentage distribution)

Residence Location 1961	High	Intermediate	Low	Unclassified [a]	Total N, all classes
Beech Creek neighborhoods	33	23	30	0	(68)
Town or neighborhoods near Beech Creek within eastern Kentucky	7	15	18	29	(42)
Outside eastern Kentucky	60	62	52	71	(161)
Total	100%	100%	100%	100%	
N =	(59)	(101)	(87)	(24)	(271)

[a] The persons in the "Unclassified" category are members of families which were not included in the original delineation by social class. There were a number of reasons these persons were not included—in some cases the families were not in the neighborhoods long; in some cases the author did not get to know enough about the families to classify them; etc. (For fuller details see Brown, 1950: 278-379).

So far as class patterns are concerned, somewhat more of the intermediate class persons left the original neighborhoods (seventy-seven percent; high-class, sixty-seven percent; low-class, seventy percent). Few high-class persons had moved to nearby areas (seven percent) compared with intermediate- and low-class persons (fifteen and eighteen percent respectively), probably because the relatively few opportunities in the rural, low-income area nearby were not sufficiently attractive to offer them any advantage over their already more favorable socioeconomic circumstances vis-a-vis their neighbors. Perhaps the most striking fact, however, is that the three groupings did not differ greatly in the proportions who had left the mountains (sixty percent of the high, sixty-two percent of the intermediate, and fifty-two percent of the low-class population). There was then, no clearly discernible selectivity in terms of social class origins in the patterns of out-migration from eastern Kentucky in these two decades.

Some of the migrants who lived in Beech Creek or nearby neighborhoods in 1961 had moved out and returned in the period from 1942 to 1961. But over forty-seven percent of the Beech Creekers once they had moved from the region never returned to live there; only about a fourth manifested some degree of indecisiveness by moving back and forth, but even half of those eventually settled outside eastern Kentucky. Individuals from low-class families were somewhat less likely to attempt to migrate from the region and those who did were less likely to be "successful" on their first attempt. Otherwise, the patterns of the three classes were strikingly similar.

In order to eliminate the bewildering maze of short, temporary moves we decided to concentrate on migrations involving six or more months' continuous residence away from eastern Kentucky (which we called "permanent moves") and, to determine differences in the out-migration patterns of persons from the three social groupings, we especially studied the Beech Creeker's *initial* "permanent" migration from eastern Kentucky.

Focusing then on those Beech Creekers who had made "permanent moves" (171 persons)[2] and on the first such move each of these migrants made, we found important class differences: forty-three percent of the low-class out-migrants moved to a nearby neighborhood compared with only twenty percent of the intermediate and five percent of the high-class migrants. For the high-class families, the pattern was either to move entirely out of the region or to remain in Beech Creek; for the low-class families, however, intervening opportunities nearby afforded additional alternatives which, for many, resulted in a "two-stage" pattern of out-migration.

Time of migration

Another significant difference among the migrants of the three classes was that they migrated at different times. From 1942 through 1947, the World War II period, over half the out-migrants were from high-class families; over two-thirds of the high-class migrants left eastern Kentucky in those years. From 1948 through 1953, the Korean War period, well over half the out-migrants were from intermediate-class families; nearly two-thirds of the intermediate-class migrants left in those years. From 1954 through 1961 over half the out-migrants were from low-class families (Table 2).

How can we account for this difference? The higher class families in Beech Creek (1942) were sensitive to and cognizant of the rapidly

TABLE 2
YEAR OF INITIAL RESIDENCE OUTSIDE EASTERN KENTUCKY AFTER 1942, BY SOCIAL CLASS ORIGIN

Year of Initial Residence Outside Eastern Kentucky		Social class origin in Beech Creek, 1942 [a] (percentage distribution)			
		High	Intermediate	Low	Total N, all classes
1942-47		68	23	23	(59)
1948-53		16	63	43	(76)
1954-61		16	14	34	(36)
	Total N =	100% (44)	100% (71)	100% (56)	(171)

[a] Those individuals who were not classified as to social class position in Beech Creek in 1942 were omitted from this and subsequent tables.

widening gap between economic circumstances in the mountains of eastern Kentucky and opportunities which existed in the industrial areas to the north. Compared with their lower class neighbors, they were already in the advantaged positions in the rural low-income area and, consequently, perceived that little could be gained through residential or occupational shifts within the region. As they saw it, upward social mobility could be affected only through out-migration. The immediately obvious method for them to enhance their own and their children's lot in life was to move elsewhere, out of the region. The lower-class families, on the other hand, because of their positions within the social class hierarchy of the Beech Creek neighborhoods, were less influenced by status differentials vis-a-vis "outsiders" and more oriented toward neighborhood and community norms in the process of formulating aspirations. Hence, lower-class families perceived that some advantages would accrue from occupation and/or residential shift within the area. Intervening opportunities nearby, in that sense, indeed did exist for them, at least until recent years.[3]

Place of destination

Though nearly three-fourths of all migrants went to Ohio with only a few going to central Kentucky, Indiana, or other states, the migrants of different classes settled in different locations (Brown, Schwarzweller and Mangalam, 1963: 60, Table 5). Of those who went to Ohio, most intermediate-class migrants settled in and around a small town in southern Ohio; low-class migrants tended to settle in the "little Kentucky" sections, or ghettos, of Dayton and Cincinnati; and finally high-class migrants concentrated in Hamilton. Place of residence, then, was apparently associated with the migrant's social class origins. We believe, however, that the primary reason for the clustering of migrants from Beech Creek in various locations is to be found in kinship ties. Since close kin tended to belong to the same class, the migrant's choice of where he would move was probably due to a combination of kinship and class factors.

Social Class Origins and The Social Structure of The Migration Process

Social class origin, we found, was also related to the social structure of the migration process. Beech Creek migrants, generally, were young persons when they moved away from the mountains. Significantly, almost a third of the migrants from high-class families were under sixteen years of age at the time compared with only about nine percent from the other two classes. Excluding these youngsters who simply accompanied parents, however, we found that migrants from high-class families tended to be older. Furthermore, adult migrants from high-class families, particularly males, tended to be more advanced in the family life cycle at the time of out-migration than those from the other classes. Although we had a relatively small number of cases, these facts suggest that the high-class pattern was a "family-uprooting" type of movement, whereas the pattern for the other social classes, especially the intermediate class, tended to follow more along the lines of a stem-family type of migration such as we discussed in an earlier paper (Brown, Schwarzweller and Mangalam, 1963). High-class families from Beech Creek moved away from the mountains as nuclear families and usually established new households as nuclear families in the areas of destination. Intermediate-class families in Beech Creek tended, however, to maintain a family homestead in the mountains, and young migrants from these families usually joined their older siblings or close kin who were already established in the areas of destination. Low-class families manifested a more diverse pattern which, we suspect, was a consequence of their ownership or nonownership of a "homestead" in the mountains.

Social Class Origins and Economic Life Chances of Beech Creek Migrants

Let us now turn to more specific consideration of the adjustment of Beech Creek migrants in the urban industrial areas, concentrating here on the patterns associated with social class origins.

First of all, we want to explore the influence of social class origins on the migrant's economic life chances in the host commmunity. In studying this relationship we must recall, as indicated above, that compared with the intermediate and lower class migrants, a larger proportion of the migrants from high-class families were youngsters when they first moved and, also important, the high-class migrants dominated, percentagewise, the first "wave" of out-migration after 1942. Migrants from high-class families in Beech Creek, therefore, had distinct advantages over migrants from the other social classes; not only the recognized ones, but also the situational advantages that followed from social class differences in the form or strategy of the migration process (whether planned, normatively organized, or accidental). Because of the nature of the research design, these considerations could not be controlled through analysis; they were taken into account, however, in our interpretation of findings.[4]

An individual migrant's social class origin in Beech Creek, we expected, would be indicative of a particular level of achievement aspiration, a particular set of value orientations, and certain kinds of social skills as well as of the possession of economic means, such as savings, all of which would be directly related to the migrant's ability to cope with the problems of adjustment encountered in the process of migration.

This expected relationship between social class origin and ability to cope with adjustment problems, we hypothesized, would be manifested in a direct relationship between the migrant's social origin in 1942 and his socioeconomic status in 1961.

In testing this hypothesis we have used: the Cornell (Danley-Ramsey) nine-item scale (Danley and Ramsey, 1959) to measure material level of living—both for individual migrants and for migrants' family households; family income; and the North-Hatt (1952) scale of occupational prestige, all of course in relation to social class origin.

We found a very high, direct relationship between the individual migrants' social class position in Beech Creek in 1942 and their level of living in the areas of destination in 1961 (Table 3).

This finding is supported by parallel analyses focusing on the family-household as a unit for analysis,[5] using family income and occupational prestige of household head as dependent variables, and social

TABLE 3

LEVEL OF LIVING OF BEECH CREEK MIGRANTS IN AREAS OF DESTINATION 1961, BY SOCIAL CLASS ORIGIN IN 1942

Level of Living 1961 (scores) [a]		Social Class of Family of Origin in Beech Creek 1942 (percentage distribution)			Total N, all classes
		High	Intermediate	Low	
High (9-7)		69	14	13	(40)
Intermediate (6-4)		28	56	33	(60)
Low (3-0)		3[b]	30	54	(44)
	Total	100%	100%	100%	
	N =	(36)	(63)	(45)	(144)

[a] The numbers in brackets refer to scores on the Cornell level of living scale.
[b] One case in low category. Percentage included only for consistency in reporting.

class origin of household head or homemaker as the independent variable. But, perhaps more important, the general pattern is supported by our observations from intensive, quasi-participant field work with a number of selected families.

On the basis of these findings, we conclude that the status hierarchy of individuals within a rural-to-urban migration system of this kind tends to maintain a reasonable degree of stability despite the seemingly disruptive process of migration. Although the system has been "upgraded" socio-economically through migration (i.e., as a collectivity), the relative positions of persons comprising the collectivity tend to be held in place.

One reason why individuals tend to maintain their social class positions relative to other individuals within a given migration system, such as the Beech Creek case, may be the differential value placed upon education by the various social classes. We found that nearly two-thirds of the migrants from high-class families had completed high school (and a third had at least some experience in college) compared with only thirteen percent of the intermediate-class migrants and only one migrant from the low-class families. Migrants who were children when they first moved to urban areas had the considerable advantage of easy access to good schools, while migrants who were reared in the mountains did not; and most of this advantage accrued to the migrants from high-class families who, in general, had moved earlier.

Social Class Origins and Patterns of Social and Psychological Adjustment

At this point in our study of Beech Creek migrants, we will delve into certain aspects of a rather intriguing and important question: how and in what sense does an extended family structure facilitate or hinder the adjustment of rural migrants? Our aim was to delineate and explore selected dimensions of the complex phenomena of social and psychological adjustment (i.e., noneconomic aspects) which, in the Beech Creek case, were relevant to migration and meaningful within the social context of the changed and changing situation in the area of destination. Consistent with our stem-family hypothesis and theoretical guidelines suggested and more fully articulated elsewhere (Brown, Schwarzweller and Mangalam, 1963), we considered the proposition that: The greater the degree of involvement with kinsfolk, the greater the social psychological adjustment of Beech Creekers to circumstances in the area of destination, under certain specified conditions. Our specific concern in this paper, of course, is with social class origin as a "conditioning variable." Hence, we want to explore the effect it has, or has had, upon the interactional patterns, orientational difficulties and sentiments of these migrants as well as to pursue the relative importance of the kinship factor in the adjustment of migrants when social class origins are taken into account. Our assumption is that the needs and the abilities to cope with problems encountered through migration differ among migrants from different class backgrounds; the validity of that assumption is our target for inquiry in this phase of our study.

Perspectives on social adjustment

In most cases, the Beech Creek migrant was not a "loner" when he arrived in the area of destination. To the contrary, he often joined kinsfolk who had migrated before him and, almost invariably, his initial residence was in the general vicinity of a number of other close kin families whom he could visit and call upon for assistance in time of need. Similarly, he in turn was often joined by other kin from Beech Creek whom he assisted.

We expect, however, that with the passage of time, and for many reasons, the migrant's need and feelings of obligation to maintain close ties with kinsfolk would become less strong, more diffuse. As the Beech Creek sociocultural system was modified or absorbed by the sociocultural system of the larger society, as the Beech Creekers' needs changed in the context of a more urban interactional system to which Beech Creekers were exposed, the interactional bonds which held branch-families together were undoubtedly changed and, in one or the other sense, weakened. Furthermore, various segments of the Beech Creek population, such as the various social classes, probably had greater or lesser need for maintaining strong interactional ties with kinsfolk and in situations where such individual need-dispositions (e.g., as derived from social class origins) were not in accord with normative pressures or could not be expressed in concrete behavioral activity, tensions were perhaps generated.

Then too we must consider that by moving from one area to another, the Beech Creeker, like other rural to urban migrants, was spatially separated from kinsfolk and friends with whom he had been intimately associated prior to moving. Those interactional ties, however, were not necessarily severed abruptly in the process of migration, although the frequency of face-to-face contact was, and had to be reduced to a considerable degree. Sentiments, i.e., normative expectations about person to person relationships, which were deeply imbedded in the migrant's personality through countless earlier socialization experiences, could not be ejected from his personality system, figuratively speaking, in one day. On the contrary, we would expect that the personality system and the normative system which tended to support it strained toward maintaining the system (i.e., the boundaries of the interactional system) as it was prior to departure from the family group and homestead in the area of origin. Behaviorally, this is manifested by frequent and regular visits back to the mountains.

In time, however, as the migrant experiences more and more of the urban world and as his needs and interests change, his interactional ties with persons in the area of origin, we expect, become less tenacious, and visiting and other forms of communication less frequent and less meaningful. It is also quite likely that various segments of a rural migrant population, such as persons from different class backgrounds, respond differently to the circumstances associated with migration. For whatever reasons, the reduction of face-to-face interaction between persons in the area of origin and persons in the area of destination, and the modification in form and content of such interactional bonds between these two subsystems, would be indicative of changes and modifications in the form of the Beech Creek stem-family system itself.

Finally, we must also consider that, with the passage of time and as the migrant comes into repeated and prolonged contact with persons native to the host community (in the neighborhood, at work, in church, etc.) such nonkin, nonmountain people may become incorporated into his pattern of interaction and frame of reference in more and more meaningful ways. The migrant's orientation toward his family and his reliance upon kinsfolk for guidance and help in crises situations, for example, may be replaced by an orientation toward and reliance upon friendship groups composed of nonkin. If nonkin with whom the migrant strikes up friendship and perhaps visiting relationships are persons from a similar sociocultural background; e.g., eastern Kentucky, then the effects of such contacts upon the migrant's personality and behavior may be negligible. Indeed, the orientations that the migrant brings with him into these confrontations may be reinforced by the interaction that ensues. But, if the nonkin persons with whom the migrant comes into repeated contact are persons indigenous to the urban community, then the likelihood is greater that a truly cross-cultural confrontation is, and has been, taking place. Such cross-cultural interaction may perform compensatory functions in the adjustment process, supplementing functions normally performed by the kinship group, and at one and the same time satisfying and changing the needs of migrants.

Dimensions of social and psychological adjustment

In view of the exploratory purposes of this study and in line with the general perspectives suggested above, we contrived a number of indicators of various aspects of interactional and psychological adjustment. For this paper, of course, conceptual definitions and validation arguments must be brief and operational procedures merely outlined.

One of our concerns is with the migrant's effective kinship group in the area of destination. "Effective kinship group" implies those persons related to him by blood or through marriage whom (in keeping with the normative standards of the Beech Creek sociocultural system) he can count on for assistance and support of one kind or another in time of crises.[6] Operationally, this group is defined as the migrant's "close kin," inclusive of parents, parents-in-law, siblings, siblings-in-law, and, if they reside outside his immediate household, his adult children. "Area of destination" implies the migrant's normal visiting community, the territory within which he can maintain a pattern of frequent visits with kinsfolk without extraordinary expenditure of time, money, or energy. This area is defined, arbitrarily, as a fifty-mile radius from the migrant's place of residence.[7]

Two indicators are utilized as parallel measures of degree of kinship involvement.[8] The first is simply a count of the number of close kin the migrant has within the area of destination. If size of branch-family network has anything to do with the adjustment of migrants, then this indicator is a useful tool for exploratory purposes. The second indicator, a more meaningful measure in many ways, is based upon the first. Frequency of visiting with each close kin in the area is converted to a yearly count; the sum divided by the total number of close kin yields an average visiting frequency. This is interpreted as an indicator of the degree of interaction the migrant maintains with kinsfolk in the area of destination and, indirectly, of the strength of familial bonds within the kinship circle.

Another concern is with the migrant's maintenance of interactional ties with the stem-family homestead in the mountains. Three indicators, two of which differ mainly in methodological terms, were employed to tap this dimension of adjustment process.[9] The first, an index of "visiting in eastern Kentucky," takes into account both the number of visits *and* duration of each visit back to the mountains during a year (i.e., a summated visiting score). The second, an index of "visiting exchanges," includes as well the number of visits (i.e., visiting units) received by the migrant in the area of destination from kinsfolk and friends living in eastern Kentucky. Visiting patterns, in both cases, are for the most part along parent-sibling, or sibling-sibling lines. But visiting is only one, albeit an important form of communication between the migrant and his relatives and friends back in the mountains. Letter writing, for example, is another kind of communication linkage that in some cases, depending upon level of literacy and other factors, may help to maintain a degree of cohesion between the branch-families and their homestead in eastern Kentucky. Hence, we developed a composite index of "letter exchanges" to explore this possibility.

We are also, of course, concerned with the migrants' social involvements with urban people and, in order to explore this dimension of interactional adjustment, two indicators were devised. The first focuses upon the degree of "social contacts with urban natives" and solicited subjective responses.[10] The second focuses upon "number of urban friends" and solicited quantifiable responses.[11] Questioning procedures, in both cases, were borrowed from Leighton's Sterling County Studies (see Hughes, et al., 1960).

In addition to these indicators of interactional patterns, six indicators of the migrants' psychological adjustment were incorporated into the research design. The term "psychological" is used to imply that these measurable attributes are indicative of certain facets of the migrant's personality structure—his state of mind, general orientation, sentiments, and the like. By observing the variability of these attributes within a multivariant analysis framework, one may discover the nature and locus of certain basic disturbances or incongruities which, by logical inference, are consequents of the migration and interactional adjustment processes.

(1) *Degree of identification with the urban locality* ("residential stability"). Rural migrants who identify more with the urban locality than with the area of origin, who feel quite satisfied with their new way of life in the host community, and who tend to think of themselves as permanent residents rather than transients in the communities to which they have migrated have made, almost by definition, a certain kind of adjustment. A simple, six-item summated attitudinal scale is utilized to measure this feeling of permanency, this particular facet of adjustment.[12]

(2) *Nostalgia for home.* Rural migrants are, symbolically speaking, caught up in and to some degree members of two sociocultural worlds. There is the social world in which they were reared—in this case, for example, Beech Creek, the mountain neighborhood, home. There is the other world—in this case, for example, Ohio, the great society, the urban community with which they and the Beech Creek sociocultural system have come into prolonged contact. A composite, four-item summated attitudinal scale is utilized to measure the migrant's feeling of attachment to the former; i.e., his longing for the old home neighborhood and an earlier way of life in the mountains. To some extent, this indicator is a polar opposite of "residential stability," though the intent here is to focus on the extreme condition of nostalgia.[13]

(3) *Expressed happiness.* Happiness is not only exceedingly difficult to know or attain, but also a very ambiguous concept to operationalize in research terms. For this study, the migrant was asked to express his own pervasive feeling about his total life situation, his own assessment of how his existential gratifications add up. The interviewing form, leading to a summary statement, followed that of Gurin et al. (1960: 22-24). Expressed happiness, nevertheless, is measured by the migrant's response to a single question; those who say they are "very happy" these days ("taking things all together") are considered "high" on the happiness scale.

(4) *Extent of worry.* The extent to which an individual worries about things as well as the sources and objects of his worries are important and sensitive indicators of certain facets of his psychological ad-

justment. Like happiness, however, "worry" is difficult to research. Again, questioning procedures are borrowed from Gurin et al. (1960: 28-30). A listing of each respondent's worries and expressed degree of concerns is utilized to construct a crude "worry index." This index is designed to measure extent of worry and does not take into account the source of such distress or involvements.

(5) *Anomia* ("normlessness and despair"). Anomia refers to an individual's state of mind with respect to his own integration into a societal structure. In that sense, the more anomic individual manifests symptoms of normlessness, hopelessness, helplessness, and the like in the face of impersonal social forces he perceives as beyond his control. The anomic individual would have very little faith in the future and in his own ability to influence the course of events in a society and would be extremely pessimistic to the point of despair. A commonly employed measure of this personality orientation, the five-item Anomia Scale developed by Srole (1956: 709-716), is used as an indicator in the present study.

(6) *Symptoms of psychological anxiety.* As Gurin and associates explain, "Specific psychological, physical, or psychosomatic symptoms have often been used as critical diagnostic indices of psychological distress, both in research on mental disturbance and in actual clinical settings." In using a symptom list as a measure of psychological disturbance, the present study leans heavily upon the work of Gurin et al. (1960: 175-187), who, in turn, have built upon the works of MacMillian (Stirling County Study, 1957), Rennie (Midtown Study, 1953), and others. As suggested from the earlier researches, a series of appropriate items was included in the interview instrument. From the results of a matrix analysis of these items (i.e., a simulated factorial analysis, in crude terms), a six-item composite indicator is constructed which, referring back to Gurin's work, is loaded somewhat on the factor "psychosomatic disposition."

Basic relationships: social class and dimensions of adjustment

We shall direct our attention here toward the main purpose of this paper, namely, to ascertain and explore the influence of social class origins upon the subsequent social and psychological adjustment of migrants in the area of destination. Other kinds of social diversity, to be sure, exist within the migrant population; those factors incorporated into our larger analysis program, though they cannot be considered directly in the present paper, are taken into account where they throw some light on the effect of social class.[14] Our first concern, at this point, is to observe the concomitance between social class background and the various indicators outlined above.

As Brown noted, differences had existed between the Beech Creek class groupings in interactional as well as in moral standards and economic means. Furthermore, as we have discovered, the original social class groupings differed in patterns of out-migration, especially in terms of where, when, and how individual migrants moved away from Beech Creek. These and other behavioral patterns and attributes associated with social class origin, we surmise, would have had some bearing upon the process of transitional adjustment of Beech Creek migrants in the area of destination. We find that class origin seems to affect three aspects of interactional adjustment (summarized in Table 4).[15]

TABLE 4

SUMMARY OF RELATIONSHIPS BETWEEN VARIOUS ASPECTS
OF INTERACTIONAL ADJUSTMENT AND
SOCIAL CLASS ORIGIN OF BEECH CREEK MIGRANTS

Interactional Adjustment Variables	*Direction of Relationship and $\bar{C}$* [a]
Size of nearby kin group	NR
Frequency of visiting nearby kin	NR
Social contacts with urban natives	.36[b]
Friendship ties with urban natives	+.26
Visiting in eastern Kentucky	NR
Visiting exchanges with eastern Kentucky	-
Letter exchanges with eastern Kentucky	-.31

[a] Where $\bar{C}$ is reported, $P < .10$; where only direction of relationship is reported, $P < .20$ but $> .10$; NR means no relationship observed. In all cases, df=2 and N=161.

[b] In this case, the distribution was not in a straight line; intermediate-class migrants scored lower than either of the other two classes.

A negative relationship between social class and degree of letter writing communication with people back in the mountains is understandable if we take into account the history of out-migration from Beech Creek. Migrants from high-class families tended to be in the earlier wave of out-migration whereas migrants from lower-class families dominated the later phases. This indicates existing differences in the stage of out-migration of the various family groups composing the social classes. Whether or not close kin members of a migrant's family group still reside in the mountains is, of course, a determinant of the number of letters that are exchanged; migrants from high-class families had fewer kinsfolk back in eastern Kentucky.

Because similar relationships are not noted between class origin and frequency of visiting in eastern Kentucky (or visiting exchanges) suggests that the concept and meaning of "family homestead" differs among the various social classes. Migrants from high-class families, as we observed during field work, often view their visits to eastern Kentucky as holiday outings rather than family reinforcement ritual; migrants from intermediate-class families, on the other hand, think of such visits more as an obligation, dictated by familistic norms. Though the pattern of frequency is similar in these cases, the content or meaning attached to such visits, we believe, differs. We would also entertain a comparable argument as to why branch-family interaction does not vary with class origin; though the kin network is not any less important for high-class migrants, the meaning of kin interaction may be quite different.

Predictably, however, the higher the social class, the more friendship ties have been established with persons "outside" of the mountain migrant community. Yet it is interesting that a larger proportion of intermediate-class migrants report "very little" contact with urban natives as compared with high- and lower-class migrants (who have greater contact). An intervening variable may be involved. We suspect that the ecological situation of intermediate-class families must be taken into account. Intermediate-class families, in general, are isolated socioculturally; they tend to reside in one or the other of the small "eastern Kentucky" settlements that have sprung up in the countryside around the great cities and industrial zones of southern Ohio. High- and lower-class migrant families, on the other hand, are more dispersed residentially and more likely to be located in the larger metropolitan centers of souther Ohio—the formet in the suburbs and the latter very often in urban slums; in either of these cases, for example, female migrants have greater opportunity to "make friends" with urban natives than do their intermediate-class counterparts.

In terms of the psychology of adjustment, our findings (summarized in Table 5) show that migrants from higher-class families are less nostalgic for "home," tend to express their feelings of happiness more positively, and are more likely to possess a sense of involvement with the larger society—an optimism about their place in the world—than migrants from lower-class families. It appears, then, that by severing or modifying certain ties with family homestead and mountain society, higher-class migrants have made or are making a relatively smoother or less stress-producing adjustment to the realities of their situation in the urban, industrial setting.

TABLE 5

SUMMARY OF RELATIONSHIPS BETWEEN VARIOUS ASPECTS OF PSYCHOLOGICAL ADJUSTMENT AND SOCIAL CLASS ORIGIN OF BEECH CREEK MIGRANTS

Psychological Adjustment Variables	Direction of Relationship and $\bar{C}$ [a]
Residential stability	NR
Nostalgia for home	-.27
Expressed happiness	+.23
Extent of worry	+
Anomia	-.39
Anxiety	[b]

[a] Where $\bar{C}$ is reported, $P < .10$; where only direction of relationship is reported, $P < .20$ but $> .10$; NR means no relationship observed. In all cases, df=2 and N=161.

[b] In this case, intermediate-class migrants scored lower than either of the other two classes; but $P > .10$.

Elaboration by Analysis: Patterns of Adjustment and the Kinship Factor

Now let us examine the patterns of relationships between various aspects of interactional and psychological adjustment in the case of each of the social class groupings of migrants (Table 6). Our aim, of course, is to interpret the meaning of these observed patterns in the light of what we know about Beech Creekers and the Beech Creek migration system.

In the case of high- and intermediate-class migrants a negative relationship emerges between extent of worry and frequency of visiting

TABLE 6

RELATIONSHIPS BETWEEN VARIOUS ASPECTS OF MIGRANTS' PSYCHOLOGICAL AND INTERACTIONAL ADJUSTMENTS, BY SOCIAL CLASS ORIGINS[a]

Psychological Adjustment Variables	Branch family		Urban ties		Stem family		
	Size of kin group	Freq. of visiting kin	Social contacts	Friendship ties	Visiting E. Ky.	Visiting exchanges	Letter exchanges

HIGHER CLASS (N = 36)

Residential stability							
Nostalgia							
Happiness							
Worry	-.32	-.44					
Anomia							
Anxiety		-.49					

INTERMEDIATE CLASS (N = 63)

Residential stability				+	-	-	-
Nostalgia		-.51			+.38	+.34	
Happiness		+.47					-.34
Worry		-.36					
Anomia			-				
Anxiety			-.47	-			+

LOWER CLASS (N = 45)

Residential stability					-.38		
Nostalgia							
Happiness		+				+	
Worry		+					
Anomia				-		-.40	-.48
Anxiety				-.39			

[a] The social class origin of seventeen migrants was not ascertainable.

nearby kin; the relationship fails to manifest itself in the case of lower-class migrants.[16] High-class migrants also show greater anxiety if they are not actively involved with a close-knit kin group.[17] Why these patterns do not hold for lower-class migrants is unclear; perhaps an aggressive confrontation with urban life (a characteristic of the high-class families) tends to generate tensions as well as to erode familistic norms. It may also be that high-class migrants view the branch-family network more as a problem-solving unit (note the negative relationship between size of kin group and extent of worry) and, for that reason, those who sense they are somewhat "alone" in the area of destination tend to worry a great deal more about various difficulties they perceive or have encountered.

Only in the case of intermediate-class migrants does a strong association appear between expressed happiness and frequency of visiting nearby kin.[18] This suggests the presence of situational or attitudinal factors which are more or less unique to the intermediate-class migrant and which, in effect, are necessary conditions for the relationship to become manifest. Compared with other migrants, intermediate-class migrants tend to be more familistic; the familistic orientation of the mountain subculture is *reinforced* by situational circumstances in the area of destination which typify the settlement pattern of the intermediate class family groups from Beech Creek. Kin interaction, then, is and remains an important determinant of happiness for intermediate-class Beech Creek migrants in particular.

With respect to the pattern of relationships between indicators of involvements with noneastern Kentuckians and various aspects of psychological adjustment, we find that only one relatioship emerges when we control on social class origin. Intermediate-class migrants who feel they have very little contact with urban natives are more likely to show symptoms of psychological anxiety.[19] This fact can be understood more clearly in the light of related findings. Female migrants generally express greater anxiety than male migrants[20] and, in their case, the less social contacts with urban natives, the greater the level of anxiety.[21] Furthermore, though intermediate-class migrants as a category express only slightly lower levels of anxiety, their level of social contacts with urban natives is considerably less than other migrants. If we are willing to make inferences from these patterns of interrelationships, we again arrive at the conclusion that the ecological situation, i.e., the relative cultural isolation of intermediate-class Beech Creek migrants, is an important factor in explaining the phenomenon in question. Female migrants from intermediate-class families are the more culturally isolated[22] and, for that reason, a larger proportion of them manifest symptoms of psychological stress. In other words, anxiety is more characteristic of female than of male migrants and it is simply a situational fact that intermediate-class female migrants in this case do not have as much opportunity for informal social contact with people from "outside" the mountain society.

It is especially interesting that no other relationships are noted (of reliable strength) between the indicators of urban social ties and psychological adjustment when social class is controlled. This aspect of interactional adjustment may be a phenomenon that has little bearing upon other aspects of migration-adjustment; the independent effect of urban social ties appears to be negligible.[23] To the extent that these indicators measure the migrants' assimilation into the informal structure of urban life, the observations made here are noteworthy for future research.

Considering the relationships between indicators of involvement with stem-family and psychological adjustment, we find that intermediate- and lower-class migrants tend to conform more to the expected pattern than higher-class migrants. Though high-class migrants, on the average, visit as much in eastern Kentucky as other migrants, such visiting behavior is not associated with any of the psychological adjustment indicators. Though high-class migrants exchange letters with people in eastern Kentucky less frequently, on the average, than lower-class migrants, those who correspond frequently with kinsfolk and friends in the mountains are not likely to be either more or less adjusted psychologically than those who do not. In short, migrants from high-class Beech Creek families, whether because of situational realities or orientational adjustments to situational realities, do not turn to the mountains in time of stress nor in their search for identity and stability; they do, however, rely to some degree upon the branch-family network as a stabilizing structure and problem-solving unit.

Stem-family ties, on the other hand, play some part in the psychological adjustment of migrants from low-class families in Beech Creek. Feelings of anomia, for example, which vary inversely with social class origin, are negatively associated with both visiting and letter exchanges by them with persons in eastern Kentucky. The greater prevalence of anomia among low-class migrants, then, may be explained by the severance or modification of interactional ties with family and friends in the mountains—a consequence of migration—coupled with an apparently strong need for such interactional ties. But we find too that those low-class migrants who have severed or modified ties with family and friends in the mountains—who visit less often in eastern Kentucky—are inclined to be more residentially stable in the urban community yet show symptoms of greater anxiety. This is not surprising if we consider that anxiety may be fostered by urban involvement, that these two aspects of adjustment— anxiety and residential instability[24]—tend to be positively associated, and that residential instability may be reinforced by interaction with stem-family. Low-class migrants were "rootless" in many respects even in the Beech Creek neighborhood situation. In their encounter with urban, industrial society, the insecurity of not having a place—a homestead—back in the mountains to which they can return in time of stress and to which they can cling as a symbolic refuge and source of identity undoubtedly contributes to their feelings of despair and hopelessness. And because they thereby assume (or must assume) an attitude of permanency in the host community, the process of alienation may have been encouraged.

In the case of intermediate-class migrants, we observe the "cushioning" effect of the stem *and* branch family network. The familistic orientation of these migrants, as pointed out earlier, tends to be reinforced by their characteristic pattern of settlement in the area of destination which itself is a consequence of the stem-family form of migration. They are happiest when actively involved with a close-knit family group; then too they are less inclined to worry about things and not as likely to experience extreme nostalgia for home and the mountain way of life.

Intermediate-class migrants, however, who for one reason or the other encounter some difficulty in adapting to the new situation, who cling to an identification with their mountain "homestead," can and do return to the mountains for "visits" and presumably for familistic and cultural reinforcement. This may be a transitional phase of the adjustment process. In any event, the stem-family structure seems to provide a "haven of safety" for those who have not been able to satisfy their interactional needs through the branch-family network. Cohesive family structures at both ends of the migration stream serve to complement each other in the reduction or alleviation of tensions resulting from the process of adjustment.

We note, nevertheless, that the level of anxiety of intermediate-class migrants varies inversely with the amount of social contact they experience with persons native to the urban area. This phenomenon, we believe, is a manifestation of the fact that these migrants (especially the women) tend to be more socially isolated by reason of their pattern of residential location. Because of generally high social interactional needs (Kentuckians like to "socialize") and the need to feel accepted by others (outside of the immediate family circle), frustration of such needs undoubtedly generates some tensions and anxieties. In other words, we believe that anxiety and feelings of social ostracism are aspects of the same syndrome and that the syndrome occurs in cases, such as these intermediate-class migrants (and especially the women), where a high need for social acceptance encounters either real, socially structured, or self-imposed barriers to social interaction.

Summarization and discussion

A brief summarization should be attempted at this point. We are aware, of course, that many of our interpretative statements are tenuous, even within a framework of empirical observations. Further verification and elaboration, both theoretical and empirical, are necessary before an explanatory model will emerge, before a developmental sequence can be posited, and before the relative importance of social class origin in the process of an individual's adjustment to situational circumstances associated with rural to urban migration can be understood. Nevertheless, let us review what we have found in the Beech Creek case.

The branch-family network, it appears, performs a supportive role for migrants from high-class families in Beech Creek by providing them with an intimate group to which they can turn for help and advice about problems before such problems become internalized in the form of psychological tensions. High-class migrants, who in many ways even prior to migration had been more committed to the value standards of middle-class America than other Beech Creekers, utilize their kinship group for much the same purposes and in much the same manner as we might expect among a "normal" American population. The notion of "family homestead" and its supportive functions seems to be inapplicable in this case. It is also plausible, of course, that those high-class migrants who are more fully committed to middle-class American norms and values and hence experience a greater degree of tension have modified their earlier close relationships with the branch-family network as a consequence of the urban encounter. In any event, if generalizations can be drawn from these observations, we must consider that "interaction with family and friends in the mountains is *not* an important factor in the psychological adjustment of migrants from high-class Beech Creek families." These migrants and their adjustment patterns, for one reason or the other, appear to represent a later phase in the migration of Beech Creek family groups.

The branch-family network, in the case of migrants from intermediate-class Beech Creek families, similarly performs an important integrative and supportive function. But, unlike the pattern for high-class migrants, involvement with persons in eastern Kentucky also shows evidence of being a stabilizing influence. Our data suggest the presence in this system of a "haven of safety;" i.e., a place with kin and friends in the mountains to which the migrant feels he can return, and indeed does, if adjustment problems are encountered in the area of destination. Such a "cushioning" bond with the stem-family homestead does not seem to exist, at least in the same form, for migrants from high-class Beech Creek families; that may be because of class differences in cultural orientation and/or differences in the stage of migration of the family groups from which the migrants originate. Furthermore, there is some reason to believe that the pattern of residential location of intermediate-class migrants affects the patterns of adjustment; familistic bonds are reinforced by social and cultural isolation from the urban context.

The complementary interplay between stem and branch-family networks, which, in the case of intermediate-class migrants, serves as a supportive framework providing stability at both or either ends of the migration stream, does not seem to exist in the case of migrants from lower-class families in Beech Creek. For the psychological adjustment of lower-class migrants appears little affected by branch-family involvements; feelings of despair and insecurity may be a normative condition among these families, and the family group, as an entity, may be unable to provide the necessary aid (economic, social, psychological) to help its members satisfy their changing needs. Furthermore, and consequently with serious implications because of their great need for economic, social, and psychological support, a "cushioning" bond with stem-family homestead such as exists for intermediate-class migrants, serves as a stabilizing factor only in a limited sense in the case of lower-class migrants. Among lower-class migrants, the transitional pattern (stabilizing functions performed by the branch-family network) appears to be absent; hence, seeking stability from an unstable situation in the mountains without complementary support from a cc¹ ·ve branch-family network can and probably does delay the transitional adjustment process.

In many ways and for a variety of reasons, then, when compared with migrants from high- and lower-class Beech Creek families, the pattern of adjustment of migrants from intermediate-class families appears to be modeled more along the lines of a stem *and* branch family system of rural to urban migration. Nevertheless, there is little doubt that the kinship factor—whether in the form of a cohesive stem-family, branch-family, or complementary stem- *and* branch-family network—enters into and affects various aspects of adjustment among all classes of migrants.

Conclusions

In the Beech Creek case over the years 1942-1961, an individual migrant's social class origins influenced, to some extent, not only when he left the mountains, where he moved, and with whom, but also his economic life chances in the area of destination and his subsequent patterns of social and psychological adjustment.[25] The latter were affected, in no small part, by social class differences in the form or strategy of migration. By the very nature of the sequence of "irrevocable decisions" and events, for example, the patterns of out-migration contributed to the maintenance and stability of the social class hierarchy within that migration system despite the seemingly disruptive phenomenon of out-migration. Even for a rural low-income population, then, social class origin seems to have validity in predicting a migrant's abilities to cope with the external environment and, consequently, his chances of enhancing his own and his family's economic well-being in an urban area.

Furthermore, we have observed throughout the Beech Creek case how the kin network functions as a stabilizing structure helping migrants satisfy their basic needs, fulfill the responsibilities of their major roles, and realize the value ends of the interactional system of which they are a part while maintaining the identity and integrity of their individual selves. Yet we have also observed that in the process of adjustment the distinctive needs of a particular segment of the migrant population (in this case, for example, the various social classes) coupled with the responsiveness and capabilities of the family-kin network to satisfy, or to provide for, or perhaps to block the attainment of those needs, determines, to some extent, the degree of tension generated within, and the personal stability of, these migrants. In short, we have noted that (and, in some respects, "how") various social classes differ in their patterns of adjustment.

The results of this study, although by no means conclusive, have, we believe, important bearing for future research and for designing more effective programs of human development in similar areas of rural poverty. Among other things, it is hoped that our findings and suggested interpretations will sensitize other researchers engaged with problems in the sociology of migration to the differential needs of and responses by various segments of a migrant population to the problems and stresses concomitant with the migration process.

NOTES

1. For an annoted bibliography of the contemporary literature on migration, and for a statement outlining the conceptual system employed in the design of this study, see Mangalam (1968).

2. When we eliminate from the 271 study population the 76 in the three classes who did not leave eastern Kentucky and also the 24 persons not put into classes in the original study, there are 171 persons left.

3. We do not claim to have established a causal relationship between social class and time of out-migration. Our discussion of the observed concomitant variation is offered as a plausible explanation consistent with our knowledge of the Beech Creek case. Nevertheless, even in this case a number of other factors *may* account for differences in the timing of out-migration; e.g., differences in age and stage in family life cycle (suggested elsewhere in this paper). Because of the relatively small number of family units included in this study, the introduction of appropriate controls was not feasible.

4. An earlier article from another study focused on the differences in economic life chances between young male migrants and non migrants from eastern Kentucky. See Schwarzweller (1964). We would have liked to consider the length of time spent in "the outside world" as a variable in the present study. This was not feasible, given the small number of individual and family units. Furthermore, the assumptions involved in the use of that variable would generate great difficulties in the interpretation of findings: the social stratification of Beech Creek (i.e., in the mountain neighborhoods) has not been a static structure over the years.

5. An important methodological correction should be mentioned. We were studying a total migration universe and a small number of cases. Our methodology, then, became greatly complicated by various kinds of unique control problems, such as the choice of a proper (i.e., valid) "unit of analysis." For example, some Beech Creekers were married to other Beech Creekers prior to out-migration, a few married after 1942, and many married persons from "outside." The problem, then, was to reconcile a comparison between social class origin, a concept or attribute linked to the family, with socioeconomic status, a similarly linked "family attribute." We attempted to resolve such difficulties by the following procedure:

Our focus of analysis shifted to the Beech Creek migrant's family-household in the area of destination. We assumed, and the assumption was checked for validity insofar as possible, that only in rare instances did a Beech Creeker marry across class lines. Where the male head of a household was included in the study population, his social class origin was ascribed to that household. Where a female homemaker had married outside the study population, her social class origin was ascribed to that household since we had no information about the social class origin of the male head; in most cases, however, the husband was an eastern Kentuckian and the assumption of comparability, we would argue, was valid. In those cases where the husband was not an eastern Kentuckian, we assumed that social class differences in origin were not very great.

6. This conception of extended family is very like that suggested by Sussman and Burchinal (1962: 320).

7. Analysis of the results of this definition shows that almost all kinsmen who have migrated are taken into account except those who are very far removed from the migrant's circle of visiting (e.g., in California, Florida, etc.).

8. Information obtained by interview was cross-checked with follow-up data accumulated over the years.

9. Because similar information was obtained from persons in the area of origin, reliability checks, through cross reference, could be and were made.

10. We asked: "In your daily life, how much contact do you have with people who are natives of this area? I mean people who were born and raised here in Ohio, Indiana, etc.—not in eastern Kentucky."

11. We asked: "About how many of these people (natives of this area) would you say were close enough friends so that you meet in one another's home?"

12. Items were selected by item-analysis from a battery of questions. The scale includes categorized responses to such questions as, "Would you prefer to live somewhere else?" and "Where would you prefer to live when too old to work?"

13. It includes such items as, "Where do you feel is really your home?" and "Where would you prefer to be buried someday?"

14. In addition to social class origins, we also considered: sex, level of schooling, level of living in the area of destination, church membership, age, and length of urban residence. Our findings in the larger program of analysis, which in no way invalidate any of the observations made here, were used to bolster the interpretations offered in this paper.

15. In the accompanying tables, all observed relationships are reported which fall within the .80 level of probability on a chi-square test of difference. One should not interpret these, in any strict sense, as tests of significance (this is a total population, not a sample). The statistical criterion is used to indicate the existence of concomitant variation among variables. The coefficient of contingency, reported only where observed relationships are at or above the .90 level, can be regarded as a crude estimate of degree of association. Since the maximum value of C depends upon the number of cells in the contingency table, a correction factor is introduced so as to approximate more closely a correlation coefficient; see McCormick (1941: 207).

16. The original relationship, for the total migrant population, is -.23.

17. The original relationship, for the total migrant population, is -.24.

18. The original relationship, for the total migrant population, is +.26.

19. The original relationship, for the total migrant population, is -.21.

20. $\bar{C} = + .49$.

21. $\bar{C} = - .34$.

22. Compared with male migrants, female migrants have fewer friendship ties with urban natives ($\bar{C} = .31$) and do not visit as much in eastern Kentucky ($\bar{C} = .22$).

23. For the total migrant population, urban social ties are negatively associated with anomia, anxiety, and nostalgia for home.

24. $\bar{C} = + .24$.

25. Whether such influences were apart from or independent of kinship and factors associated with kinship was not demonstrated. In the Beech Creek case social class and kinship are, in empirical terms, inseparable.

REFERENCES

BROWN, J. S. (1952a) The Farm Family in a Kentucky Mountain Neighborhood. Lexington: University of Kentucky Agricultural Experimental Station, Bulletin 587 (August).
--- (1952b) The Family Group in a Kentucky Mountain Farming Community. Lexington: University of Kentucky Agricultural Experimental Station, Bulletin 588 (June).
--- (1952c) "The conjugal family and the extended family group." American Sociological Review 17 (June).
--- (1951) "Social class, intermarriage, and church membership in a Kentucky community." American Journal of Sociology 57 (November): 233.
--- (1950) "The social organization of an isolated Kentucky mountain neighborhood." Unpublished Ph.D. dissertation, Harvard University.
BROWN, J. S. and G. A. HILLARY, JR. (1962) "The great migration: 1940-1960." In T. R. Ford (ed.) The Southern Appalachian Region: A Survey. Lexington: University of Kentucky Press.
BROWN, J. S., H. K. SCHWARZWELLER and J. J. MANGALAM (1963) "Kentucky mountain migration and the stem-family: an American variation on a theme by LePlay." Rural Sociology 28 (March): 66.
DANLEY, R. A. and C. E. RAMSEY (1959) Standardization and Application of a Level of Living Scale for Farm and Nonfarm Families. Ithaca: Cornell University Agricultural Experimental Station, Memoir 362 (July).
DAVID, A. and J. DOLLARD (1940) Children of Bondage. Washington, D. C.: American Council of Education.
GURIN, G., J. VEROFF and S. FELD (1960) Americans View Their Mental Health. New York: Basic Books.
HUGHES, C. C., M. TREMBLAY, R. W. RAPOPORT and A. H. LEIGHTON (1960) People of Cove and Woodlot. New York: Basic Books.
MANGALAM, J. J. (1968) Human Migration. Lexington: University of Kentucky Press.
McCORMICK, T. C. (1941) Elementary Social Statistics. New York: McGraw-Hill.
NORTH and HATT (1952) "Jobs and occupations: a popular evaluation." Pp. 411-426 in R. Bendix and S. M. Lipset (eds.) Class, Status, and Power. Glencoe, Ill.: The Free Press.
PARSONS, T. (1940) "An analytical approach to the theory of social stratification." American Journal of Sociology 45 (May): 850.
SCHWARZWELLER, H. K. (1964) "Education, migration, and economic life chances of male entrants to the labor force from a low-income rural area." Rural Sociology 29 (June).
SCHWARZWELLER, H. K. and J. S. BROWN (1967) "Social class origins, rural-urban migration, and economic life chances: a case study. Rural Sociology 32 (March): 5-19.
SROLE, L. (1956) "Social integration and certain corollaries: an exploratory study." American Sociological Review 21 (December): 709-716.
SUSSMAN, M. S. and L. BURCHINAL (1962) "Parental aid to married children: implications for family functioning." Marriage and Family Living 24 (November): 320.

Chapter 6

Differential Experience Paths of Rural Migrants to the City

ROBERT C. HANSON
and OZZIE G. SIMMONS

The migration of rural people to cities all over the world creates serious problems for action agencies concerned about the successful integration of new residents into the urban community. It is obvious that after a few years in the city, the variety of adjustment outcomes among the migrants is startling, ranging from complete dependence on the city's welfare programs, through marginal survival, to thriving growth and exemplary citizenship. While we all agree that action programs should be based on knowledge of the situation, action agencies are more likely to establish programs on the basis of observed needs rather than on knowledge of the adjustment process. One agency, for example, may set up a skill training program based on the observation that many rural migrants arrive in the city lacking formal education or marketable work skills. The agency assumes that such migrants will need help if they are to make a successful adjustment.[1] But we know that other migrants, with similar entering attributes, do make successful adjustments. Another agency may set up a legal aid clinic, after observing that many rural migrants get into financial trouble and other legal difficulties. Here the agency assumes that adjustment depends on emergent phenomena, that

Author's Note: *The project, "Urbanization of the Migrant: Processes and Outcomes," was supported by the National Institute of Mental Health Grant No. MN 09208. We wish to thank Fu-Chin Shih and Geraldine Macdonald for assistance in the analysis and preparation of data presented here. This research was conducted in the Program of Research on Social Processes, Institute of Behavioral Science, University of Colorado.*

is, on what happens after the migrant settles into the life of the city. No doubt both agencies provide needed services. But it would be useful to know how some unskilled migrants make a successful adjustment despite the disadvantages they come with, and what kinds of migrants, through what processes, get into financial and legal trouble.

In this paper we illustrate a method for examining the adjustment process. The urban experiences of four groups of migrants provide the data for the analysis. The composition of the groups is held constant while we observe what happened to each group on a series of variables describing their urban experiences over time. Such an examination yields an interpretation of some of the conditions and processes producing the variety of adjustment outcomes evident in the case histories of rural migrants to the city.

Method

We classify migrants at the beginning (e.g., on some arrival attributes) and at the end of a period of time in the city (e.g., on an indicator of adjustment). Then we examine what happened over this period of time for the different comparison groups. The results suggest an interpretation of the intervening steps between the starting point and the outcome. This inductive procedure helps us find out what happens to different types of migrants. Once the processes are understood, a later stage of inquiry can be deductive. Then we need not classify migrants by outcome types; rather, differential experience paths generated by the operation of a theoretical model yield the alternative outcomes of groups classified only by their entering attributes.[2]

Time Trend Data

The procedures used to collect data on the experiences of rural migrants to the city have been described elsewhere (Hanson and Simmons, 1968: 152-158). The data base for the generation of time trends can be described briefly. Detailed case history interviews were obtained from sixty-six Spanish-American rural migrants to the city of Denver. Their experience in the city ranged from two to seven years with an average of about five years. After conceptualizing a model which simulates the interaction of the migrants with the opportunities they confront in the city, a detailed code book was prepared. Coders then were able to describe the changing situations of each of the sixty-six migrants, month by month, for their total urban experience. The month-by-month data on each migrant (a total of 3,415 man-months) were then punched on IBM cards. A series of factor analyses were undertaken to reduce the large number of concrete variables to a smaller number of theoretical concepts and indexes. Then typical demographic variables (such as age, family size, and socioeconomic status) and other constructed variables drawn from the factor analyses (such as a "family cycle index" and a "degree of rootedness" index), a total of eighty-four, were standardized to values between .00 and .99. These time-oriented data were organized for two kinds of analyses: (a) multiple regression analyses to produce predictive equations needed in the simulation model, and (b) time trend analyses which reveal social processes characteristic of variously defined groups of migrants.[3]

In the time series program the variables were ordered month by month for the first forty-eight months of each migrant's experience in the city.[4] For each run, the program provides for the definition of up to four comparison groups. The composition of groups is then held constant while behavior on up to four other subsequent variables is followed through time.[5]

Figure 1 illustrates the kind of output available from the program. We have divided the migrants into two groups based on a measure of their socioeconomic status at the time they arrived in the city. Then these two groups of migrants, thirty-five who were at or above the mean on socioeconomic status versus thirty-one who were below the mean, were fixed while a variable called financial independence, which is the complement of the amount of welfare help received, is followed over time for the two groups. It is clear from the figure that the advantaged

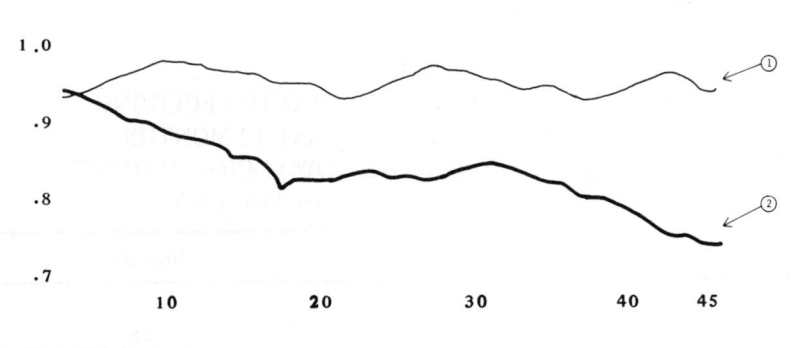

(1) High SES ⩾ 0.46, 1st month N = 35
(2) Low SES < 0.46, 1st month N = 31

Figure 1. FINANCIAL INDEPENDENCE (I.E., COMPLEMENT OF WELFARE HELP RECEIVED) FOR MIGRANTS CLASSIFIED BY SOCIOECONOMIC STATUS ON ARRIVAL IN THE CITY (SMOOTHED BY A MOVING AVERAGE OF ORDER SIX)

migrants, those with higher socioeconomic status, remained far more financially independent than the disadvantaged migrants. The lower status group received welfare help early and continued to receive more welfare help over time than did the higher status group.

In the higher status group, however, ten out of the thirty-five received welfare help at some time during the thirty-sixth to forty-eighth months (or during the last twelve months prior to their interviews). Among the thirty-one persons in the lower status group, on the other hand, fourteen of the thirty-one managed to stay financially independent during the comparable period. As shown in Table 1, there are twenty-five persons who arrived with higher socioeconomic statuses and who received no welfare help at all during the last twelve of forty-eight months in the city. For present purposes we will label this group "The Thrivers." The ten persons who arrived in an advantaged position but failed to remain completely financially independent during the last twelve months are called "The Stumblers." The third group of fourteen migrants who arrived in the city in a disadvantaged status but managed to stay off the welfare rolls in the last twelve months are labelled "The Strugglers." And finally, the seventeen migrants who arrived in a disadvantaged status and who ended up receiving welfare help during the last twelve months are called "The Losers."

Analysis of the time trends of these four groups provides plausible answers to the questions: (a) Why do some advantaged migrants end up on welfare? (b) How do some disadvantaged migrants manage to stay off welfare?

TABLE 1
THE DISTRIBUTION OF MIGRANTS RECEIVING
WELFARE HELP IN THE LAST 12 MONTHS
CLASSIFIED BY HIGH AND LOW SOCIOECONOMIC
STATUS ON ARRIVAL IN THE CITY

	Low SES	*High SES*
No Welfare Help	14	25
Welfare Help	17	10
	31	35

The Consequences of Unstable Employment

Figure 2 shows the trends of the four groups on an index of employment. (The standardized index takes into account whether the migrant was fully, partially, or not at all employed this month, plus the number of months totally unemployed since the last job.) Within six months, the Thrivers establish a high level, stable degree of employment. In contrast, from the very beginning the Stumblers exhibit an irregular, fluctuating, unstable pattern of employment.

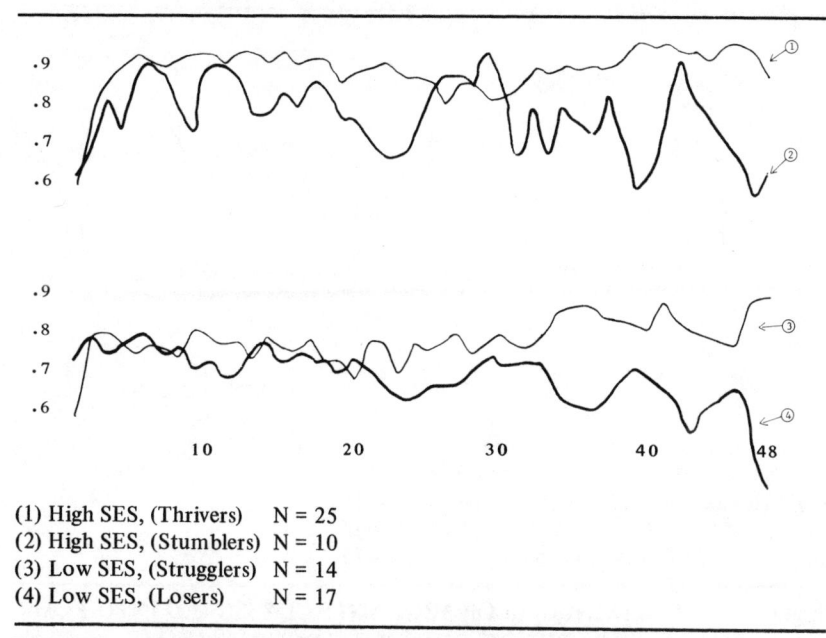

(1) High SES, (Thrivers)   N = 25
(2) High SES, (Stumblers) N = 10
(3) Low SES, (Strugglers) N = 14
(4) Low SES, (Losers)   N = 17

Figure 2. DEGREE OF EMPLOYMENT OF MIGRANTS CLASSIFIED INTO FOUR TYPES OF OUTCOME GROUPS

There is also an evident contrast between the Strugglers and the Losers. The Strugglers establish a slowly but steadily rising trend of employment whereas the trend for the Losers is one of gentle but steady decline.

The final outcomes for the four groups are, of course, not independent from our initial classification of groups based on welfare help in the last twelve months. The relative lack of employment among the Stumblers and Strugglers at the end determines, in part, their need for welfare assistance at that time.

One of the consequences of the different employment trends is reflected in Figure 3, which shows the monthly earned income among

[150] BEHAVIOR IN NEW ENVIRONMENTS

the four groups. The Thrivers, by the end of the first six months, are at the top in earned income and remain at a consistently high level. Reflecting their employment trend, the Strugglers show a slight and fairly steady improvement in income over time. In contrast, the Losers start, within the first year, a steadily declining trend ending up with very little monthly earned income after four years in the city. Finally, the Stumblers exhibit the same cyclical and fluctuating pattern characteristic of their degree of employment.

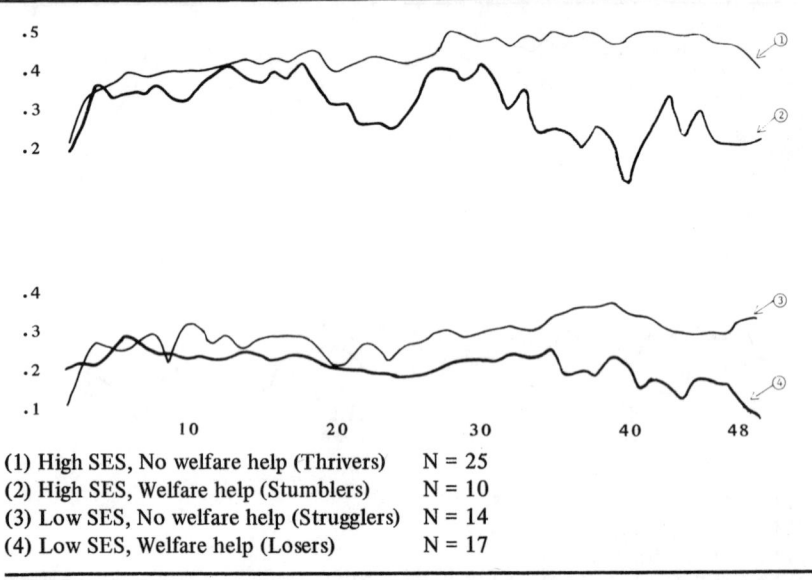

(1) High SES, No welfare help (Thrivers)   N = 25
(2) High SES, Welfare help (Stumblers)   N = 10
(3) Low SES, No welfare help (Strugglers) N = 14
(4) Low SES, Welfare help (Losers)   N = 17

Figure 3. EARNED INCOME OF MIGRANTS CLASSIFIED INTO FOUR TYPES OF OUTCOME GROUPS

We expect, of course, that employment and income patterns are so fundamental to survival in the city that the consequences of these patterns will appear in other measures of economic well-being or of integration into the social structures of the city. A fair indicator of the standard of living among the four groups is shown in Figure 4. The curves trace over time the monthly average standardized value of an index of total expenses. The index includes the monthly cost of rent or mortgage payments, food and general living expenses, current purchase payments, leisure activity expenditures, and medical and legal costs.

The fairly steady increase in spending over time is apparent for both the Thrivers and Strugglers. Again, the early high, later low and generally unstable, fluctuating path of the Stumblers is obvious. Finally, the Losers exhibit a fairly stable, low level of total expenditures with a slight downward trend appearing in the third year. The generally higher values for both the Thrivers and Stumblers suggests that amount of customary spending is partly determined by socioeconomic status at the time of arrival in the city.

Figure 5 shows the month-by-month averages on a "rootedness" index, an indicator of the degree of stable integration of a migrant

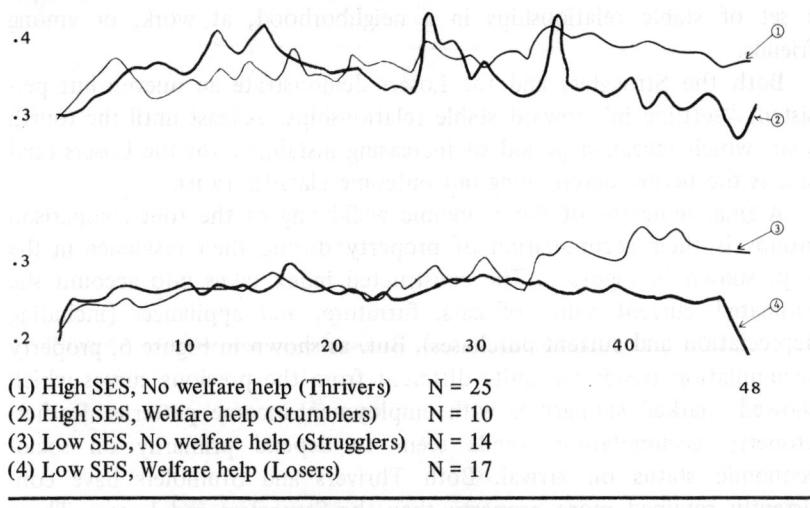

(1) High SES, No welfare help (Thrivers)   N = 25
(2) High SES, Welfare help (Stumblers)   N = 10
(3) Low SES, No welfare help (Strugglers) N = 14
(4) Low SES, Welfare help (Losers)   N = 17

Figure 4. TOTAL EXPENSES OF MIGRANTS CLASSIFIED INTO FOUR TYPES OF OUTCOME GROUPS

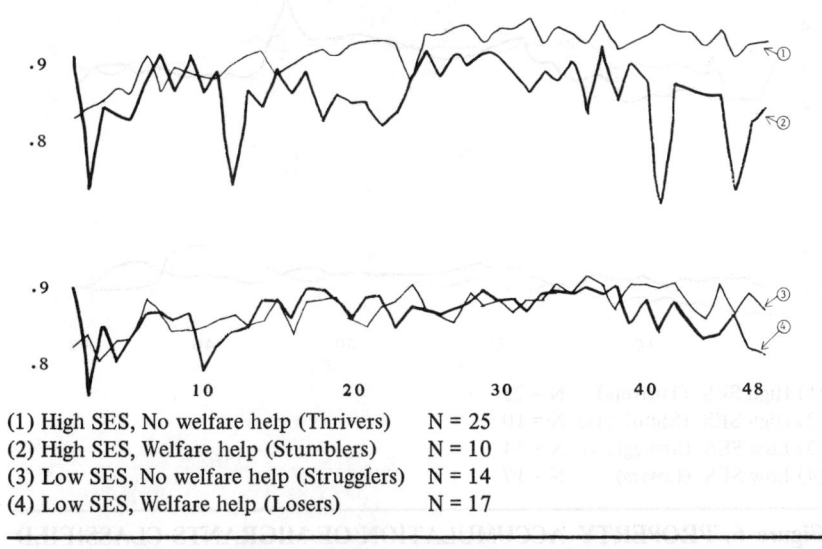

(1) High SES, No welfare help (Thrivers)   N = 25
(2) High SES, Welfare help (Stumblers)   N = 10
(3) Low SES, No welfare help (Strugglers) N = 14
(4) Low SES, Welfare help (Losers)   N = 17

Figure 5. ROOTEDNESS INDEX OF MIGRANTS CLASSIFIED INTO FOUR TYPES OF OUTCOME GROUPS

[152] BEHAVIOR IN NEW ENVIRONMENTS

within social structures of the city. The index takes into account recent turnovers in housing, employment, and friendships: the more stable the situation of the migrant, the higher the index.

The trends of the Thrivers and Stumblers clearly reflect their employment-income patterns: a persistent improvement over time for the Thrivers and a highly unstable, fluctuating pattern for the Stumblers. After four years in the city, as a group the Stumblers have not achieved a set of stable relationships in a neighborhood, at work, or among friends.

Both the Strugglers and the Losers demonstrate an uneven but persistent "settling in" toward stable relationships, at least until the fourth year, which reveals a period of increasing instability for the Losers (and also is the period determining our outcome classification).

A final indicator of the economic well-being of the four comparison groups is their accumulation of property during their residence in the city, shown in Figure 6. The constructed index takes into account the estimated current value of cars, furniture, and appliances (including depreciation and current purchases). But, as shown in Figure 6, property accumulation trends are quite different from the previous curves which showed marked similarities with employment-income patterns. Rather, property accumulation trends seem to depend primarily on socioeconomic status on arrival. Both Thrivers and Stumblers have consistently retained more property than the Strugglers and Losers. There

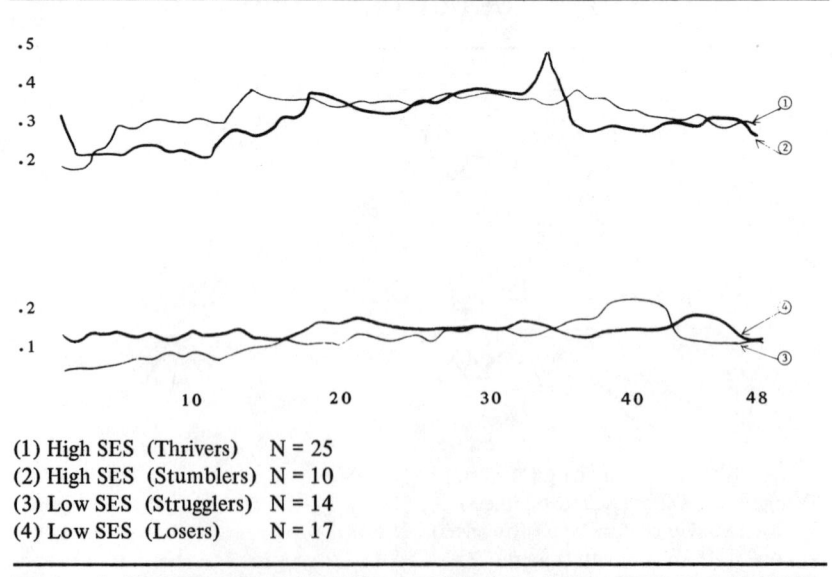

(1) High SES (Thrivers)   N = 25
(2) High SES (Stumblers)   N = 10
(3) Low SES (Strugglers) N = 14
(4) Low SES (Losers)   N = 17

Figure 6. PROPERTY ACCUMULATION OF MIGRANTS CLASSIFIED INTO FOUR TYPES OF OUTCOME GROUPS

appears to be little difference between groups in the same socioeconomic status level.

The consequences of unstable employment may be summed up as follows. Among migrants entering the city at the same level of socioeconomic status, dramatic differences emerge between groups depending on their employment experience. Stable employment, for either high or low status groups, leads to slowly increasing earned income, a persistent rise in the standard of living (as measured by total monthly expenditures), and stable integration into the city's social structures (as indicated by a "rootedness" index). For migrants relatively advantaged on arrival in the city, unstable employment produces an erratic, fluctuating income pattern which, in turn, leads to a highly variable spending pattern (and thus, perhaps, to property repossessions, bankruptcy or legal problems, as will be seen later), and to a failure to achieve stable social relationships in the city. Employment instability is the main factor differentiating the experience paths of higher status migrants who ended up receiving welfare help (the Stumblers) from those higher status migrants who remained financially independent after four years in the city (the Thrivers).

The Consequences of Misfortune Experiences

There are other possible explanations for welfare assistance besides the lack of income arising from unemployment. For example, a family may need outside financial help when the breadwinner is in jail or in the hospital, or to cover the cost of an operation for a family member and so on. These and other misfortunes sap the resources of the family so that the family may then require assistance from welfare agencies in the city.

Figures 7 and 8 provide an indication of the frequency of legal trouble and of the economic severity of legal and medical troubles among the four comparison groups. Figure 7 shows the proportion of each group having some incident of legal trouble each month, where an incident of legal trouble was defined as any one of sixteen specified kinds of trouble (e.g., picked up by the police, fined, jailed, court appearance for any reason) or any other action whereby the migrant or a member of his family was judged by the coders to have been in legal trouble. It is obvious that higher proportions of the Stumblers and Losers faced legal difficulties of some kind for more months (especially during the fourth year when the welfare assistance classification was determined), than either the Thrivers or Strugglers.

When the total cost of illness and legal trouble is plotted month by month, as in Figure 8, no essential differences are apparent among

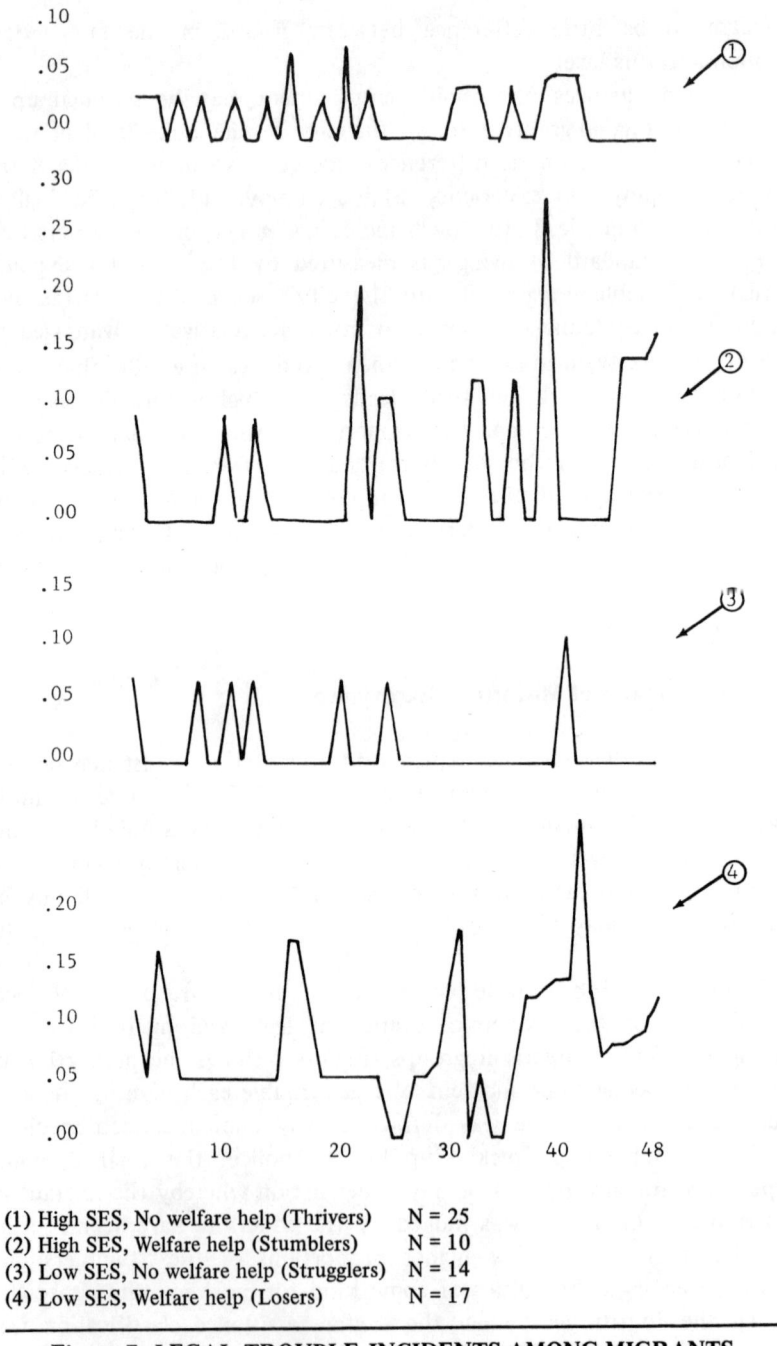

(1) High SES, No welfare help (Thrivers)   N = 25
(2) High SES, Welfare help (Stumblers)   N = 10
(3) Low SES, No welfare help (Strugglers)   N = 14
(4) Low SES, Welfare help (Losers)   N = 17

Figure 7. LEGAL TROUBLE INCIDENTS AMONG MIGRANTS CLASSIFIED INTO FOUR TYPES OF OUTCOME GROUPS

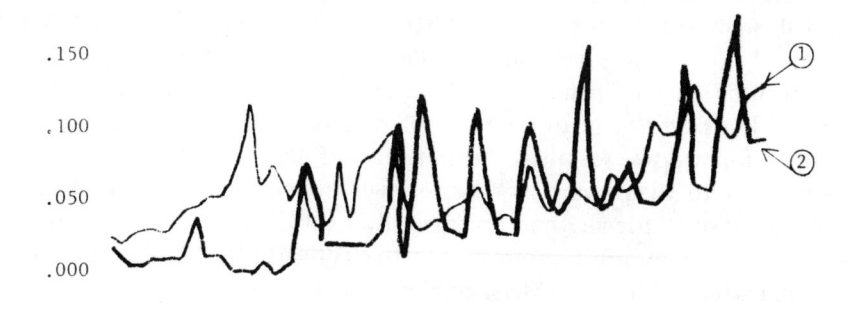

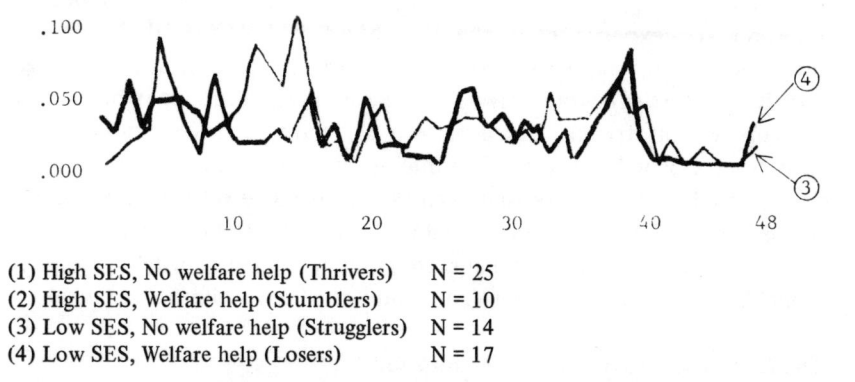

Figure 8. MISFORTUNE EXPENSE OF MIGRANTS CLASSIFIED INTO FOUR TYPES OF OUTCOME GROUPS

(1) High SES, No welfare help (Thrivers)   N = 25
(2) High SES, Welfare help (Stumblers)   N = 10
(3) Low SES, No welfare help (Strugglers)   N = 14
(4) Low SES, Welfare help (Losers)   N = 17

groups at the same socioeconomic status level. All groups demonstrate erratic patterns (especially the Stumblers), but the general trend for both Thrivers and Stumblers is toward an ever increasing burden of medical and legal costs, whereas the level of cost for the lower status groups remains low if not decreasing slightly over the four-year period. The Thrivers and Strugglers experience few incidents of legal difficulties, but the Thrivers show rising misfortune expenses (which includes medical bills) while the Strugglers do not. The Stumblers and Losers experience more legal trouble, but the Stumblers show rising misfortune expenses while the Losers do not.

These results, and those which follow, create a puzzle which no doubt allows for several alternative interpretations of the different adjustment outcomes among the four groups. The somewhat complicated argument developed here starts with the question: how can we account for the relatively high incidence of legal problems, increasing misfortune expenses, and general instability pattern of the Stumblers which leads eventually to their use of welfare assistance? In the figures which follow, we see that the downfall of the Stumblers cannot be attributed to their living in "unsafe, high trouble risk" neighborhoods, nor to deviant friends, nor to especially heavy drinking. We therefore conclude that in direct contrast with the Thrivers, the unstable employment pattern of the Stumblers, combined with their pattern of relatively high spending, causes their legal difficulties (which we assume must be primarily economic—repossessions, bankruptcy, garnishment of wages, etc., though to prove this assumption we would have to decompose the index of legal trouble), and that their increasing misfortune expenses (including hospitalization and medical care associated with births) depends, in part, on their rapidly expanding families throughout their residence in the city. In the course of presenting the evidence to interpret the experience of the Stumblers, we observe their obvious differences from the Strugglers (who appear to be a relatively young, mostly unmarried group living in poor neighborhoods where they develop deviant friends and a pattern of relatively heavy drinking, but who have escaped many legal problems thus far), and from the Losers (older, established impoverished families who move into poor neighborhoods and are in misfortune trouble from the beginning due to poor health on arrival, which perhaps contributes to their unstable employment, later legal problems, and consistent need for welfare assistance).

The Consequences of Poor Neighborhood, Deviant Friends, and Heavy Drinking

Figure 9 demonstrates that selection of neighborhoods to live in after arrival in the city is very much a function of initial socioeconomic status. Neighborhood conditions ranged in a five-category index from "very unsafe" and "poor neighborhood" to "a pleasant and safe residential neighborhood." The trends of the Thrivers and Strugglers indicate slow but fairly steady improvement, while the trends of the Stumblers and Losers are somewhat curvilinear—slow initial improvement with later decline. The differences between status groups remain about the same at the end as at the beginning (i.e., the Losers start low and remain low, the Thrivers high and remain high, while the Stumblers and Strugglers begin to converge because of the decline of the Stumblers and rise of the Strugglers).

.6

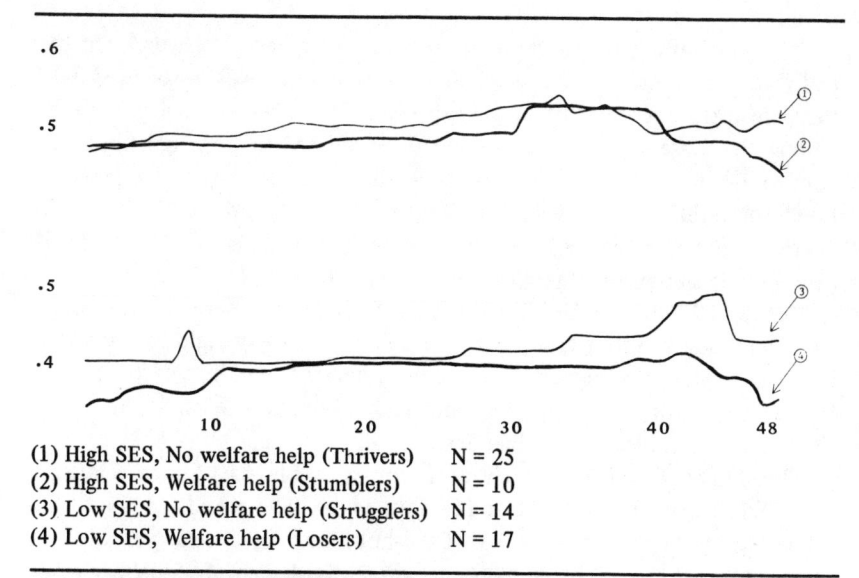

.5

.5

.4

10 20 30 40 48

(1) High SES, No welfare help (Thrivers)   N = 25
(2) High SES, Welfare help (Stumblers)   N = 10
(3) Low SES, No welfare help (Strugglers)   N = 14
(4) Low SES, Welfare help (Losers)   N = 17

Figure 9. NEIGHBORHOOD CONDITIONS OF MIGRANTS CLASSIFIED INTO FOUR TYPES OF OUTCOME GROUPS

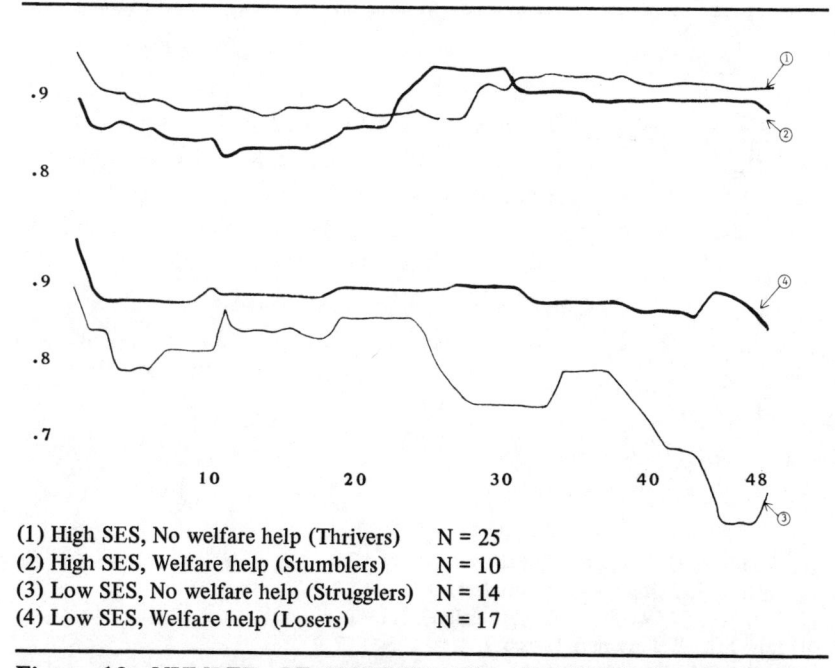

.9

.8

.9

.8

.7

10 20 30 40 48

(1) High SES, No welfare help (Thrivers)   N = 25
(2) High SES, Welfare help (Stumblers)   N = 10
(3) Low SES, No welfare help (Strugglers)   N = 14
(4) Low SES, Welfare help (Losers)   N = 17

Figure 10. NUMBER OF NONDEVIANT FRIENDS OF MIGRANTS CLASSIFIED INTO FOUR TYPES OF OUTCOME GROUPS

[158] BEHAVIOR IN NEW ENVIRONMENTS

It is obvious that adjustment outcome differences between Thrivers and Stumblers cannot be attributed to consequences associated with living in different types of neighborhoods. The incidence of legal trouble among the Losers might be thought to be related to poor neighborhood conditions, where we would expect an increased risk of being jailed or fined for public drunkenness, brawling, traffic violations, etc. Their legal trouble, however, is not related to deviant friends nor to "heavy" drinking, as shown by the curves in Figures 10 and 11.

Figure 10 shows the standardized average number of nondeviant friends among the four groups of migrants for the forty-eight months. The lower the value, the higher the number of deviant friends, where a deviant friend was defined by the respondent himself as a friend who either fights a lot, drinks a lot, or has been in jail.

The Thrivers and Losers show a similar stable pattern of relatively few deviant friends throughout the four years. The trends of the Stumblers and Strugglers are similar for about the first two years; then, surprisingly, the Stumblers begin to drop off deviant friendships while the Strugglers pick up more and more deviant friends over time.

The degree of sobriety among the four groups follows a somewhat similar pattern, as shown in Figure 11. Degree of sobriety is an index

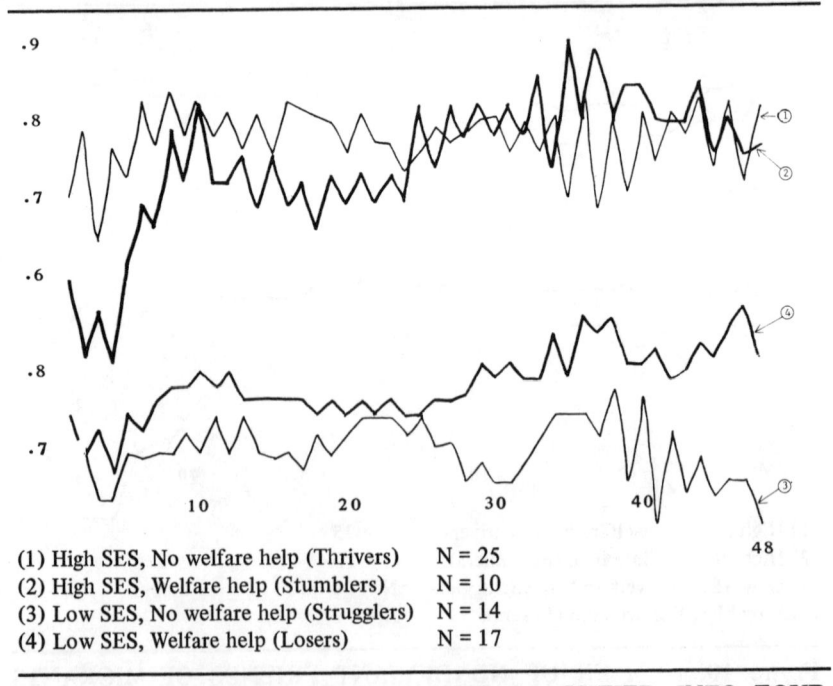

(1) High SES, No welfare help (Thrivers)   N = 25
(2) High SES, Welfare help (Stumblers)   N = 10
(3) Low SES, No welfare help (Strugglers) N = 14
(4) Low SES, Welfare help (Losers)   N = 17

Figure 11. SOBRIETY OF MIGRANTS CLASSIFIED INTO FOUR TYPES OF OUTCOME GROUPS

based on the number of times the respondent was drunk in recent months. The trends of the Thrivers and Losers indicate an early and high-level pattern of sobriety. The evident trend of the Losers toward more sobriety rather than toward more heavy drinking clearly eliminates the hypothesis that their adjustment problems are bound up with the consequences of heavy drinking.

Obviously the Stumblers were the heaviest drinkers during the first year in the city, but they increase their level of sobriety over time, whereas the Strugglers show a tendency toward more heavy drinking, thus showing patterns similar to the deviant friendship patterns in the two groups.

Contrary to possible theoretical expectations, the adjustment problems of the Stumblers and Losers do not appear associated with "bad companions" or "problem drinking." Also, while the trends for the Thrivers turn out as might be expected—to live in better neighborhoods with few deviant friends and with a fairly high level of sobriety—a different pattern emerges for the Strugglers. They start out in poor neighborhoods, and despite their slow general improvement in neighborhood conditions, they demonstrate a friendship and drinking pattern more consistent with a forecast of future trouble rather than with continuing easy adjustment to city life. The deviant friend and heavy drinking pattern of the Strugglers suggests that the composition of this group may be quite different from the others. This hypothesis is borne out in the next section.

Consequences of Rapid Family Expansion and Poor Health

A family cycle index based on current marital status, size of the family, and number of teenagers in the family is plotted month by month in Figure 12. The value of the index is lowest for unmarried men, increases for young marrieds and as births increase family size, and reaches high levels when the family moves into a later established phase with teenaged children.

It is obvious that the composition of the Strugglers is quite different from the three other groups. Many of the members were single or recently married when they arrived in the city. Family expansion does not occur as a strong trend until about two years after arrival. The Losers, in contrast, came into the city with relatively larger, older, established families which slowly continued to expand.

The trends for the Thrivers and Stumblers appear quite similar. Both groups are composed mainly of families in the rapidly expanding phase. The Stumblers, then, in contrast with the Strugglers, faced the costs of

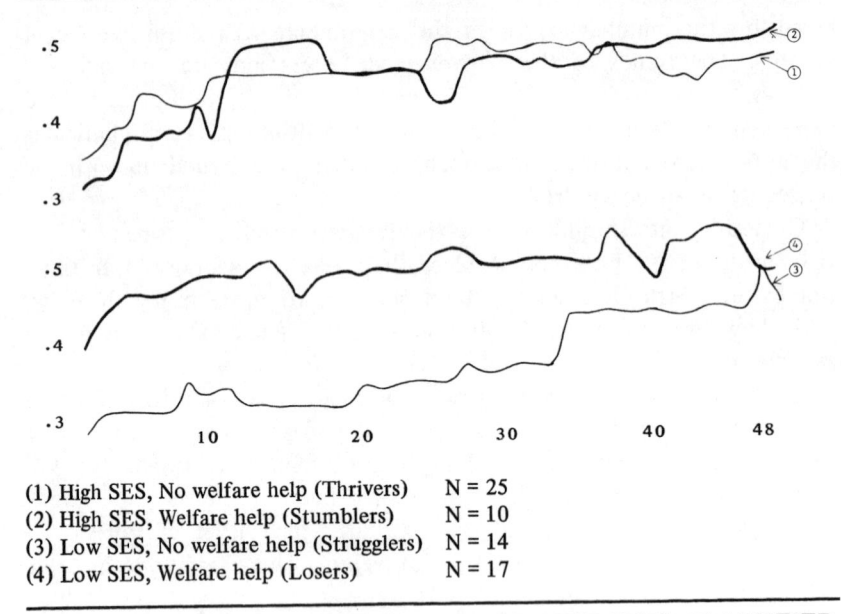

(1) High SES, No welfare help (Thrivers)   N = 25
(2) High SES, Welfare help (Stumblers)   N = 10
(3) Low SES, No welfare help (Strugglers) N = 14
(4) Low SES, Welfare help (Losers)   N = 17

Figure 12. FAMILY CYCLE STATUS OF MIGRANTS CLASSIFIED INTO FOUR TYPES OF OUTCOME GROUPS

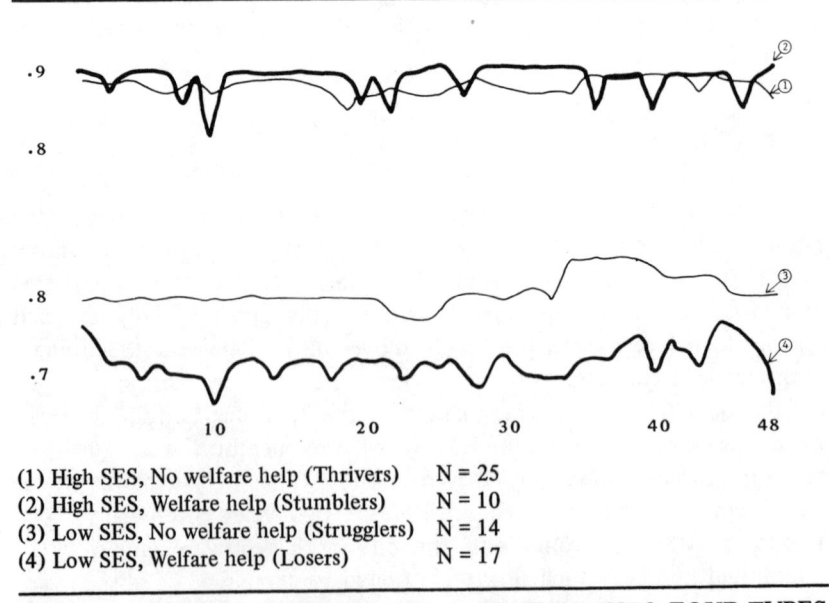

(1) High SES, No welfare help (Thrivers)   N = 25
(2) High SES, Welfare help (Stumblers)   N = 10
(3) Low SES, No welfare help (Strugglers) N = 14
(4) Low SES, Welfare help (Losers)   N = 17

Figure 13. HEALTH OF MIGRANTS CLASSIFIED INTO FOUR TYPES OF OUTCOME GROUPS

supporting extra family members (reflected in their spending and misfortune expenses curves), and, in contrast with the Thrivers, lacked the stable employment and the increasing amount of income needed to avoid dependence on welfare assistance.

Finally, a dramatic difference in the two lower socioeconomic status groups is observed on an indicator of the migrant's relative "good health." The standardized index derives from a combination of the coder's judgments of a migrant's current state of physical health, whether he was physically able to work or not, and his score on a nine-point mental health scale, built into the index as a personal constant. Figure 13 shows that the average Loser arrived in the city in poor health and remained in a relatively poor state of health throughout his urban residence, in obvious contrast with the average member of the other three groups. There can be little doubt that poor health contributed both to their high unemployment rates and to their need for welfare assistance, especially when we have inferred that the Loser group consists mainly of large, already established families.

An Interpretation of the Processes of Rural Migrant Adjustment to the City

It is now appropriate to summarize our conclusions about the characteristic attributes and trends of the four comparison groups, and to propose an interpretation of the adjustment processes of rural migrants to the city. Summary characterizations appear in Table 2.

One major conclusion generated from the summary table is that, among migrants who arrive in the city with an above average socioeconomic status and in good health, unsuccessful adjustment by part of the group cannot be predicted from arrival attributes, but depends on emergent phenomena such as employment and misfortune experience. But among migrants who arrive in the city with a below average economic status, two arrival attributes bear on probable later unsuccessful adjustment: state of health and phase in the family cycle. Healthy, single men or breadwinners in newly-formed families are more likely to maintain financial independence than older migrants who arrive in relatively poor health and who have established, large families to support. The high probability of the need for welfare assistance for the latter group can be predicted from the beginning, because likely experiences of unstable employment and low income follow from the family's arrival characteristics.

Healthy migrants with above average skills and education are likely to integrate into the city within a period of about six to twelve months. They will have accumulated property prior to their arrival, and they

TABLE 2
SUMMARY CHARACTERIZATIONS OF RURAL MIGRANT ADJUSTMENT OUTCOME GROUPS: ON ARRIVAL ATTRIBUTES AND CITY EXPERIENCE TRENDS

	Thrivers	Stumblers	Strugglers	Losers
Attributes On Arrival				
SES	above average	above average	below average	below average
Property Accumulation	above average	above average	below average	below average
Phase in Family Cycle	middle, expanding	middle, expanding	beginning	late
Health	good	good	average	poor
City Experience Trends				
Employment	high level	fluctuating	increasing	declining
Income	high level	fluctuating	increasing	declining
Total Spending	high, increasing	high, unstable	low, increasing	low, stable
Property Accumulation	high	high	low	low
Legal Trouble Incidents	low	high	low	high
Misfortune Expenses	increasing	increasing	low	low
Neighborhood Conditions	good, gets better	good, gets worse	poor, gets better	poor
Friends	nondeviant	increasingly nondeviant	increasingly deviant	nondeviant
Drinking	light	heavy at first, then lighter	heavy	light
Rootedness	high level	low, fluctuating	average	average

move into relatively good neighborhoods. If they settle into a stable job providing steady income, their adjustment to the city appears easy. They expand their standard of living, they can absorb increasing costs of medical care for their expanding families, and they are able to establish a stable residence and stable groups of nondeviant friends.

How does it happen that some of the migrants who arrive in the city in a relatively advantaged position—with higher socioeconomic status, in good health, and with a substantial accumulation of property—end up needing welfare assistance after three years in the city? In our interpretation of the process of adjustment, the primary experience factor is unstable employment. Some otherwise advantaged migrants become locked into a cyclical, seasonal, fluctuating pattern of employment. Unstable employment produces uncertain income. A rapidly expanding family and a customary high level of expenditures for consumer goods produces high average monthly total expenses. But since income is unstable, these migrants run into financial problems leading to relatively frequent legal difficulties—repossessions, bankruptcy, and relatively high misfortune expenses. The instability in jobs and income determines other instabilities: frequent changes of residence and a high turnover in friendships resulting in a failure of these migrants to integrate successfully into the social structures of the city. Consequently, in times of financial trouble, these migrants turn to welfare agencies for assistance. For the average migrant who falls into this failure pattern, his adjustment problems do not appear to derive from poor neighborhood conditions, deviant companions, or problem drinking.

How do some migrants who arrive in the city in a relatively disadvantaged socioeconomic status succeed in maintaining financial independence whereas others end up needing welfare assistance? Of crucial importance from the very beginning are the migrant's health and burden of family responsibilities. Among those who are relatively less educated and skilled, the single, healthy, younger men are more likely to find more stable employment and to earn enough income to support themselves without a high level of total expenses and without welfare assistance. But they are also likely to move into a rooming house environment in poor neighborhoods where it is easy to develop friendships with other men in a similar situation and to spend time drinking with them in bars or at dances. As they marry and establish households, they move to slightly better neighborhoods, begin to raise their level of expenditures and to accumulate furniture, appliances, and other property. With steady employment, fairly stable residence and friendships can be established. But the process of marriage and creation of new families makes for a certain instability in residence and friendships. For the group, therefore, a high level of integration into the city has not yet been achieved. Also, the leisure patterns of these single men or young couples includes the probability of relatively heavy drinking and of developing friendships with other single men or young couples with similar tendencies, that is, to drink, to get into fights, and, thus, perhaps, to land in jail.

There are others in the low socioeconomic status group who arrive disadvantaged because of relatively poor health, with larger, established families to support and having low education and skill levels. They must move into the worst neighborhoods and are least likely to find steady employment from the beginning. The consequences are that earned income is low and decreases over time, property does not accumulate, total expenses remain low, families remain in bad neighborhoods, and welfare assistance is required to support the survival of the family. We infer that the relatively high incidence of legal trouble among this group is due primarily to financial problems such as failure to pay rent or bills, and to juvenile delinquency problems associated with the poor neighborhoods in which they live. The legal trouble incidents do not appear to depend on problem drinking nor on deviant companions among this group of older males with established families. What integration into the city is obtained is negative: stability in poor neighborhoods and increasing dependence on welfare.

Action Implications

In the ideal case, if all rural migrants were received into the city at a central orientation point, only a relative few would require immediate attention: those in poor health, and/or the less educated and less skilled older men with large families to support. Their "experience paths" should be guided and observed from the beginning. The healthy, younger men or newly-formed families are likely to avoid welfare assistance, regardless of skill level, on the basis of their ability to find steady employment (assuming, of course, that there are openings available, if searched for). The majority of those arriving with above average socioeconomic status are also likely to move into good neighborhoods and, with steady income, to begin the process of integration into the social structures of the city. But, ideally, the process should be monitored and helpful steps taken for those who run into misfortunes. Otherwise, some of the advantaged migrants will probably blunder into financial binds and associated legal difficulties, because of unstable employment combined with increasing expenditures for consumer goods and medical costs associated with rapidly expanding families. What these families need most is stable employment with sufficient income to meet their needs. Secondly, they probably need advice on money management and family planning. The goal is to dampen the highly fluctuating pattern of unemployment and job changes, residence changes, and friendship changes which contribute to their failure to integrate into the city.

For most rural migrants to the city, the key to the adjustment process is stable employment which produces a steady income which must then be managed properly. But the typical public or private employment agency provides little help for the migrant needing *stable* employment. Perhaps a working answer is some partial intervention into the network of informal social relations the migrant soon encounters as he develops friendships and acquaintances. A "gatekeeper structure," consisting of recognized ethnic group leaders, could be organized to provide information to incoming migrants (Kurtz, 1966). Ideally, personal friendship or acquaintance linkages would connect the leaders, who have access to reliable and sound information at the top, with the migrants newly arrived in a neighborhood or job. Then the experience path of the newcomer would be monitored by a friend or acquaintance tied into the gatekeeper structure. When misfortunes arise, or a job is lost, the migrant would have access, through his friend, to the person or persons at higher levels of the structure who have the appropriate knowledge for handling the problem.

NOTES

1. That poorly educated, unskilled migrants *are* more likely to have trouble adjusting to the city has been clearly demonstrated in many studies. See Beijer (1963). The process of decline is shown in a number of time trend graphs presented in Hanson, Simmons and McPhee (1968a). Also note Figure 1 in this paper where financial independence is used as an indicator of adjustment for two groups differing in socioeconomic status on arrival in the city.

2. The model was described in a paper presented at the annual meetings of the American Sociological Association, San Francisco (1967). Illustrations of some model concepts are also presented in Simmons, Hanson and Potter (forthcoming).

3. A description of the standardization and analysis procedures, with illustrative output, is available in Hanson, Simmons and McPhee (1968b).

4. Of course, a migrant with less than forty-eight months of urban experience drops out of the analysis when his data run out. The loss of these cases creates some instability in the trend curves in the latter part of the forty-eight-month period, especially if the comparison group was composed of a small number of cases in the beginning.

5. For each comparison group, each month, the program computes the mean for continuous variables or a percentage for binary variables. The final output from the computer is a table presenting the month-by-month statistics and a figure which presents the statistics in the form of time trend curves. The program was written by Richard Jones of the Institute of Behavioral Science.

REFERENCES

BEIJER, G. (1963) Rural Migrants in Urban Setting. The Hague: Martinus Nijhoff.
HANSON, R. C. and O. G. SIMMONS (1968) "The role path: a concept and procedure for studying migration to urban communities." Human Organization 27 (Summer): 152-158.
--- O. G. SIMMONS and W. N. McPHEE (1968a) "Time trend analyses of the urban experiences of rural migrants to the city." Paper presented to the Conference on Adaptation to Change sponsored by the Foundation's Fund for

Research in Psychiatry. San Juan, Puerto Rico (June).
——— O. G. SIMMONS and W. N. McPHEE (1968b) "Quantitative analyses of the urban experiences of Spanish-American migrants." Pp. 65-83 in Proceedings of the 1968 Annual Spring Meeting, American Ethnological Society. Detroit: American Ethnological Society.
———, O. G. SIMMONS, W. N. McPHEE, R. J. POTTER, and J. J. WANDERER (1967) "A simulation model of urbanization processes." Paper presented at the annual meetings of the American Sociological Association, San Francisco (August).
KURTZ, N. R. (1966) "Gatekeepers in the process of acculturation." Ph.D. dissertation, Univ. of Colorado.
SIMMONS, O. G., R. C. HANSON and R. J. POTTER (1969) "The rural migrant in the urban world of work." In the Proceedings of the Eleventh Interamerican Congress of Psychology, Mexico City (December).

Chapter 7

The Economic Absorption and Cultural Integration of Inmigrant Workers:
Characteristics of the Individual Versus the Nature of the System

LYLE W. SHANNON

GROUP IDENTITIES AND THE PROCESS OF ABSORPTION AND INTEGRATION

People may be thought of as occupying social spaces in the ongoing social system of which they are a part. For idiosyncratic reasons or for reasons related to their position in various groups or segments of the larger society, people decide to migrate, to change their spatial and social places. When migrants make such a move or series of moves they are more or less faced with the task of becoming absorbed into the economy and integrated into the culture of either the larger society of the host community or into a subsociety with its own more or less distinctive subculture. Whatever their decision, the new social spaces they occupy will initially be based to some extent on their individual

---

Author's Note: *The Racine study of inmigrant labor to which this paper refers was conducted with the support of the Research Committee of the University of Wisconsin's Graduate School, a grant of the National Institutes of Health (Project RG 5342, RG 9980, GM 10919, and CH 00042), the National Science Foundation, the Urban Research Committee of the University of Wisconsin, and the Ford Foundation grant to the University of Wisconsin. From 1958 to 1962 the project was located at the University of Wisconsin, from 1962 at the University of Iowa. Support for analysis of the data continued from NIH until 1965. The project has since been supported by the College of Liberal Arts and the Division of Extension Services at the University of Iowa. The 1960 sample consisted of 236 Mexican Americans, 284 Anglos, and 280 Negroes.*

characteristics and various fortuitous circumstances, but more probably on the overall position of others in the social groups of which they are a part or on the societal segments with which they are identified. Some studies have concentrated on the characteristics of the individual in determining the extent to which migrants are absorbed and integrated while others have emphasized the influence of group identities and the organization of society. Neither is wholly correct. The salient question is which combinations of individual characteristics (socially acquired as part of past experience) and group identities determine the level at which the migrant is initially absorbed and the rate at which he ultimately moves upward in the social system of the host community, whether it be in a subsociety and subcultural group or in the larger society.

While human and organizational factors determine the level of living of persons in urban-industrial societies, aspects of the natural environment were of greater importance in hunting and gathering societies. Primitive migrants lived within an ecosystem in which the main problem was to cope with the challenge of their physical environment, the flora and fauna of their geographical location. Modern day migrants are part of a system within which people must learn to cope with a pleasant or unpleasant, facilitating or unfacilitating, social and organizational environment in which the most powerful determinants are neither vegetative, nor animal, nor meteorological. Inmigrants (migrants who have settled in an urban area) are interacting with each other and with persons who are more fully integrated into one segment or another of a society that is in transition from rural to urban-industrial.

Among those with a professional interest in migration research have been some who are concerned with the "push" that makes men move and others with the attraction of a place that "pulls" migrants to it. Some have described the social system from which migrants departed while others have been concerned with their adjustment in the new society. In some research the characteristics of the individual are hypothesized to be the determinant of the decision to move and the determinant of the form and degree of adjustment in the new or host community, while in others the organization of the society into which the individual is being economically absorbed and culturally integrated is emphasized. The designs of these studies, however, have too frequently called for no more than a description of the characteristics of migrants at *one point or another in time* while on the move. They fail to adequately describe the experiential chains, or *chains of events rather than single factors,* that account for decisions to move. They fail to help us understand how some people are successfully absorbed into the economy of host communities and integrated into the larger society while others are not.

To efficiently predict the course of absorption and integration we must know which individual characteristics or group identities increase or decrease life chances (opportunities for and probabilities of upward or downward movement) within the social system the migrant has left and into which he intends to move or has moved. Our strategy should be to study migrants as occupants of social spaces who may or may not be engaged in behavior which they believe will change their relative position in the total system. We must remember that the migrant may exercise to a limited extent the option of attempting absorption and integration into the inmigrant community, a subsociety in itself, or into the larger community and larger society, but that the success of his initial attempt to enter and of his later progression into either society depends on how the society is organized, how the migrant perceives himself, how he has formulated his short-term and long-term goals, and on a variety of fortuitous circumstances. Finally, we must take into account the fact that inmigrants modify the social environment of which they become a part, thus affecting the process of adjustment for those who follow at a later period in time.

Race and Ethnicity, Religion, and Sex and the Possibility of Subcultures

In this report, based on a study of Mexican Americans, Negroes, and Anglos, we are concerned with the interrelationship of race and ethnicity, religion, and sex—how combinations of these factors may be related to the generation or appearance of social and social psychological characteristics which will influence the rate of economic absorption and cultural integration of inmigrants.

Though the parameters of race, ethnic, religious, and sexual subcultures are neither mutually inclusive nor exclusive of a universe of attributes, participation in or identification with any of these subcultures would influence the weight of a person's group identity in determining the position or positions assigned to him in any society or subsociety. To be specific, in the Racine data we had Protestant and Catholic Anglos, Negro Protestants, and a Mexican American group that was predominantly Catholic but with a sufficient number of Protestants to be included in the analysis. These groups were further partitioned by sex. We hypothesized that the combination of Catholicism with the Mexican American subculture (folk or peasant) would produce a different world view or attitude toward life than when combined with the Anglo (urban-industrial, blue-collar) subculture.[1] If religion in itself played a role in determining world view, all other variables controlled, then Anglo Protestants would differ from Anglo Catholics.

Having included years of formal education among the measurable influences on world view, we hypothesized that the more educated persons within each religious group would have an active, independent world view and the less educated a more passive, fatalistic view. This world view of the educated probably does not square with how things really work in all segments of society. While middle and upper socioeconomic status Anglos may well be able to manipulate their social-environment to their advantage, not everyone who has acquired an active world view is in a position to do so, particularly if he belongs to a minority racial or ethnic group.

We hypothesized that Catholic influence on world view, as it has been traditionally perceived and especially in reference to the world view of Mexican Americans, would be more likely than would Protestant influence to overshadow the effect of education or more accurately, we hypothesized that traditional Catholic influence would have a different direction from education while Protestant influence would be in the same direction as education. If it is, in fact, a combination of Catholicism and Mexican traditions rather than one or the other that is the determinant of the world view of the Mexican American, then Anglo Catholics and Protestants with similar levels of education should have world views that differ from each other less than from the world views of Mexican Americans. Furthermore, if it is a combination of tradition and religion, then years of education, level of aspiration for children, level of living, and occupational level should have a different pattern of correlations with world view for Mexican American Catholics than for persons outside this subculture. Protestant and Catholic Anglos should also differ in this respect but to a lesser degree.

If the Protestant Negro subculture emphasizes rewards at some future date rather than at the present, just as the traditional Mexican American Catholic subculture, then the correlation between education (or other variables) and world view among Protestant Negroes would be more similar to that of Catholic Mexican Americans than to either Protestant or Catholic Anglos.

To summarize, extreme differences in world view would be found between male Protestant Anglos and female Catholic Mexican Americans. All other variables would have their extremes in these groups as well. Between the opposite polar types other groups would be ranked in the following order: Mexican American Catholic males, Mexican American Protestant females, Mexican American Protestant males, Negro Protestant females, Negro Protestant males, Anglo Catholic females, Anglo Catholic males, and Anglo Protestant females. Education would have an increasingly high correlation with world view in each subcategory commencing with Mexican American Catholic females and extending to Anglo Protestant males. The pattern of correlations between world view and other variables would differ from one racial or ethnic and religious subcategory to another but exactly how was not hypothesized. The data, in a sense, constitute the quantitative ethnology of ethnic and racial, religious, and sexual subgroups.

The Relationship of Each of the Variables to Race and Ethnicity, Religion, and Sex

In the Racine study, the process of economic absorption was more narrowly defined than was that of cultural integration. Economic absorption involves not just securing work but becoming a part of the regularly employed labor force at a level consistent with one's capabilities and the capabilities of others at every level or position in the economic institution. Cultural integration refers to integration into the whole gamut of institutional life. It involves the transformation of values, the acquisition of new behavioral patterns, and social participation beyond one's own primary group, although not necessarily in this order.

When the consequences of people in interaction as members of groups in the larger society are represented by a correlation diagram in which economic absorption and cultural integration (it is unnecessary to assume that absorption and integration are goals of either all inmigrants or all persons in the larger society) increase with education, work experience, association with appropriate role models, and level of aspiration (some of the assumed determinants of success in the larger society), most persons fall along either a fairly straight or a slightly curved line. Although the Racine data showed this general pattern, the correlations for the combined Racine sample of Anglos, Mexican Americans, and Negroes were based on the fact that almost any measure of absorption and integration or of the independent determining variables produced high Anglo scores, intermediate Negro scores, and low Mexican American scores. Therefore, the statistically significant relationships appearing in the larger sample tended either to decrease, disappear, or change in direction when race and ethnicity were controlled.[2]

The measures or variables with which we are concerned in this report are world view, educational level, level of aspiration for children, level of living, and occupational level. World view and level of aspiration for children are utilized as measures of cultural integration, i.e., integration into the larger Anglo society. Educational level is an example of an hypothesized determinant of success in the larger society, while level of living and occupational level are taken as measures of economic absorption. Data on the five basic variables are presented in Table 1 in percentages, for Mexican Americans, Anglos, and Negroes by religion and sex.

World View

Anglos had significantly more active world views than Negroes and Mexican Americans, but Negroes did not have significantly more active world views than Mexican Americans.[3] The Mexican American males had more active world views than did the females. Negro Protestant males had more active world views than did the Negro Protestant females, but not significantly so. Catholic Anglos had slightly less active world views than did Protestant Anglos but the difference was not significant. Among both Catholic and Protestant Anglos the males had more active world views than did the females, although the difference was statistically significant among only the Catholics. Anglo Protestant females had more active world views than did Anglo Catholic females; the males were almost identical in world view.

But, as we had hypothesized, differences in world view were both large and statistically significant when based on race and ethnicity, or race and ethnicity combined with religion and sex as shown in Table 2. For example, the difference between Mexican American Catholics and Anglo Protestants was represented by a correlation of .4179 and the difference between Mexican American Catholic females and Anglo Protestant females by a correlation of .4417. Largest of all was the difference between Mexican American Catholic females and Anglo Protestant males, represented by a correlation of .5330.[4] Sex and religion, in themselves, do not appear to be important determinants of world view since Mexican American/Anglo differences above were represented by a correlation of .3992 and Negro/Anglo differences by .3298.

Educational Level

On the whole, the greatest differences to be found between groups were those based on education. Anglos were by far the best educated with the Negroes second and the Mexican Americans last, with less difference between Anglos and Negroes than between Negroes and Mexican Americans. Eighty-three percent of the Anglos had nine or more years of education while 83 percent of the Mexican Americans had eight years of education or less. The difference was statistically significant at the .001 level between each of these groups as shown in Table 3.

Differences between Mexican Americans and Anglos were greatest between females; the largest contrast, as measured by the coefficient of

TABLE 1

RELATIONSHIP OF SELECTED VARIABLES TO RACE OR ETHNICITY, RELIGION AND SEX: EACH VARIABLE DICHOTOMIZED AND DISTRIBUTION SHOWN IN PERCENTAGES

	World View		Educational Level		Level of Aspiration for Children		Level of Living		Occupational Level	
	Active	Passive	8 Years or Less	9 Years or More	Low	High	Low	High	Low	High
MEXICAN AMERICAN										
CATHOLIC	37	63	83	17	79	21	68	32	84	16
Male	37	63	83	17	80	20	67	33	83	17
Female	45	55	83	17	70	30	73	27	82	18
PROTESTANT	30	70	82	18	88	12	60	40	85	15
Male	38	62	88	12	71	29	82	18	88	12
Female	44	56	89	11	78	22	78	22	78	22
	29	71	87	13	63	37	88	12	100	0
NEGRO										
PROTESTANT	44	56	43	57	59	41	73	27	73	27
Male	44	56	43	57	59	41	73	27	73	27
Female	48	52	50	50	58	42	72	28	72	28
	40	60	37	63	60	40	73	27	75	25
ANGLO										
CATHOLIC	77	23	17	83	42	58	11	89	30	70
Male	73	27	15	85	40	60	13	87	30	70
Female	82	18	20	80	46	54	15	85	25	75
PROTESTANT	64	36	11	89	34	66	11	89	34	66
Male	79	21	18	82	44	56	9	91	30	70
Female	84	16	24	76	42	58	11	89	32	68
	74	26	12	88	47	53	6	94	29	71

TABLE 2
THE RELATION OF WORLD VIEW TO RACE OR ETHNICITY, RELIGION AND SEX

	Mexican-American	Negro	Anglo
Race and Ethnicity Controlled			
World View × Religion	NS .0028	*** .3298	NS .0704
World View × Race and Ethnicity		Mexican/Negro	Mexican/Anglo
	NS .0726		*** .3992

	Prot.	Cath.	Prot.	Cath.	Prot.	Cath.
Religion Controlled						
World View × Race and Ethnicity	NS .0801	* [blank]	*** .3399	* .2567	*** .2778	*** .3394
World View Cross Race and Religion	MexC/AngP *** .4179		MexC/NegP NS .0738	MexP/AngC * .2567	AngC/NegP *** .2600	

	Mexican-American	Mexican/Negro	Negro	Negro/Anglo	Anglo	Mexican/Anglo
Race and Ethnicity Controlled						
World View × Sex	°° .1566	° .1562	NS .0803		* .1521	° .2031

	Prot.	Cath.	Prot.	Cath.	Prot.	Cath.
Race, Ethnicity and Religion Controlled	Male Female	Male Female	Male Female	Male Female	Male Female	Male Female
World View × Sex	NS .0377	° .1134	NS .0803	*** .3510	*** .4025	*** .4047

	Protestant		Catholic		Protestant		Catholic		Protestant		Catholic	
	Male [AngM/NegF	Female AngF/NegM]	Male [MexM/AngF	Female MexF/AngM]	Male [MexM/NegF	Female MexF/NegM]	Male [MexM/AngF	Female MexF/AngM]	Male [MexP/AngC	Female MP/AC]	Male [MexF/NegM	Female MP/AC]
Religion and Sex Controlled												
World View × Race and Ethnicity	NS .1363	NS .0523	NS .2019	*** .3115	*** .3521	*** .3270	.1875	*** .4930	*** .2938	°° .2762	*** .3583	*** .3332

	Protestant		Catholic		Protestant		Catholic		Protestant		Catholic	
	Male [MexC/NegP	Female MC/NP]	Male [AngMc/NegP	Female AC/NP]	Male [MexC/AngP	Female MC/AP]	Male [MexP/AngC	Female MP/AC]				
Religion Controlled												
World View × Sex	*** .4210	*** .2557	NS .0196	** .3698	*** .1118	*** .4417	*** .3198	NS .0888				
Cross Race and Ethnicity	*** .4210	*** .2557	NS .0196	** .3698	*** .1118	*** .4417	*** .3198	NS .0888				

	Male [ACM/NPF	Female ACF/NPM]	Male [MCM/APF	Female MCF/AMP]	Male [MCM/NPF	Female MCF/NPM]	Male [MPF/ACM	Female ACF/MPM]				
Sex Controlled												
World View × Religion	NS .0374	° .1467	*** .3115	** .2170	*** .3962	*** .4417	*** .3198	*** .2295				
Cross Race and Ethnicity												
World View × Race	*** .3736	*** .1408	** .2975	** .5330	.0422	.1907	** .4093	.1408				
Cross Religion and Sex												

Box contains r, coefficient of correlation and key to level of significance for Chi Sq. [blank] = no cases or not sufficient for computations. Level of significance indicated as follows: *** = .001, ** = .01, * = .02, °° = .05, ° = .10.

TABLE 3
THE RELATION OF EDUCATIONAL LEVEL OF RESPONDENT TO RACE OR ETHNICITY, RELIGION AND SEX

Race and Ethnicity Controlled		Mexican-American		Negro	Anglo
Education X Religion		NS .0379		*** .2919	NS .0379
		Mexican/Negro		Negro/Anglo	Mexican/Anglo
Education X Race and Ethnicity		*** .4097			*** .6651
		Prot.	Cath.	Prot.	Cath.
Religion Controlled		*** .2165		*** .2622	*** .4842
Education X Race and Ethnicity		MexC/AngP	MexC/NegP	MexF/AngC	
Education Cross Race and Religion		*** .6476		*** .4024	*** .5820
				AngC/NegP	
				*** .2718	
		Mexican-American		Negro	Anglo
Race and Ethnicity Controlled		NS .0137		°° .1345	°° .1489
Education X Sex					
		Prot.	Cath.	Prot.	Cath.
Race, Ethnicity and Religion Controlled		NS .0212	NS .0122	°° .1345	NS .1081	NS .1245
Education X Sex					
	Male	Female	Male	Female	Male	Female	Male	Female
Sex Controlled	***	*		***	***	***	***	***
Education X Race and Ethnicity	.3532	.2369		.4619	.2936	.2745	.6117	.7176
	Mexican-American/Negro	Mexican-American/Anglo
	Protestant	Catholic	Protestant	Catholic
	Male	Female	Male	Female	Male	Female	Male	Female
Religion and Sex Controlled	[AngM/NegF AngM/NegM]	[MexM/AngF MexM/AngM]		
Education X Race and Ethnicity	° .1945	*** .3962			*** .2623	*** .2745	*** .4268	*** .5669	*** .6219	*** .6882
	Protestant	Catholic	Protestant	Catholic
	Male	Female	Male	Female	Male	Female	Male	Female
Religion Controlled	[AngM/NegF AngM/NegM]	[MexM/AngF MexM/AngF]	[MexM/AngF MexM/AngM]	[MexF/AngM MexF/NegM]
Education X Sex	° .1276	*** .5940			*** .4010		NS .4930		** .2560	° .1780
Cross Race and Ethnicity	Male	Female	Male	Female	Male	Female	Male	Female
	[MexC/NegP AngC/NegP]	[MCM/APF MCF/APM]	[MexC/AngP AC/NP]	[MexC/AngP MC/AP]	[MexP/AngC MC/NP]	[MexP/AngC MP/AC]
Sex Controlled	*** .3442	*** .4582	*** .3508	*** .7122	*** .5948	*** .2632	*** .7007	*** .7030	*** .5398	*** .6311
Education X Religion										
Cross Race and Ethnicity	[ACM/NPF ACF/NPM]	[MCM/NPF MCF/APM]	[MCM/NPF MCF/APM]	[MCM/NPF MCF/ACM]	[MPF/ACM ACF/MPM]
Education X Race	°° .1663	*** .3791			*** .5837		*** .4619	*** .3347	*** .5129	*** .6566
Cross Religion and Sex										

Box contains r, coefficients of correlation and key to level of significance for Chi Sq. ▭ = no cases or not sufficient for computations. Level of significance indicated as follows:
*** = .001, ** = .01, * = .02, °° = .05, ° = .10.

correlation, was .7176 between Mexican American females and Anglo females. Other Mexican American versus Anglo contrasts were: between Mexican American Catholic males and Anglo Protestant females, .7122; between Mexican American Catholic females and Anglo Catholic females, .6882; between Mexican American Catholic males and Anglo Catholic females, .7030; and between Mexican American Catholic females and Anglo Protestant females, .7007.

There was no significant difference within either the Mexican American or the Anglo group based on religion when educational level was dichotomized at eight years or less and nine years or more.[5] Anglo males, both Catholic and Protestant, had less education than the females but not significantly so. Among the Negroes, males had significantly less education than females.

Level of Aspiration for Children

A scale measuring level of aspiration for children was based on responses to questions about education and occupation.[6] Anglos had the highest level of aspiration for their children followed by the Negroes, with the Mexican Americans at the lowest point; all differences were statistically significant, but did not compare in size with differences based on education. The Mexican American Catholics had somewhat lower levels of aspiration for their children than did the Mexican American Protestants; the Anglo Protestants had lower levels of aspiration for their children than did the Anglo Catholics, but not significantly so. The difference between Mexican Americans and Anglos was therefore greatest among the Catholics, being even greater than that between Mexican American Catholics and Anglo Protestants. The Mexican American Catholic females had significantly lower levels of aspiration for their children than did the males; Protestant males had lower levels of aspiration than did the females. Although the differences were sizable, they were statistically significant only among the Catholics. Within both the Anglo Protestant and Catholic groups the difference on a basis of sex was the opposite of that among the Mexican Americans, but not statistically significant—Anglo Catholic males and Protestant females had the lower levels of aspiration for their children. Among the Negroes, females had only slightly lower levels of aspiration for their children than did males.

The highest aspiring group consisted of Anglo Catholic females and the lowest of Mexican American Catholic females, producing a coefficient of correlation of .5588, which represented the greatest difference between groups on level of aspiration for children. It is very clear that the large difference between Anglo and Mexican American Catholic females influenced every Mexican American and Anglo comparison that included either Mexican American or Anglo Catholic females.

Level of Living

The level of living or the possessions scale was the only scale on which Mexican Americans and Anglos were not at opposite extremes.[7] Anglos had the highest level of living followed by the Mexican Americans and Negroes, with little difference between the Mexican Americans and Negroes. Among the Mexican Americans, Catholics had a higher level of living than did Protestants, whereas among the Anglos, Protestants were higher than Catholics, but neither difference was significant. The highest level of living scores were obtained in homes where Anglo Protestant females were interviewed and the lowest level of living scores were obtained in Mexican American Protestant female interviews.

High correlations were obtained in every instance involving a contrast in the level of living of Negroes and Anglos or Mexican Americans and Anglos. For example, Negro and Anglo females produced a correlation of .6782 and Negro and Anglo Protestant females produced a correlation of .6433. Mexican American Catholic males and Anglo Protestant females were correlated .6672.

Occupational Level

Anglos were at the highest occupational levels, with Negroes next and Mexican Americans at the lowest level.[8] Each group was significantly higher than the other but with greater differences between Anglos and Negroes than between Negroes and Mexican Americans. Within the Mexican American group, Catholics and Protestants had similar levels of living but the Catholics were somewhat higher. Catholics and Protestants had almost identical occupational levels among the Anglos.

In homes where the male was interviewed males tended to have higher occupational levels than in homes where the female was interviewed, but the difference was in no case significant. The data suggested that there might well have been a tendency on the part of males to describe their jobs in such a manner that they were coded at a higher level than when the same job was described by the female spouse.

The greatest difference in occupational level was between Mexican Americans and Anglos, particularly when the correlation involved religion and sex as well.

*Recapitulation of the Relationship of Variables
to Race and Ethnicity, Religion, and Sex*

In summarizing the data just described we first note that every Mexican American versus Anglo comparison was statistically significant and that three of the five had correlations of .5000 or more. Every Negro versus Anglo correlation was also statistically significant and one was above the .5000 level. Although three of the five Mexican American versus Negro correlations were statistically significant, none of the correlations were above .5000.

Among the cross-race and -ethnicity comparisons and cross-religion and -sex comparisons, those involving Catholic Mexican American males and Catholic Anglo females, Mexican American Catholics and Anglo Protestants (both male and female), or Mexican American Catholics and Anglo Protestants, sizable correlations of statistical significance were generated following the pattern of Mexican American versus Anglo correlations without controls.[9]

The Relationship of World View to Other Variables with Controls for Religion and Sex

Complexity as a Consequence of Sophistication

There are basically two alternative approaches to explaining the relation of world view to other variables. World view could be regarded as the determinant of how much education a person will manage to obtain, the level of aspiration that he will have for his children, the occupational level at which he will be employed, and the level at which he will be living. Equally plausible is that years of education, occupational level, level of living, and the level of aspiration for children (all in interaction within a social context) will shape a person's total view of the world. In either case, education is closely related to world view.

Within this framework we could expect that some who have been well educated will in the process have acquired a world view that serves as a rationalization for their success in life—that whatever success they may have achieved was due to their own efforts. Theirs is an active, individualistic world view and one which also includes the notion that others who make the effort can be equally successful in manipulating their environment. On the other hand, the highly sophisticated Anglo may say that he has what he has because of the way that the system works—that it is more or less initial advantage and/or lucky breaks that have enabled him to move upward in the system.

Those who have relatively little in the way of material goods, status, and education may also rationalize their general lack of that which is sought in the larger society by blaming it on how the system works or on external forces which determine what one will have in life. The less educated Mexican American, for example, may have high aspirations for his children but be willing to settle for less because the forces that determine how much education a child will receive and how far a child will move upward occupationally have been defined as insurmountable or almost entirely beyond his control. For some these forces are believed to be supernatural and for others they are simply fortuitous or chance. The less-educated Negro may share the same more or less fatalistic view but his explanation is based on the way that the white power structure functions to prevent upward movement for some people, but facilitates upward movement for others, principally Anglos.

To recapitulate, it is possible for Anglos, Mexican Americans, and Negroes with similar but somewhat different views to have the same score on the world view scale. Some believe that society is organized in such a way as to make it possible for certain categories of persons to achieve more readily than others while others believe that chance or powerful external forces of a mystical nature dictate their lives—but all tend to be passive and fatalistic.

Since Protestant Anglos and Catholic Mexican Americans have tended to be at opposite extremes and since there are few sophisticated Anglos, we would expect relatively high Protestant Anglo correlations between world view and other variables, particularly education or occupational level. The pattern of correlations for unsophisticated Catholic Anglos should be similar to that for Protestant Anglos. The next lowest set of correlations should be for Protestant Negroes, the assumption being that some have a passive world view related to traditional patterns of socialization and education but that others have a passive world view for somewhat the same reason as the sophisticated Anglos. The correlations between world view and education and other variables for Mexican Americans should be lower than any others on the assumption that their Catholic Mexican American background is the determinant of their world view rather than education, occupational level, or other variables.

There is also the possibility that none of the Anglo correlations will be very high since one kind of Anglo will cancel out another type of Anglo or that world view is relatively independent of the variables with which we have compared it. Let us turn to the data presented in Table 4.

World View and Education

When the total sample of Mexican Americans, Anglos, and Negroes was considered there was a statistically significant correlation of .3531 between world view and education, a correlation based to a large extent

TABLE 4
RELATIONSHIP OF WORLD VIEW TO EDUCATIONAL LEVEL, LEVEL OF ASPIRATION FOR CHILDREN, OCCUPATIONAL LEVEL AND LEVEL OF LIVING, CONTROLLING FOR RACE OR ETHNICITY, SEX AND RELIGION

	Educational Level	Level of Aspiration for Children	Occupational Level	Level of Living
MEXICAN AMERICAN CATHOLIC				
Male	.2324c	.1802a	.0610	.2921d
Female	.2320b	.1959a	.0788	.1043
NEGRO PROTESTANT				
Male	.1693a	.1395	.0977	.2520d
Female	.1374	.1904b	.1517	.1048
ANGLO				
CATHOLIC				
Male	.4187c	.1402	.3780c	.1713
Female	.1950	.0355	.0424	.2337
PROTESTANT				
Male	.3892d	.1967	.2478a	.1006
Female	.2473a	.1828	.2126	.1457

NOTE: r_4 coefficients of correlation of world view and selected variables significant at the following levels:

$^a = \chi^2$ significant at .10 level

$^b = \chi^2$ significant at .05 level

$^c = \chi^2$ significant at .02 level

$^d = \chi^2$ significant at .01

on the fact that 57 percent of the Mexican Americans were in the low education and passive world view cell, and 69 percent of the Anglos were in the opposite high education and active world view cell in a two-by-two table. In other words, differences in the distribution of the marginals and opposing modal cells for each of the samples resulted in a relatively high correlation between world view and education before the introduction of race and ethnic or religious controls. Before controls for religion and sex were introduced the correlations between world view and education were .3040 for Anglos, .1875 for Negroes, and .2423 for Mexican Americans.

After introducing controls relevant to the hypothesis of the existence of subcultural groups based on race and ethnicity, religion, and sex, we see that Anglo Protestant males have a correlation of .3892 between world view and education, Anglo Catholic males .4187, and Anglo Protestant females .2473; all are statistically significant. Catholic females do not have a statistically significant correlation between education and world view. Anglo Protestant females have essentially the same correlations as do Mexican American Catholic females and males. But none of the correlations between world view and education were sufficiently high, even when statistically significant, to account for much of the variation in world view. We must conclude that although Anglos, Negroes, and Mexican Americans differ in their education and world views, the interrelationship of these variables within each ethnic group, religious group, and sex category gives some but not much support to the more detailed subcultural hypothesis.

World View and Level of Aspiration for Children

Level of aspiration for children was correlated with world view in the total sample in somewhat the same manner as was education. The correlation for the total sample was .2240, for Anglos .1287, Negroes .1657, and Mexican Americans .2249; all were statistically significant. About half of the Mexican Americans had passive world views and low levels of aspiration for their children, and about half of the Anglos had active world views and high levels of aspiration for their children. More of the Negroes were in the passive world view and low aspiration for children cell than in any other cell, but proportionately more Negroes were in the passive world view and high aspirations for children cell than either Anglos or Mexican Americans, suggesting that Negroes may have high aspirations for their children even though they have a view of the world that is considerably different from that which is held by Anglos with high aspirations for their children.

When the Anglos were partialled by religion and sex, the Protestant correlations were higher than the Catholic, but in no case was level of aspiration for children significantly related to world view. The best explanation for these low correlations is the number of Anglos in the active world view and low level of aspiration for children cell who did not have specific occupational levels for their children but simply stated that they would "leave it up to the children." The Anglos may have meant that they would leave it up to their children if they selected an appropriate occupational level. Because they did not express their high aspirations for children in precisely the same manner as did the Mexican Americans and Negroes, the number of Anglos who would be in the active world view and high aspiration for children cells was reduced and the number in the active world view and low aspirations for children was increased.

Although there were statistically significant relationships between world view and level of aspiration for children among the Mexican Americans, the magnitude of these correlations was essentially the same as for the Anglo Protestants. The Negroes differed from the other groups only in that the female correlation was higher than the male and was statistically significant.

World View and Occupational Level

Turning to the relationship of occupational level to world view, the total sample had a statistically significant correlation of .2207, attributable in part to the fact that about 70 percent of the Anglos were in the active world view and high occupational level cell with a correlation of .1395 and about 40 percent of the Mexican Americans were in the opposing passive world view and low occupational level cell with a correlation of .1362. There was practically no relationship between world view and occupational level for the Negroes.

Both Catholic and Protestant Anglo males had world views that were significantly related to occupational level and in the direction hypothesized, i.e., that an active world view was associated with high occupational level. Others showed either less or in some cases practically no relationship between world view and occupational level.

The data suggest that the Anglo who has been absorbed into the economy at a relatively high level not only has a different world view than does the Negro and Mexican American but also that world view among Anglo males varies directly with the degree of success in the world of work. This was also found to be true in an earlier paper where income was controlled within each race and ethnic group; world view varied significantly within groups with the greatest variation being between low and high income Anglos. To be explicit, 59 percent of the low income Anglos versus 87 percent of the high income Anglos had an active world view. Among the Negroes, 40 percent of those with low incomes had an active world view as compared to 58 percent of those with high income. Among the Mexican Americans, 31 percent of those with low income but 57 percent of those with high income had an active world view (Shannon, 1968: 34-64).

World View and Level of Living

The relationship between level of living and world view showed an overall correlation of .2972, based to a considerable extent on the distribution of the two variables within each racial and ethnic group. Seventy percent of the Anglos fell in the high level of living and active world view cell while 46 percent of the Mexican Americans and 43 percent of the Negroes fell in the low level of living and passive world view cell. The Anglo correlation was .1586, the Negro correlation was .1572, and the Mexican American correlation was .0988, the latter not statistically significant.

When all controls were introduced, the only significant correlations found between level of living and world view were for Negro Protestant males and Mexican American Catholic males, although all others were at least .1000 and ranged upward to .2337.

SUMMARY AND CONCLUSIONS

Two quite opposite approaches to the study of migrants are possible: (a) studies of the characteristics of migrants as individuals in their places of origin and in the host communities in which they have settled, in places of origin compared with those who stayed, and in the host community compared with long-time residents, and (b) studies of migrants as persons occupying social spaces in ongoing social systems. Studies of the first type tend to be static in character while those of the latter are processual and dynamic. The former tend to culminate in descriptive and comparative reports that are time-bound while the latter attempt to explain the workings of the system—the complex processes by which the migrant comes to be in the social spaces that he occupies.

Previous analyses of the data had already shown that race and ethnicity are important determinants of the social space that a migrant occupies in his community of origin and in the host community. While the existence of more or less distinctive subcultures for Mexican Americans and Negroes was demonstrated by different response patterns for samples of each of these groups as compared to Anglos and by differences in patterns of the interrelationship of the variables in a correlation matrix for each of the groups, race and ethnicity did not account for all variation in responses to the questions. Although income and other characteristics of the inmigrants and long-time residents were also related to world view, more variation was explained in any characteristic or dimension of the total sample by race and ethnicity than by any other variable.

Further analysis raised the question of whether or not the addition of religion and sex as variables for control might not result in separation of the larger sample into yet more diverse subcultural groups with sufficiently distinctive characteristics as to have consequences for their absorption and integration. Might not the combination of being a Catholic, being socialized in the Mexican American tradition, and being a female result in maximum differences in world view and other characteristics as contrasted with being Protestant, Anglo, and male? And would the intercorrelations of world view and other variables differ within groups on any basis that could be predicted from existent theory about the interrelationship of world view, education, and measures of position within the larger society and its subgroups?

To answer the first question, Anglo Protestant males were at the extreme active end of the world view scale and Mexican American Catholic females were at the passive end of the scale, accompanied by Mexican American Protestant females. Although Anglo Catholics and Protestants were at the most favorable end of the scale and Mexican Americans were usually at the lowest end of the scale, neither religious nor sex differences followed a systematic pattern for other variables.

As for the second question, world view was significantly correlated (none of the correlations were very high) with three of the four variables for one group only—the Mexican American Catholic males. Overall the correlations of the greatest magnitude were for Anglo Catholic males; the lowest correlations were for Anglo Catholic and Negro Protestant females. Thus, with controls for race and ethnicity, religion, and sex, the hypothesis that the highest correlations between world view and other variables would be in the Anglo group (Protestant males) and the lowest correlations in the Mexican American group (Catholic females) with other groups falling between the extremes in a systematic pattern, must be rejected. (It could be noted that Anglo Catholic and Protestant males had the highest correlations between world view and education or occupation but if we were to continue this "male" pattern, Mexican Americans and Negroes would fall in an order incompatible with the hypothesis.)

The general idea of subcultures being represented by differences in the interrelationship of variables is not rejected although the specific pattern of differences hypothesized for the groups observed was not present. If each of the variables, although related to race and ethnicity, does not vary by religion and sex in either the more or the less systematic fashion that was suggested, nor in any other readily discernible pattern, is it possible that these conjoint relationships do not increase the visibility of various inmigrant groups and as a consequence the probability of occupying different social spaces than they would occupy simply on a basis of their racial and ethnic identities?

NOTES

1. For examples of earlier writings on the relationship of religion to the economic system, see Weber (1930), Sombart (1913), Tawney (1926). More recently the relationship of religious identification to attitudes and behavior pertinent to the present report have been dealt with in the following: Lenski (1961), Mack, Murphy and Yellin (1956) and Mayer and Sharp (1962).

2. See Krass, Peterson and Shannon (1966), Shannon and Krass (1963) and Shannon and Morgan (1966).

3. As we pretested our questions it became apparent that world view consisted of three facets: a person's perception of his own manipulative power versus the organization of the society or some more powerful determinant, his time perspective as oriented toward the present versus the future, and his hierarchy of values involving individual achievement against the ties of the group. A seven-item Guttman scale was developed from the following questions:

(1) Not many things in life are worth the sacrifice of moving away from your family.

(2) The secret of happiness is not expecting too much and being content with what comes your way.

(3) The best job to have is one where you are part of a group all working together, even if you don't get much individual credit.

(4) Planning only makes a person unhappy, since your plans hardly ever work out anyway.

(5) Nowadays, with world conditions the way they are, the wise person lives for today and lets tomorrow take care of itself.

(6) Not many things in life are worth the sacrifice of moving away from your friends.

(7) When a man is born, the success he is going to have is not already in the cards; each makes his own fate.

Respondents with the most individualistic orientation, or "active" respondents, did not agree with the first six statements on the scale but did agree with the last statement. The next most active respondents, or the next most individualistic oriented, agreed with the first statement but did not agree with the statements below it except for the last statement, and so on with each type of respondent down to the most fatalistic and group-oriented, or "passive" respondents who agreed with every statement except the last, with which they disagreed.

To be acceptable, a scale must have a coefficient of reproducibility of .9000. This scale has a coefficient of reproducibility of .9011 and a minimum coefficient of reproducibility of .7125. The lower the minimum coefficient of reproducibility, the greater the improvement of the scale over marginal reproducibility.

4. The use of the r_4 coefficient of correlation as an index of difference rather than a measure of association may seem unusual to the reader. Actually, it is employed as a measure of the association of variables with race and ethnicity, religion and sex—the larger the association then the greater the *difference* between groups on the variables referred to. Rather than simply use any one of a number

of available measures of the significance of differences between groups we have used a measure of the relationships and χ^2, a test of the significance of the difference. This measure of association will then make it possible to compare this set of relationships with the relationship of world view scores to other variables such as educational level and level of living.

5. Better educated respondents were defined as those with nine or more years of education and less educated respondents as those with less than nine years of education—a meaningful cutting point for urban-industrial persons with backgrounds similar to those of our respondents. Anglos had significantly more education than Negroes, and Negroes had significantly more education than Mexican Americans.

When Mexican Americans were dichotomized at four years of education or less and five or more, rather than at eight years, the Protestants were considerably less educated than the Catholics. Forty-five percent of the Catholics as compared to 65 percent of the Protestants had four years or less of education, while the difference was only five percentage points when dichotomized at eight years of education. Similarly, the Mexican American males were less educated among both the Catholics and the Protestants, particularly among the Protestants.

6. Persons with the highest level of aspiration for their children wanted college for them, would only be satisfied if their children went through college, thought it would be financially possible to send their children through college, and wanted their children to be professionals.

The questions utilized in constructing the level of aspiration for children scale were as follows:

(1) About how much schooling would you (have) like(d) your children to have?

(2) It is sometimes hard to tell how things will actually work out. If things (had) turned out that your children completed junior high school (9th grade), and then went to work, would you (have been) be satisfied or dissatisfied? high school? two years of college? college degree?

(3) You can't always tell about the way things will work out. Here are some statements. Tell me, as far as you can see, which statement will come the closest to one that you would agree with: (1) for a person in my financial position, it will be practically impossible to keep my children in school past the 9th grade; (2) 12th grade; (3) to put my children through college.

(4) Is there any special line of work you would like any of your children to go into? (If they are no longer in school, read: Was there any line of work you would have liked your children to go into?)

Although this scale does not quite meet the minimum standard of reproducibility, having a coefficient of reproducibility of .8984 and a minimum coefficient of reproducibility of .6675, the difference between the coefficient of reproducibility and the minimum coefficient of reproducibility was so great that it was decided to accept the scale.

7. A level of living score was constructed from whether or not respondents possessed six items; the most frequently possessed was refrigerator, and in descending order, washing machine, telephone, sewing machine, fabric rug in front room, and 1957 or newer model car. The level of living scale had a coefficient of reproducibility of .9413 and a minimum coefficient of reproducibility of .7600.

8. Male respondents or the spouses of female respondents were classified according to occupational level into seven categories ranging from professional to agricultural labor.

9. In order to obtain better distributions for the purpose of showing the relationship of one variable to another within racial and ethnic and religious groups, different cutting points were utilized in some cases in the next section of this report than were used in Table 1. An attempt was made to select cutting points that would give us fairly even marginals for most of the variables and questions, or at least not skew the marginals to such an extent that the coefficients of correlation would be distorted.

REFERENCES

KRASS, E. M., C. PETERSON and L. W. SHANNON (1966) "Differential association, cultural integration, and economic absorption among Mexican-Americans and Negroes in northern industrial community." Southwestern Social Science Q. 47: 239-252.

LENSKI, G. (1961) The Religious Factor. New York: Doubleday.

MACK, R. W., R. J. MURPHY and S. YELLIN (1956) "The Protestant ethic, level of aspiration, and social mobility: an empirical test." American Sociological R. 21 (June): 295-300.

MAYER, A. J. and H. SHARP (1962) "Religious preference and worldly success." American Sociological R. 27 (April): 218-227.

SHANNON, L. W. (1968) "The study of migrants as members of social systems." Pp. 34-64 in Proceedings of the 1968 Annual Spring Meetings of the American Ethnological Society. Seattle: Univ. of Washington Press.

――― and P. MORGAN (1966) "The predictions of economic absorption and cultural integration among Mexican-Americans, Negroes and Anglos in a northern industrial community." Human Organization 25: 154-162.

――― and E. M. KRASS (1963) "The urban adjustment of migrants: the relationship of education to occupation and total family income." Pacific Sociological R. 6: 37-42.

SOMBART, W. (1913) The Jews and Modern Capitalism (M. Epstein, translator). London: F. Unwin.

TAWNEY, R. H. (1926) Religion and the Rise of Capitalism. New York: Harcourt Brace.

WEBER, M. (1930) The Protestant Ethic and the Spirit of Capitalism (Talcott Parsons, translator). New York: Charles Scribner.

Chapter 8

Coping with Urbanity: *The Case of the Recent Migrant to Santiago de Chile*

FRED B. WAISANEN

The dramatically accelerating urbanization process in Latin America in recent decades is one facet, and a crucial one, of the broad and complex issue of modernization. In this context, the more specific phenomenon of internal migration introduces questions about antecedents and consequences, actors and social systems, losses and gains, over short term and long. The present paper focuses upon some consequences of internal migration, with specific focus upon attitudinal bases of participation in urban Chilean life.

The literature abounds in value-laden (and perhaps unanswerable) questions about Latin American population shifts (see, for example, Beyer, 1967 and Hauser, 1961). Arguments have been advanced that the

---

Author's Note: *The author is a professor in the Department of Sociology, the Department of Communication, and the International Communication Institute, at Michigan State University. He received significant support from these units of affiliation, as well as from International Programs of Michigan State University and from the Midwest Universities Consortium for International Activities, Inc. Guillermo Briones, Professor of Sociology at the University of Chile, shared direction for the larger study from which the data for this paper were obtained, and his collaboration is gratefully acknowledged. The project participation of Jerome T. Durlak is also appreciated.*

volume of migration to the large urban centers has been too high, leading to a drain from rural areas of younger, more innovative and achievement-oriented (and thus more potentially productive) manpower. Simultaneously, arguments run, the new migrants, lacking vital occupational and social skills, tax the ameliorative resources of the city. To the extent that the rural-urban migrants move into culturally and ecologically marginal urban slums (callampas, favelas, barriadas, etc.), consequential problems may be intensified and are certainly made more visible.

In 1962, thirty-four percent of the population of Santiago de Chile was made up of migrants, with about a quarter million people (approximately ten percent of the total urban population) living in *callampas*, sharing the deprivation pattern endemic to slum residence in Latin America (Rogler, 1967). In addition to low educational and income levels, the salient characteristics of the callampas include peripheral location and segregation from the urban nucleus, single-room dwelling, lack of sanitary facilities, and low degree of interfamilial ties (Caplow, 1964: 162-163). Irrespective of individual and intrafamilial orientation, the normative structure and the attitude configurations of the residents are rooted in traditionalism.

The Research Problem

An earlier paper by Briones and the present author (1967) reported data supportive of the hypothesis that conditions of ecological marginality (defined as slum residence) depress aspiration levels and relate to a more "materialistic" orientation toward education. These effects held with education and income controlled. The present paper has a similar focus upon the depressant effect of the callampas, with a particular concern (at dependent variable level) for an attitude configuration of modernity.

The present work stems from a theoretical concern for the nature of actor-social system relationships and the identification process. There are two social systems at issue (i.e., the callampas and Santiago) and the normative structures of the two systems differ in theoretically relevant ways. At the core of these contrasts may be norms that specify maintenance and enhancement. In general, the contrasts are those that differentiate traditional and modern behavioral modes—with the callampas system being the more traditional.

Given the physical proximity of the two systems, the significant normative contrasts force recognition of impediments to the idea diffusion process. The very presence of the enclaves and their normative differentiation from the larger urban system testify to the presence of barriers to acculturation and integration. Among the themes in the barrier configuration are (a) lack of interest, (b) lack of skills and (c) lack of acceptability.

Lack of interest refers primarily to the system of primary (or more salient) commitment, and may involve, on the one hand, confident and gratifying participation in one system (the callampas), and on the other hand, fear and distrust of the other system (Santiago).

The skill barrier may be operative even if interest in escape is high. More is involved here than occupational expertise; coping in the urban systems involves a cognitive complex that includes role taking ability (particularly, ability to take the role of the impersonal other), ability to generalize and to perceive cause-effect relationships. Most generally, these skills relate to the condition of functional literacy.

Lack of acceptability may be less a lack of the actor who aspires to participation in the modern system and more to patterns of exclusion and membership criteria carried in the normative structure of the city.

These barriers, i.e., low levels of interest, skill and acceptability of an actor, can be understood only in the context of general social processes that produce and maintain these barriers. The general social processes that touch the barrier issue most directly appear to be identification, education and stratification. An actor's interest in participation in another social system represents a reordering of reference groups, which is a shift in patterns of identification; cognitive and interactive skills are acquired in classrooms or in other mobility experiences; and acceptability is dependent upon ranking and admission criteria at issue in the stratification process.

This processual underlayment to the barrier problem appears to be particularly relevant in the callampas-Santiago comparison. Identification, education and stratification either facilitate or produce barriers to intersystemic contact, (i.e., contact between Santiago and the callampas) thus impeding the diffusion process that modernization requires. Consequently, the central hypothesis asserts: Residents of the callampas will show lower modernity scores than the residents of Santiago, controlling for education and income. Secondarily, we expect that education will show effects independent of income and place of residence, and that income will show effects independent of education and place of residence.

The Variables and Their Instrumentation

We have neither data nor present inclination to grapple with the problem of what it means to be modern.[1] Whatever the core configuration be, it would appear to be in fit with the literature to assume that there are contrasts with traditional modes, and that among the aspects of that contrast might be self-perceptions of (1) ability, (2) opportunity, (3) influence, (4) innovativeness, (5) sociability, and (6) confidence in the future. More specifically, modern man perceives that he is able to significantly affect his life trajectory and that the system provides opportunity for self-enhancement. He perceives less that he is subordinate to the system and more that he exerts influence upon it. He searches for new ideas and practices and recognizes that the maximization of self-esteem is dependent upon their adoption. He views the future without crippling anxiety, and with confidence that social ascent will be an output of energy and skill inputs.

These indicators, individually and collectively, constitute the dependent variable. Although we do not and cannot argue that these variables represent the essence of attitudinal modernity, we do assume that they are interrelated components of a modern attitude system. The zero-order correlations reported in Table 1 provide some support of the interrelatedness, if not the representativeness, of the six indicators.

TABLE 1
INTER-ITEM CORRELATIONS OF THE DEPENDENT MEASURES

Variable	2	3	4	5	6
1. Ability	.46	.26	.26	.43	.32
2. Opportunity		.26	.14	.34	.27
3. Influence			.35	.48	.26
4. Innovativeness				.39	.21
5. Sociability					.24
6. Optimism					

Instrumentation

The independent and control variables. Place of residence was an aspect of sample design. Education was measured by last year of schooling completed, and income was expressed as total monthly income of subject. The monetary unit in Chile is the Escudo; at the time of data collection the exchange ratio was 3.6 Escudos to $1.00 U.S.A.

The dependent variables. We used modifications of the Cantril ladder scale to tap each of the indicators of modernity.[2] In this technique, the interviewer handed the subject a card containing a simple picture of a 10-step ladder. The steps of the ladder were numbered from 0 to 10, with step 10 at the top of the ladder. When asking a question, the interviewer pointed to the ladder, moving his finger rapidly up and down the steps. The specific items were asked as follows:

(1) Perceived ability. "At the top of the ladder stands a person who has the personal capacity to make his life happier. At the bottom of the ladder stands a person with very little personal capacity to make his life happier. Where do you stand on the ladder now?

(2) Perceived opportunity. "At the top of the ladder stands a person who has all the opportunities to do anything he wants to do. At the bottom is someone who doesn't have any opportunities or chances to do anything he wants to do. Where do you stand on this ladder?"

(3) Perceived influence. "At the top of the ladder stands a person who has very much influence over people at work, with neighbors, friends and others. At the bottom of the ladder is a person with no influence over others. On what step do you think you stand right now?"

(4) Perceived innovativeness. "At the top of the ladder stands a person who wants to do new things all the time. He wants life to be exciting and always changing, although this may make life quite troublesome. At the bottom is a person who wants a very steady and unchanging life. On what step do you think you are at the present time?"

(5) Perceived sociability. "At the top of the ladder stands a person who likes other people very much. He has many friends and is liked very much by his neighbors, people at work, etc. At the bottom of the ladder is a person who doesn't like other people and who is not liked by others. On what step would you say you are at the present time?"

(6) Optimism. "At the top of the ladder stands a person who is completely free of worry about the future, who feels confident and unworried. At the bottom of the ladder is a person who feels very insecure. On what step of the ladder do you stand right now?"

Field Methodology

The variables utilized and the data reported here come from a larger study on communication and migration in Chile,[3] and are derived from two samples, (a) Gran Santiago and (b) the callampas. The Gran Santiago subjects represent a subsample selected from a larger sample of 1540 noncallampas migrants drawn by the Institute of Economics of the

University of Chile. This sample totals 160 subjects. The callampas subjects were randomly selected from a recent census of thirty callampas on the peripheries of Santiago, and the 108 interviewees correspond closely by demographic characteristics to the census means. In each sample heads of households were interviewed.

The interviewers were students at the University of Chile and were trained at the Institute for Economic Planning at the University. Interview time averaged seventy-five minutes, and the interviewing was completed in December, 1964.

Analysis

Given place of residence as the independent variable, with education and income as controls, the three dichotomously treated variables produce an eight-cell table. A separate table was constructed for each of the six indicators of modern cognitive style (the dependent variable), with mean values along an eleven-point scale entered into the appropriate cell. Each table thus provides four "tests" of the effect of place of residence upon the particular indicator of modernity, or a total of twenty-four tests. The pattern of means and comparison possibilities is shown in Figure 1.

Place of Residence	Income	Lower	Higher
Callampas	lower	$\bar{x}_1$	$\bar{x}_5$
	higher	$\bar{x}_3$	$\bar{x}_7$
Santiago	lower	$\bar{x}_2$	$\bar{x}_6$
	higher	$\bar{x}_4$	$\bar{x}_8$

FIGURE 1: MEANS FOR DEPENDENT VARIABLE BY PLACE OF RESIDENCE, INCOME AND EDUCATION

The effect of place of residence, with income and education controlled can be examined by comparison of $\bar{x}_1 - \bar{x}_2$, $\bar{x}_3 - \bar{x}_4$, $\bar{x}_5 - \bar{x}_6$ and $\bar{x}_7 - \bar{x}_8$. The effect of education, with place of residence and income controlled, can be assessed by comparison of $\bar{x}_1 - \bar{x}_5$, $\bar{x}_3 - \bar{x}_7$, $\bar{x}_2 - \bar{x}_6$, etc. Finally, one can attend to the effect of income, with place of residence and education controlled, by comparison of $\bar{x}_1 - \bar{x}_3$, $\bar{x}_2 - \bar{x}_4$, $\bar{x}_5 - \bar{x}_7$, etc.

If place of residence is unrelated to modern cognitive styles, then the differences between the relevant means should show random patterns of differences, i.e., as often "negative" as "positive." Using the binomial distribution, one can establish the probability of any predominant effect, given twenty-four trials and probability of .50 for either positive or negative effect (see Mosteller, 1961:241-283).

Findings

The relevant means are given in Table 2. The comparison of means by place of residence, with income and education controlled, produces nineteen "successes" out of twenty-four trials, a success being a difference in hypothesized direction, i.e., with Santiago residents showing higher mean scores on the dependent measures. Three of the four exceptions come from low education, low income groups. The probability of chance occurrence of nineteen successful predictions out of twenty-four trials (at $p = .5$) is .003, enabling rejection of the null hypothesis.

Education and income show comparable effects. In the case of education, controlled by place of residence and income, eighteen out of twenty-four trials show the hypothesized effect. Income, controlled by place of residence and education, produces successful predictions in nineteen out of twenty-four trials. The probability of occurrence in each case (.008 and .003, respectively) enables rejection of the null hypothesis.

The data suggest the possibility of interaction effects in the three variables at issue. For example, the summary data in Table 3 suggest that place of residence may have greater impact for higher income groups than for lower. (Among higher income subjects, there are eleven successes out of twelve, as against eight out of twelve for the lower income group.) Similarly, education appears to have greater effect in the callampas than in Santiago, while income shows a greater effect in Santiago than in the callampas.

Some preliminary notions of the differential effects of education and income can be gleaned by comparing low education, low income Santiago subjects with various patterns of callampas subjects.

First, comparing high education, high income callampas with low education, low income Santiago produces five out of six comparisons favoring the callampas, suggesting that education and income tend to overcome the depressant effect of callampas residence. Moreover, education alone tends to overcome the callampas effect. When high education, low income callampas subjects are compared with low education, low income Santiago subjects, the callampas means are greater (i.e., more "modern") in five out of six comparisons.[4]

TABLE 2

MEAN SCORES ON DEPENDENT VARIABLES BY RESIDENCE, INCOME AND EDUCATION

Dependent Variable[a]	Residence	Income[b]	Educ. 0-3 Yrs.	Educ. 4 Yrs. or More
1. Ability	Callampas	Lower	6.05 (32)	5.78 (25)
		Higher	4.00 (10)	5.38 (13)
	Santiago	Lower	5.62 (16)	6.15 (42)
		Higher	7.45 (10)	7.46 (67)
2. Opportunity	Callampas	Lower	4.41 (33)	4.95 (29)
		Higher	4.60 (10)	5.19 (13)
	Santiago	Lower	4.47 (16)	5.10 (42)
		Higher	6.80 (10)	5.90 (69)
3. Influence	Callampas	Lower	4.18 (31)	4.78 (23)
		Higher	3.41 (11)	5.46 (13)
	Santiago	Lower	4.77 (15)	5.08 (42)
		Higher	6.00 (10)	6.27 (69)
4. Innovativeness	Callampas	Lower	4.78 (26)	4.62 (24)
		Higher	5.75 (10)	6.54 (13)
	Santiago	Lower	4.22 (16)	5.12 (42)
		Higher	6.61 (9)	6.16 (67)
5. Sociability	Callampas	Lower	5.47 (33)	5.76 (25)
		Higher	4.15 (10)	6.27 (13)
	Santiago	Lower	5.90 (16)	6.25 (42)
		Higher	7.10 (10)	6.91 (69)
6. Optimism	Callampas	Lower	4.29 (34)	4.62 (25)
		Higher	2.20 (10)	4.96 (13)
	Santiago	Lower	3.44 (16)	3.92 (32)
		Higher	5.80 (10)	5.57 (69)

[a] In each case, the coding was from "0" (less modern), to "10" (more modern).
[b] In each case, lower income represents 150 Escudos or less per month, higher income 151 Escudos or more.

The mean comparisons that are contrary to expectation appear to be distributed unevenly among the six indicators of the dependent variable.

TABLE 3

EFFECT OF PLACE OF RESIDENCE, EDUCATION AND INCOME ON MODERNITY

Effect of	Controlling for	No. of Tests	No. "Successes"	p.
1. Place of Residence	Education, Income	24	19	.003
2. Education	Place of Residence, Income	24	18	.008
3. Income	Education, Place of Residence	24	19	.003
4. POR - High Income	Education	12	11	.003
5. POR - Low Income	Education	12	8	.121
6. POR - High Educ.	Income	12	10	.016
7. POR - Low Educ.	Income	12	9	.054
8. Educ. - Santiago	Income	12	8	.121
9. Educ. - Callampas	Income	12	10	.016
10. Income - Santiago	Education	12	12	.00+
11. Income - Callampas	Education	12	7	.193

In the twenty-four comparisons relevant to effect of place of residence, for example, variables 2, 4 and 5 show particularly strong patterns of hypothesis support. Indeed, these three dependent variable indicators produce only four failures out of thirty-six comparisons made in assessing the effect of place of residence, education and income. By contrast, variables 1, 4 and 6 produce twelve failures in thirty-six predictions. In this post hoc context, place of residence, education and income may have their most vital impact upon perceived influence (need-achievement), perceived sociability (need-affiliation), and perception of opportunity to continue influence and sociability into the future.

The possibility that other variables intrude upon these matters is, of course, both possible and probable. The purification of relationships such as those reported here is neither easy to do nor easy to defend. We did, however, examine the possibility that age and length of urban residence may have an effect upon the callampas-Santiago comparison, the Santiago residents being older and having lived in the urban setting for a longer time than callampas residents. The results of this analysis were encouraging. Holding education, income and place of residence constant, neither age nor years of urban residence produced a correlation with any one of the six dependent measures that was significantly different from zero.

Summary

Data on educational achievement, income level and selected aspects of an attitudinal configuration of modernity were collected from 108 residents of the callampas (slums) and from 160 residents of the city of Santiago de Chile. The subjects were all randomly selected heads of households.

Six indicators of attitudinal modernity, viz, self perception of ability, opportunity, influence, innovativeness, sociability and confidence in the future, were treated collectively to enable a number of tests of the effect of place of residence, education and income upon the dependent attitudinal variables. The effect of each was as hypothesized and statistically significant by the binomial distribution.

These data provide no new dimensions to the issues of social ascent and participation in the larger society. To the degree that the independent variables point to housing, schooling and jobs, they are in agreement with both reason and the much publicized demands of reformers and politicians.

However, the effects do not appear to be uniform. In the callampas, education is strongly related to modern attitudes, while income is not. Conversely, in Santiago, income is strongly associated with modernity and education is not. When comparison of mean scores on the dependent variables is made between low education, low income Santiago subjects and high education, high income callampas subjects, the callampas means are generally higher. Thus education and income tend to overcome the depressant effect of slum residence. When high education, low income callampas residents are compared with low education, low income Santiago subjects, the callampas modernity scores are similarly higher. It may be, therefore, that education alone can overcome the depressant effect at issue.

That education is a powerful stimulant to modernization is beyond argument; but the theoretically relevant components of the education-modernization dynamic are less well established. Mobility processes are certainly involved. Within a framework of two social systems that differ by normative structure and value orientations, the school introduces theoretical and practical models of alternative behavioral modes; there is, thusly, an awareness function in education that is critically intersystemic. Similarly, education must involve self-evaluative processes. In the context of modernization, education emphasizes that social ascent is desirable and possible, thus stimulating identification with urbanity and dissociation from tradition.

The stimulant value of income may be of a different order and operative at a different level. Income is, of course, more a consequence of coping skills than an antecedent of them. Within an education-to-occupation-to-income model, however, income provides the base for perceptions of distributive justice (or injustice) and can thus function to stimulate change.

The significance of place of residence stems from identification with the social system of primary socialization. As age, participation, rank and esteem in a social system increase, role circumscription increases, opportunity for intersystemic contact decreases, and identification with the system is intensified. When this happens in the callampas its residents become immobilized due to resignation, entrapment, or choice.

NOTES

1. See Smith and Inkeles (1967) and Waisanen and Durlak (1967).

The emphasis here is on individual modernity, as in Smith and Inkeles. There are several options in operationalization; one can use demographic, behavioral or (as we did) attitudinal indicators. We are, however, less concerned with the question of what the essence of modernity is and more concerned with an exploration or a possible depressant effect of place of residence upon some representative attitudinal dimensions of modernity.

2. For a recent discussion of the ladder technique see Cantril (1965: 22-26).

3. A study supported by (1) Programa Interamericano de Informacion Popular, a cooperative project of the Institute of Agricultural Sciences of the Organization of American States and the American International Association and (2) the Institute of Economic Planning of the University of Chile.

4. While the probability of five "successes" out of six events is not significantly different from chance at customary levels (p = .094), the pattern is interesting enough to warrant further research.

REFERENCES

BEYER, G. H. [ed.] (1967) The Urban Explosion in Latin America. Ithaca: Cornell University Press.

BRIONES, G. and F. B. WAISANEN (1967) "Aspiraciones educationales, modernizacion y integracion urbana." America Latina 10 (October-December): 3-21. Also (1969) pp. 252-264 in Meadows and Mizruchi (eds.) Urbanism, Urbanization and Change. Reading, Mass.: Addison-Wesley.

CANTRIL, H. (1965) Pattern of Human Concerns. New Brunswick: Rutgers University Press.

CAPLOW, T. et al. (1964) The Urban Ambiance. Totowa, N.J.: Bedminster.

HAUSER, P. M. [ed.] (1961) Urbanization in Latin America. New York: International Document Service.

MOSTELLER, F. et al. (1961) Probability with Statistical Inference. Reading, Mass.: Addison-Wesley.

ROGLER, L. H. (1967) "Slum neighborhoods in Latin America." Journal of Inter-American Studies 9 (October): 507-528.

SMITH, D. H. and A. INKELES (1967) "The OM scale: a comparative socio-psychological measure of individual modernity." Sociometry 29 (December): 353-377.

WAISANEN, F. B. and J. T. DURLAK (1967) "Mass media use, information source evaluation, and perception of self and nation." Public Opinion Quarterly 31 (Fall): 399-406.

PART III

THE SOCIOCULTURE
and
INDIVIDUAL
BEHAVIOR

Chapter 9

Context and Behavior: *A Social Area Study of New York City*

ELMER L. STRUENING, STANLEY LEHMANN and JUDITH G. RABKIN

Whether seeking to explain or treat identified pathology, researchers and practitioners in the mainstreams of medicine, psychiatry and psychoanalysis have traditionally regarded the individual as their focal unit of study. Following the medical model, illness was conceptualized as a foreign agent that existed within the patient, under his skin. Isolated individuals opposed this view, but in general the relationship between the individual and his environment was largely ignored.

Within the past several years, an increasing number of writers interested in behavior and behavior disorders have sought to modify the basic assumptions implicated in the clinical or medical model of illness. They come from such diverse backgrounds as psychoanalysis, general systems theory, contract psychology, social biology and ecology. These men

---

Author's Note: *This study was supported by Grant No. 5 R11 MH 02308-01 from the National Institute of Mental Health. Portions of this paper were presented at the April, 1968, meeting of the Eastern Psychological Association and at the April, 1968, meeting of the National Conference on Mental Health Statistics. We are grateful to the many members of the Research Division of Lincoln Hospital Mental Health Services who participated in the collections and processing of the data of this paper. We thank Dr. Jack Cohen for consultation and Mrs. Mattie Jones for typing the manuscript.*

subscribe either explicitly or by implication to the position that psychopathology is not "subdermal" but develops in a context larger than the individual. Instead of focusing on factors within the individual that lead to psychiatric difficulty, the subject of study becomes the transactions between the individual and his environment.

In the earlier stages of the community mental health movement, attention was largely devoted to other individuals in the patient's environment, but there has been increasing attention directed at prevailing environmental conditions and characteristics. In line with this trend, mental health workers have become interested in the sociological studies of urban typology and particularly in the method of social area analysis in relation to disordered behavior in large metropolitan areas.

Recognition that the social context might have equal or greater relevance than individual characteristics in predicting behavior has led to a change of treatment strategy as well as focus. Today many community action programs (e.g. Mobilization for Youth or Lincoln Hospital's Neighborhood Service Centers) adopt an "advocate role," fighting for the individual in order to change his social context, and replacing the traditional psychotherapeutic goal of only changing the individual. In health and mental health services, the clinical or medical model introduced to nineteenth century medicine is being increasingly displaced by the public health model. The latter conceives of illness not as a collection of independent and discrete entities like measles and diphtheria, but as interchangeable and shifting manifestations of disorder best dealt with in the context of social and environmental factors.

Despite these widespread changes in theory and action characterizing mental health and other community programs across the nation, little study has been devoted to the comparative merits of the various methods of intervention. While the massive influx of federal aid provided under the Community Mental Health Centers Act of 1963 assures widespread acceptance of the public health model in psychiatry, little is known about the techniques that will most effectively fulfill current goals. Many mental health organizations throughout the country will soon be responsible, for the first time in their history, for the mental health or social competence of entire catchment area populations rather than just those who actively and spontaneously seek treatment. These changes of emphasis and structure have profound implications for the planning and implementation of community mental health programs, and also create an urgent need for their systematic evaluation. Using methods derived from social area analysis, we have become involved in the problems of evaluation and planning of such programs, as well as the analysis of underlying assumptions regarding the effect of environment on its inhabitants.

Specifically, one aim of the present study was to develop a method of accurately describing the social characteristics of specific urban areas in terms meaningful to the planners of community health programs. A second aim was to evaluate the effect of environments on the social behavior of their inhabitants, and a third was to develop procedures for analyzing the above information to determine the most effective means of intervention and to compare the efficacy of alternative approaches along these lines.

Ideally, an integrated and comprehensive information system describing the catchment area population and its environment should be available as one basis for planning a network of mental health services. This would include a series of population, environmental and behavioral variables and their interrelationships to facilitate the design and implementation of community programs. After this preliminary phase of establishing an information system, a research project should be inaugurated to evaluate the effectiveness of existing programs. This would entail measurement of change of variables presumably influenced by mental health programs with other influences held constant. The present paper describes the first step in this plan: the establishment of an information system.

The set of procedures we designed to meet these research aims is largely derived from work in urban sociology, and more specifically, social area analysis. Urban sociology is a twentieth century specialization developed by Park and Burgess at the University of Chicago, concerned with the spatial distribution of social problems and the formulation of urban typologies. Mapping techniques were used to reveal such spatial distributions, relying largely on census data for basic information. The early theories of urban typology included Burgess' concentric zones hypothesis, Hoyt's sector hypothesis, and Harris' multinucleation hypothesis. Although these typologies generated considerable research, flaws in underlying assumptions precluded their validation, and interest in this approach declined. More recently, the studies of Shevky (1949, 1955), Tryon (1955, 1967) and others in the analysis of social areas have provided a theoretical basis and specific methods for delineating the social characteristics of areas. These investigators have been less interested in spatial distributions, and define social structure solely in terms of the relationships among characteristics.

Shevky and his colleagues used 1940 census data for Los Angeles to develop three indexes—social rank, urbanization, and segregation—in terms of which census tracts could be scored. Tracts with similar patterns of scores are regarded as representing a particular type of social area. Tryon, working from 1940 census data for San Fransisco, derived three indexes by cluster analysis of census items, which he labelled socio-economic achievement, family life and assimilation. These are roughly equivalent to the indexes that Shevky worked out on theoretical grounds. Hadden and Borgatta (1965) did a factor analysis of thirty-two census tract items and obtained three stable factors that applied to different regions of the United States: socioeconomic status, suburb factor and disorganization-deprivation. The first two are clearly comparable to the indexes of Shevky and Tryon, and since the item with the highest loading on disorganization-deprivation is the percentage of Negroes, this factor is also similar to Shevky's segregation and Tryon's assimilation index.

Social area analysis has been adopted by psychologists as a method for studying the relation between environmental and behavioral variables. Until now, it has most often been used in the study of juvenile delinquency (Chein, 1963; Cartwright and Howard, 1966). In the current study, we adopted its general framework to study the relations between environment and a variety of social, psychiatric and physical problems. This method rests on the assumption that any large geographical area can be divided into two or more subareas to be observed and described just as people or objects may be described. These geographical subareas or units ideally have the following characteristics:

(1) Homogeneity in terms of the variables observed so that their measurement will acurately describe the area. For example, a unit may be delineated so that it includes very poor and very rich people. If one is interested in financial status, the statistic of median income in such an area does not accurately describe the population contained in it.

(2) Stable boundaries to permit the measurement of change over time on a given set of measures. Since census data is often used, there are characteristically ten-year intervals between measurements.

(3) A sufficiently large population in each unit of observation to yield reliable observations.

There must also be enough units of observation to insure stable or replicable results and relationships. The reliability and validity of the measures chosen for study should be relatively invariant across the units. Since the variables they are intended to describe may have different statistical distributions, both simple and complex relationships among these measures must be considered.

Application to New York City Health Areas

In the course of reviewing possible sources of data for New York City, it became apparent that public and private agencies use many different units of observation, often based on heterogeneous areas, making comparative and correlational studies difficult if not impossible. To mitigate the effect of such problems, we chose the health area, as defined by the New York City Department of Health, as our unit of observation. Since each health area consists of two or three census tracts, we could describe each health area in terms of the indices available in census data, as well as epidemiological data compiled by New York City agencies that are based on health areas.

We chose twenty-five types of variables describing social and behavioral characteristics, most of which were measured at two points in time. This study reports findings for nineteen of them, each representing a single occasion. In order to have our data in comparable form, we used frequency of occurrence (with two exceptions) to quantify our variables; this was calculated separately for each of the health areas. We classified these nineteen variables into two groups. The first, derived primarily from census data, was regarded as descriptive of the *environment*, and included such variables as unemployment, overcrowding and the number of residents who lived abroad in 1955. The latter variable was highly correlated with the number of Puerto Ricans in the population while also reflecting migration rates. The other group of variables describes the *behavior, health and mental health* of populations, including number of premature births and number of psychiatric hospitalizations. The latter group of measures was conceived as outcome variables presumably linked to environmental conditions.

Thus, we selected nineteen variables to work with intensively; of these, nine were regarded as predictor variables describing environmental conditions. The other ten were regarded as outcome variables describing behavior, health and mental health, presumably representing certain crucial outcomes of living determined to some extent by environmental conditions.

(A) *Environmental (predictor) variables* (from 1960 census data unless otherwise noted)

(1) number of unemployed people 14 years or older

(2) number of overcrowded housing units

(3) number of people living in the South in 1955 (mostly Negro)

(4) number of people living abroad in 1955 (mostly Puerto Rican)

(5) median family income

(6) number of people 14 years or older who are divorced or separated

(7) median years of education completed by people 25 years or older

(8) number of people on welfare funds (1962)

(9) number of people in the health area

(B) *Behavioral (outcome) variables* (from 1960 data unless otherwise noted)

   (1) number of admissions to state and city mental hospitals (1961)
   (2) number of out-of-wedlock births (1962)
   (3) number of infant deaths under one year
   (4) number of births under 2501 grams (premature)
   (5) number of births with prenatal care beginning in ninth month
   (6) number of births with prenatal care beginning in seventh or eighth month
   (7) number of cases of reported venereal disease
   (8) number of reported homicides
   (9) number of reported suicides
  (10) number of reported arrests in the age group 7-20

Method

Our objective was to investigate the degree to which two sets of predictor variables would explain differences among health areas on the ten outcome variables listed above. To determine the magnitude of the relationships among all nine predictor variables describing health area environments and the ten outcome variables, a nineteen by nineteen product moment correlation matrix was computed for each borough. To estimate the degree to which differences among health areas could be explained by describing environmental conditions, multiple correlations of the nine predictor variables (including health area population size) with each of the outcome variables were computed. These coefficients are presented under the symbol Rp in Tables 1 and 2 for the Bronx and Brooklyn respectively. Subsequent computation partialled out the effect of health area population size.[1] These corrected coefficients are presented in column two of the same tables under the symbol Rc.

The two variables estimating the frequency of recent migration from the South and from abroad comprise the second set of predictor variables. The multiple correlation coefficients indicating the degree of relationship between these two predictor variables and each of the ten outcome variables are presented in column four of Tables 1 and 2 under the symbol Rm.

The sixth column of the two tables, symbolized by Rc^2-Rm^2, indicates the percentage of outcome variance explained by the health area population size-corrected set of nine predictor variables minus the percentage of outcome variance explained by the two migration variables in weighted linear combination. This column simply indicates the degree of advantage provided by the nine-variable predictor set over the two variable set.

TABLE 1

THE PREDICTION OF THE TEN OUTCOME VARIABLES
BOROUGH OF THE BRONX
N=63 HEALTH AREAS

Outcome Variables	Rp	Rc	Rc^2	Rm	Rm^2	Rc^2-Rm^2
Hospitalization for Mental illness	91	88	77	77	59	18
Out-of-wedlock births	97	97	95	88	78	17
Infant deaths	81	80	65	71	51	14
Premature births	93	92	85	89	79	6
Prenatal care (9th month only)	95	95	89	85	73	16
Prenatal care (7th-8th month)	97	97	94	91	83	11
Reported V.D.	97	97	93	81	65	28
Reported homicides	68	68	46	66	43	3
Reported suicides	48	37	14	13	02	12
Reported arrests	94	94	88	85	72	16

Rp: Multiple correlation of nine predictor variables, including health area population size, with outcome variables.

Rc: Multiple correlation of nine predictor variables with outcome variables, corrected for health area population size as described in the text.

Rc^2: The above value squared.

Rm: Multiple correlation of the two migration variables (from the South; from abroad) with the ten outcome variables.

Rm^2: The above value squared.

Rc^2-Rm^2: The percentage of outcome variable variance explained by the nine predictor variables (corrected for health area population size) minus the percentage explained by the two migration variables.

Decimals omitted; all values in .XX format.

TABLE 2

THE PREDICTION OF THE TEN OUTCOME VARIABLES
BOROUGH OF BROOKLYN
N=113 HEALTH AREAS

Outcome Variables	Rp	Rc	Rc^2	Rm	Rm^2	Rc^2-Rm^2
Hospitalization for mental illness	92	90	81	81	66	15
Out-of-wedlock births	97	97	94	91	84	10
Infant deaths	93	93	86	85	72	14
Premature births	91	89	80	82	68	12
Prenatal care (9th month only)	96	96	91	89	79	12
Prenatal care (7th-8th month)	97	97	93	89	79	14
Reported V.D.	96	96	92	89	79	13
Reported homicides	82	82	67	74	55	12
Reported suicides	46	28	08	22	05	03
Reported arrests	92	92	84	75	56	18

Rp: Multiple correlation of nine predictor variables, including health area population size, with outcome variables.

Rc: Multiple correlation of nine predictor variables with outcome variables, corrected for health area population size as described in the text.

Rc^2: The above value squared.

Rm: Multiple correlation of the two migration variables (from the South; from abroad) with the ten outcome variables.

Rm^2: The above value squared.

Rc^2-Rm^2: The percentage of outcome variable variance explained by the nine predictor variables (corrected for health area population size) minus the percentage explained by the two migration variables.

Decimals omitted; all values in .XX format.

Results

By referring to Tables 1 and 2, it can be seen that the degree of relationship between the nine predictor variables and the outcome variable is consistently high for both boroughs, except for the measure of suicides. In both Bronx and Brooklyn data, six of the ten corrected multiple correlations (Rc) exceed .90, indicating a strong relationship between environmental conditions and behavior and health. Thus in predicting such factors as premature births, infrequency of use of medical facilities, venereal disease, out-of-wedlock births and reported arrests, more than three-quarters of the variance is accounted for by environmental parameters. High if less spectacular correlations were also found between number of homicides, infant deaths and hospitalizations for mental illness, and environmental variables.

Relationships among the nine predictor variables are moderately high. For the Bronx they have a mean correlation of .53, and for Brooklyn, .49. This suggests that the number of variables in our predictor battery could be reduced without substantially reducing its power. To examine this tentative conclusion, multiple correlations between two of the nine predictor variables—the number of recent migrants from the South and from abroad—and the ten outcome variables were computed for each of the two boroughs. The coefficients are presented in column 4 of Tables 1 and 2 under the symbol, Rm.

In the data of both Bronx and Brooklyn at least six of the Rm multiple correlations exceed .80, indicating that the extent of migration into health areas from the South and abroad is a reasonably accurate predictor of differences among these health areas on certain measures of health, mental health and behavior. It is worth noting, however, by comparing the values under Rc with those under Rm, that the predictive power of the nine-variables battery is considerably greater than that of the two migration variables. This advantage is expressed in terms of the additional percentage of outcome variance accounted for by the nine-variable battery in the sixth column of the Tables under $Rc^2 - Rm^2$.

In order to identify the most powerful predictor batteries we plan to systematically compare the efficiency of different combinations of variables in predicting a variety of outcome variables in the five boroughs of New York City. Based on data we have at this point, it appears that the role of a predictor variable varies considerably as a function of the particular outcome variable it is to predict and of the borough in which it was observed.

Discussion

Description of a given health area in terms of the nineteen environmental and outcome variables has considerable meaning to the health and mental health planner. The number of admissions to mental hospitals provides one estimate of the manpower needed to contend with such related problems as helping former mental patients adjust to community living. The number of pregnancies receiving medical care only in the third trimester reflects the nature of attitudes toward the use of health facilities and almost certainly the quality and style of service available to these people. The extent of migration from southern rural areas where medical and social services were seldom available offers one basis for planning educational programs.

Such variables also describe the type of stimuli which constantly impinge upon the people of an area. As Tryon (1955) points out, even if an individual is different from the majority of people living in his neighborhood and is therefore inadequately described by averaged statistics and demographic data, we still have specific knowledge of the social situation in which he lives. As we have found, description of environmental variables reveals the level of social competence, nature of the opportunity structure, extent of antisocial behavior and family stability characterizing a given area. Certainly all of these factors play a role in a resident's level of aspiration, self-esteem and pattern of behavior.

The question of the generality of our results is of great interest to us. In view of the variety of possible factors influencing the observed relationships between predictor and outcome variables, the similarity of the obtained multiple correlation coefficients for the two boroughs is striking. Seven of the ten uncorrected r's varied no more than two points between boroughs. We cannot be sure that the same findings will emerge elsewhere, although we expect them from Queens and possibly Manhattan, perhaps with some shift in the relative weights of predictor variables. But of course this is an empirical matter to be resolved through continued study.

Assuming our results are in fact general, it is worth thinking about their implications for planning a community health program. Is it possible to improve the mental health of people living in miserable physical and social settings without also improving those environmental conditions to which behavior appears to be so powerfully linked? The answer to this question is probably an emphatic NO and we suspect that a number of community mental health directors are keenly and anxiously aware of the odds against them unless the more fundamental social and economic problems of their catchment areas are also considered. By creating, for example, the Model Cities program, the federal government is indicating the need for comprehensive services focused on both environmental conditions and their consequences in order to eliminate slums.

Whatever may be the best strategy for improving the functioning of communities, our interest is in the development of a set of procedures for evaluating alternate models of intervention. The twenty multiple correlation coefficients presented here hardly do justice to the richness of the data at our disposal. There are three additional kinds of analysis we plan to undertake. The first concerns the classification of health areas into types in terms of their similarity on both predictor and outcome variables, following the procedures of Tryon or other profile analytic methods. With similar health areas thus identified, changes under the influence of different conditions or programs could then be compared at regular intervals over a period of time. As mental health programs become activated in our great cities, changes in *similar* social areas with *different* programs could be compared on a variety of relevant predictor and outcome variables. Deviant social areas of a city—those changing abruptly in one way or another—could be identified and studied to determine those factors contributing to rapid deterioration or growth.

A second form of data analysis we intend to pursue concerns interborough comparisons in order to determine differential effictiveness of predictor variables and the social significance thus entailed. For example, the variable showing number of southern Negroes who came to New York City within the last five years is a far more powerful predictor of social and health problems in the Booklyn data than it is in the Bronx. In contrast, recent migration from Puerto Rico is much more strongly related to maladaptive behavior in the Bronx, suggesting that the Puerto Rican is at a greater disadvantage here than in Brooklyn. Quite obviously, more detailed study of the interrelations between minority group membership and other predictor variables is needed before reasonably valid interpretations are drawn, but these preliminary observations suggest the value of this kind of analysis.

The third approach to data analysis that we are planning concerns the relationship between our predictor variables and a more comprehensive sample of behavioral measures presumably associated with social conditions. The range of measures could be as broad as the programs they were to evaluate. For example, by describing the social areas of a city in terms of the reading levels or academic growth scores of seventh grade students living in each social area, relationships between school performance and the neighborhood environment could be estimated. By acquiring information from established agencies and by gathering additional data through independent studies, each social area within a city could be comprehensively described.

The kind of data needed for the accurate and efficient description of urban areas is, unfortunately, often difficult to obtain. It seems to us, however, that the amount of money currently spent by most cities in gathering information by a variety of agencies using different and incompatible units of observation is probably adequate to sponsor a single data-gathering agency. Its task would be the coordination of relevant information for effective program planning and evaluation. Such an agency has been extablished for the entire state of Maryland (Gorwitz et al., 1963; Bahn, 1960; Gorwitz, 1967), so that it would seem equally feasible for urban use.

While the foregoing procedures may seem complex, we submit that such an approach is necessary to develop an empirically based urban sociology and to evaluate successfully the proliferating attempts at influencing the health and mental health of communities.

NOTE

1. Rc was computed by using the following formula:

$$Rc = \sqrt{\frac{Rp^2 - r^2}{1 - r^2}}$$

where:
Rp is the multiple correlation of the nine predictor variables, including health area population size, with a given outcome variable.
  r is the relationship of health area population size to a given outcome variable.
  Rc is the health area's size-corrected multiple correlation coefficient, indicating the degree of relationship between the weighted linear combination of the predictor variables and the outcome variable.

REFERENCES

BAHN, A. K. (1960) "The development of an effective statistical system in mental illness." American Journal of Psychiatry 116: 798-800.
CARTWRIGHT, D. and K. HOWARD (1966) "Multivariate analysis of gang delinquency: I. ecological influences." Multivariate Behavioral Research 1:321-371.
CHEIN, I. (1963) "Some epidemiological vectors of delinquency and its control: outline of a project." Research Center for Human Relations, New York University (unpub. mimeo.).
GORWITZ, K. (1967) "Mental health and mental illness in Maryland: a report from the Maryland Department of Mental Hygiene." Maryland Department of Mental Health (unpub. mimeo.).
GORWITZ, K., A. K. BAHN, C. CHANDLER and W. MARTIN (1963) "Planned uses of a statewide psychiatric register for aiding mental health in the community." American Journal of Orthopsychiatry 33: 494-500.
HADDEN, J. K. and E. F. BORGATTA (1965) American Cities: Their Social Characteristics. Chicago: Rand McNally.
HOYT, H. (1939) The Structure and Growth of Residential Neighborhoods in American Cities. Washington, D. C.: United States Government Printing Office.
SHEVKY, E. and W. BELL (1955) Social Area Analysis. Stanford: Stanford University Press.

SHEVKY, E. and M. WILLIAMS (1949) The Social Areas of Los Angeles: Analysis and Topography. Berkeley and Los Angeles: University of California Press.

TRYON, R. C. (1967) "Predicting group differences in cluster analysis: the social area problem." Multivariate Behavioral Research 2: 453-475.

——— (1955) Identification of Social Areas by Cluster Analysis. Berkeley and Los Angeles: University of California Press.

Chapter **10**

Migration and Ethnic Membership in Relation to Social Problems

ELMER L. STRUENING, JUDITH G. RABKIN, and HARRIS B. PECK

Despite the variety of studies designed to investigate the relationship between migration and mental illness, emergence of consistent and clearcut results were long impeded by a variety of methodological problems. Gradually, however, investigators have come to agree on common definitions and measures of the concepts of migration and mental illness so that the more recent studies can be compared, samples can be combined, and some general conclusions can be tentatively drawn. Before summarizing the findings available to date, some of the methodological and procedural difficulties will be reviewed to facilitate evaluation of the studies conducted in this field.

The most common methods of approaching the study of migration and mental illness are those of case-finding in the general community (e.g. Tietzse, 1942; Srole et al., 1962) and analysis of residence records of hospitalized mental patients (e.g. Malzberg and Lee, 1956; Lazarus et al., 1963). The former method seeks to estimate the rates of illness in a given community regardless of the number of people seeking treatment. As Hammer and Leacock (1961) point out, community surveys circumvent some of the problems involved in the use of treatment figures, but provide questionable data since the more intensive the survey, the

---

Author's Note: *This paper was prepared for the conference on "Migration and Behavioral Deviance," November 4-8, 1968, Dorado Beach, Puerto Rico. This study was supported by Grant No. 5 R11 MH 02308, Public Health Service, National Institute of Mental Health.*

higher the obtained rates. The majority of investigators employ the second of these methods: analysis of the prior residential histories of psychiatric patients at the time of their first admission to state hospitals.

Before 1940, over 200 studies in this area were conducted, but they were so varied in their definition of migration and mental illness, and their sample characteristics and methods of measurement, that few results could be compared or replicated. In 1940 the national census introduced a specific item asking for the respondent's place of residence five years prior to the census. It has subsequently been suggested that there is a negative correlation between migrant's length of stay and rates of hospitalizations occurring within a year of the actual change of residence (Wilson et al., 1965; Malzberg and Lee, 1956). Although it may be debated that the five-year interval is too short or too long this census item has been widely adopted for practical reasons as the criterion for defining migration.

In addition to the issue of time elapsed since migration, a further problem concerns the geographical distance covered. While 20 percent of the American population moves annually, most of these moves are within the same county. Shryock (1964) reports that less than one-sixth of these moves are across state lines. Although the absolute distance and changes in life style (e.g. from rural to urban) are not controlled for, state boundaries are often used as the criterion for differentiating between mobile and migrant populations for practical reasons. In general then, with some important exceptions, most American investigators define migration as a move across state lines within the past five years.

Progress has also been made regarding the definition of mental illness. Early studies ignored the variable of age at the time of the patient's first admission to a mental hospital, thereby often invalidating the obtained results. As Goldhamer and Marshall (1953) have made clear, admission rates not controlled for age can and often do lead to spurious and distorted findings; this is particularly true in studies of migration where the age of the migrant or immigrant differs consistently from that of the general population, being concentrated in the younger adult brackets where mental illness rates are highest in all groups. This source of error contributed to the early reports that immigrants had far higher rates of mental illness than native Americans.

Sample selection is another potential source of difficulty in this area. While most studies include only patients in public psychiatric hospitals because these records are most readily available, others include patients in private hospitals, as well as clinic outpatients and/or those in private treatment. There is also the question of interpreting mental illness rates based on hospitalization figures. As Goffman (1964) and others have pointed out, hospitalization is characteristically based on "contingencies" such as socioeconomic status, visibility of the offense, proximity to a mental hospital, and availability of interested relatives. The use of hospitalization rates as an index of mental illness in a larger population consequently provides results that are not altogether straightforward.

A related problem is selection of the base population to which migrants are compared. While it is possible to compare migrants to the "population of origin"—that is, the group they left behind, it is far more common to compare migrants to the "population of destination," the group which they join. Ideally, the base population should exclude all those who entered or left a given region within a stated time interval (Thomlinson, 1962). Even if this definition is adopted, certain difficulties remain. For example, Guttentag and Denmark (1965) have pointed out that Southern Negro migrants as a group display psychiatric characteristics typically regarded as pathological, in comparison to Northern Negroes as well as whites. In their study they found that a group of Southern Negroes in northern psychiatric hospitals did not differ significantly in terms of formal Rorschach characteristics from a sample of Southern Negroes in Northern communities. However, both of these migrants groups did differ significantly from Piotrowski's normative sample in the direction of showing greater pathology. Guttentag and Denmark claim that Southern Negro migrants who are brought to psychiatric admitting offices are being inappropriately compared to a fundamentally different base population which significantly increases the risk of their psychiatric hospitalization regardless of actual extent of psychiatric impairment. This point, relevant as it seems to be, is not dealt with by most investigators in this field.

Another methodological problem that deserves attention is that of interpretation of rate differences between migrant and native populations if indeed they occur. It has proved remarkably difficult to demonstrate unequivocally that group differences are specifically attributable to migration rather than intervening variables such as social class, race, or speed of transition. There is finally the question of whether migrants comprise a group that is sufficiently homogeneous to permit the emergence of clearcut differences from other population groups. While many of these problems have been handled quite effectively in the studies to be noted below, it will be helpful to keep them in mind in considering inconsistencies and discrepancies that have been reported in this literature.

Review of the Literature

During the last fifty years there have been perhaps four overlapping theoretical positions regarding the relationship between migration and mental illness, and although adherents of all of them can still be found, their popularity has tended to be successive rather than simultaneous. The first position is that immigrants are disturbed before they start to move. The second position asserts that the process of migration itself precipitates psychiatric difficulties, and the third seeks to explain higher rates of mental illness in migrants in terms of demographic characteristics which make them dissimilar to the base population to which they are compared. The fourth position emphasizes the interaction that occurs between the migrant and the receiving society, considering such phenomena as pressure for assimilation and culture shock which appear at the interface of the two groups as they confront each other. The first position, based almost entirely on studies of immigrants to the United States, hold that mental disorders precipitate immigration, that those whose psychological adjustment has been marginal or unstable to begin with are more apt to immigrate. This explanation included a racial component to the effect that immigrants came from inferior stock. Such conclusions were typically based on poorly designed studies showing rates of hospitalization for immigrants two or three times higher than those of the native-born.

Two long-term programs of investigation in the 1930s and 1940s, led by Ødegaard in Norway and Malzberg and Lee in New York, seemed to rule out racial explanation and instead supported the position that the process of migration itself generated sufficient stress to lead to the development of mental illness in hitherto intact people. Working independently, they demonstrated that when the variables of age and sex are controlled, most of the differences in hospitalization rates for immigrants and natives could be accounted for. Over the past thirty years, Ødegaard has examined rates of mental illness of native Norwegians, those who came here and stayed, those who came here and returned to Norway, and internal migration within Norway. His findings discredited the notions that higher immigrant rates of mental illness were attributable to their ethnic origin and also that unstable people are more apt to immigrate. He found that Norwegians in Minnesota had significantly higher rates of hospitalization than those in Norway, and that female immigrants had higher rates than males. Since it was seldom the females who initiated the act of immigration, he concluded that immigrants were not self-selected in terms of a predisposition to mental illness.

Malzberg and Lee (1956) and Malzberg (1962) studied both immigrants from abroad and migrants from other parts of the United States in their analyses of mental illness in the various groups admitted to all psychiatric hospitals in New York State. Like Ødegaard, they found that relative to natives, foreign-born females had higher admission rates, although there was no significant differences for males. Malzberg and Lee were perhaps the first to report that, in the United States, migrants have first-admission rates that are appreciably higher than those of immigrants as well as nonmigrant natives. Using 1940 census data and the census definition of migration (having lived outside the state five years previously), they found that migrants, both white and black, from urban as well as rural points of origin, had higher admission rates at all ages. In 1962 Malzberg reported additional data from the 1950 census supporting his earlier findings. It may be noted in passing that Malzberg's "native" population included an unknown percentage of migrants who had arrived in the state more than five years ago.

Subsequent work in this area has confirmed the observation that migrants in America have higher hospitalization rates than either foreign-born or nonmigrant native-born. Several studies were designed to emphasize the variable of migration as such in accounting for the group differences reported, although in their analyses the investigators noted other variables that also differentiated the groups under study. Their main interest, however, was the comparison of hospital admission rates for the different groups. Lazarus et al. (1963: 41), studying hospitalization rates in New York, Ohio and California in 1950 and defining migrant status in terms of state of birth, reported that "migration per se is apparently a major determinant of admission to mental hospitals for all classes of natives." Lee (1963), using New York State data, again found that age-specific rates of hospitalization for migrants were considerably higher than those of nonmigrants, with immigrant rates being intermediate between them. Locke and Duvall (1964) reported studies of both 1950 and 1960 census data in relation to admissions to Ohio public mental hospitals. Their first study showed results like those of Lazarus and Lee, while their analysis of 1960 census data actually showed *lower* rates for immigrants than either of the native-born groups; migrants continued to have the highest rates. Whether this signifies a widespread change in admission rates for immigrants is unclear since few other published studies utilize the most recent census data.

Defining migration as movement between sections of North Carolina as well as across state lines, Keeler and Vitols (1963) found that 40 percent of the new schizophrenic admissions to the state hospital serving Negroes had a history of migration preceding admission. This in itself is ten times the rate of migration for nonhospitalized Negroes in North Carolina. Of the migrant patients, the authors believed that 65 percent had not been schizophrenic preceding migration, while the rest were apparently schizophrenic before leaving home. The authors emphasize that migrants do not constitute a homogeneous population and should not be treated as such in research designs.

All of these studies were designed to control for the variables of age, sex and race and dealt only with first admissions, since Malzberg and Lee demonstrated the importance of doing so. As analyses of rate differentials progressed, it became increasingly clear that other intervening variables were also related to the obtained differences in hospitalization rates found for migrants, immigrants and nonmigrant natives. In fact, several investigators in the field became doubtful that migration itself was the major distinguishing characteristic of migrant groups, and gradually moved toward a third theoretical point of view; that the higher admission rates of migrants are largely due to intervening variables such as race, education, occupation, social class and residence. Findings related to these variables will be reviewed in turn.

As investigators of the effects of social change altered their focus from immigrant groups to native American migrants, it became apparent that the variable of racial membership was very much involved in their work. During the last thirty years, most migration within the United States has consisted of rural Southern Negroes moving to North and West urban centers in search of better living and employment conditions. This sector of the migrant population is specifically known as "immigrants." As early as 1944, Malzberg found that the much higher admission rates for Negroes in New York State were almost exclusively due to the presence of Southern Negroes who had migrated north. (See Malzberg's essay in Klineberg, 1944.) Lazarus et al. (1963) found that nonwhite migrants had highest hospitalization rates, with nonwhite nonmigrants, white migrants, and white nonmigrants following in that order. Race evidently was the variable most strongly associated with hospitalization, and the authors conclude that "color is more important than migration status in our statewide patterns of differentials." While hospitalization rates for Negroes are consistently higher than those of whites regardless of migration status, it has also been established that migrants have higher hospitalization rates within each racial group. The only clear exception to this general finding is the work of Kleiner and Parker (1960, 1965) working in Philadelphia, who have repeatedly reported lower rates of mental illness for Southern-born than native Negroes. They have also found that Negroes migrating from other Northern states to Philadelphia had the highest rates of all. One possible explanation for this striking divergence of findings about the interaction of migration and race concerns the nature of the sample. Migrants are here defined somewhat unusually as those who spent the majority of their first seventeen years of life outside of Philadelphia; this group probably differs from other research samples defined by the census item pertaining to residence within the past five years, which is the more commonly used criterion. It is also possible that Southern Negroes who migrate to Philadelphia are in some ways unlike those who travel further North or West to larger industrial centers. In sum, even if its influence is not entirely clear, the variable of racial membership appears to be important in studies of migration.

Less conclusive evidence has been reported for the variables of social class, occupation and education. Lee (1963) found that standardization of New York State hospitalization data for education and occupation did not greatly alter conclusions based on age-specific rate alone. Locke and Duvall (1964) found negative correlations between admission rates and education and occupation. The variable of social class (often measured in terms of occupation as well as income) was found by Srole et al. (1962) to account for most of the differences in psychiatric impairment ratings between various white ethnic groups in New York City. Clark (cited in Murphy, 1965) showed that the same overall relationship between foreign- and native-born held up within occupational categories, with the former having higher hospitalization rates in each category. In short, as Murphy (1965) observed, the effects of social class, education and occupation have less of an effect than might have been expected in analysis of hospitalization rates for migrant and nonmigrant groups, but these variables apparently do account for a little of the differences between them.

In general, for all the intervening variables here noted, there still remains an association between migration and hospitalization rates which cannot be entirely accounted for by peculiarities of the migrant group. Instead of continuing to study such variables, increasing attention is now being devoted to the physical and cultural contexts within which migrants live, rather than demographic or psychological characteristics of the migrants themselves. This can be thought of as the fourth point of view about migration and mental illness. In a sense, this change of focus follows the trend in psychotherapy treatment and research away from emphasis on the individual (in this case the migrant) as the unit of study, in favor of extended surveillance of the individual within his social and environmental context. This represents basically an adoption of the general systems theory and model in the study of behavior. With respect to migrants, the two major areas to be studied following this approach are the environmental characteristics of migrants at origin and destination, and the cultural implications and consequences of the process of migration.

Residence at origin and destination have long been regarded as relevant variables in the study of mental illness rates for migrant and immigrant groups. It has been generally assumed that the social change of migration would be greater for those moving from rural to urban environments (the converse occurs too seldom to warrant analysis).

Supporting evidence is provided by Srole et al. (1962). These investigators initially expected new immigrants in their sample to have higher prevalence rates of mental illness. When this was not found, they decided that they had a new kind of immigrant in their sample: people who moved from urban setting to urban setting rather than from rural to urban. When first-generation immigrants were classified as coming from these two sources, the expected differences for rural-urban immigrants were obtained.

Rural or urban origin is not significant as such, but rather serves as a basis for measuring the amount of social change in environmental climate and life style experienced by the migrant. Similarly, the type of city at destination is also relevant. Ødegaard found, for example, that Norwegian migrants as a group had lower hospitalization rates than nonmigrants in all rural and urban areas except for Oslo, where migrants from both rural and urban origins had higher rates than nonmigrants. Since Oslo is the only Norwegian city at all comparable to American large cities, this finding may have bearing on the study of American migrants. Finally, while it has been found that American migrants moving from rural to urban centers have higher hospitalization rates than nonmigrants, the same is also true of migrants moving from one rural area to another. The change from rural to urban context therefore does not account entirely for the differences between migrant and nonmigrant groups.

The variable of residence at destination has been approached from both an ecological and cultural point of view. Faris and Dunham (1939) found that areas of Chicago with a high percent of home ownership had lower hospital admission rates, while the reverse was true of those areas with many transients as measured by percent of hotel and lodging-house residents. Gerard and Houston (1953) found that the concentration of schizophrenics in poorer areas of the city of Worcester was due to the large number of recent single migrants entering such areas, and that the variable of migratory status or single occupancy rather than that of poverty seemed to account for the observed high rates of mental illness.

Different implications about the role of slums for city newcomers are suggested by Fried (1964, 1966), who conceptualizes migration as a crisis of transition. Noting the tendency of migrants to move to slums populated by others of their own kind, Fried sees this as a means of minimizing changes in cultural and social living styles. By seeking out compatriots, migrants and immigrants find a certain increment of security and shelter in familiar patterns of family relationships, religious and recreational activities, enabling them to devote more attention to adapting to urban occupational demands. Based on this analysis, it would seem that migrants settling in urban areas with a high proportion of earlier migrants from similar backgrounds would adjust more easily than those moving to areas with few other migrants or where the other migrants came from a different cultural origin.

H. B. M. Murphy also emphasizes cultural context in evaluating the risks and problems that migrants encounter upon arrival at their destination. Impressed with the findings that areas with large immigrant populations show lower rates of psychiatric hospitalization, as in Israel and Singapore, he has delineated two factors which seem to influence migrant hospitalization rates. The first concerns group membership in the new environment. When migrants can easily join an existing social group similar to the one they came from, the stresses entailed in the move are reduced or prevented. This is essentially Fried's point regarding the role of ethnic slums. As supporting evidence, Murphy (1965) cites the observation that in Canada, Chinese have the lowest hospitalization rates of all minority groups in British Columbia where they constitute a large group. But in Ontario, where they are scattered about in "penny numbers," they have the highest hospitalization rates of all ethnic minorities. It is not clear whether group membership as such is responsible for these differences, or whether the major factor is reduction of culture conflict. Murphy thinks it is the former, referring to his study of different dialectal groups of Chinese migrating to Singapore. Although there is no culture conflict between any Chinese, he found that hospitalization rates for the various dialectal groups were inversely related to their size in the community.

The second major factor that Murphy has pointed out as an influence on ease of adjustment for migrants is the amount of pressure in the receiving society to assimilate, to "join the melting pot." In some regions and countries such as the United States and Australia, newcomers are expected to quickly resemble everybody else. The countries have reported higher hospitalization rates for newcomers than those which accept or encourage the maintenance of ties to the original culture, as in French Canada or among the Singapore Chinese.

In summary, almost all studies show higher rates of mental illness as measured by hospitalization rates for migrant or immigrant groups than for the nonmigrant native-born. Some aspect of the transition from one society to another apparently constitutes a hazardous situation leading to increased risk of psychiatric hospitalization. There seems to be mounting evidence attesting to the notion that the development of mental illness in migrant groups is not so much due to qualities inherent in the migrants, but rather to the interaction between migrants and the societies that receive them. Stress seems to be most intense for those migrants whose original culture differs radically from the new, where

[226] BEHAVIOR IN NEW ENVIRONMENTS

there are no available familiar groups to join and thus modify the intensity of the cultural change, and where the receiving society most actively emphasizes rapid assimilation.

Purpose of the Study

The present study examined the role of migration from a broader and somewhat different perspective than those characterizing the work reviewed above. It was reasoned that adaptation to a new social context by migrants would result not only in high rates of mental illness but also in an increased incidence of physical illness and deviant or disruptive social behavior. Therefore a more extensive range of social problems or difficulties in adaptation were considered. Among those included were measures of infant mortality, premature birth, family disruption, economic difficulty, severe mental illness and antisocial behavior.

Migrants are defined, in census terms, as residents who have moved into a given area within the past five years. In this study migrants of two ethnic groups, Negroes and Puerto Ricans, were considered. Within the Negro group only migrants from Southern states were included; the Puerto Rican group consisted of migrants from abroad (usually Puerto Rico).

The major purpose of this study was to examine the influence of migration on specified social, health and mental health problems listed above, using the health areas[1] of two boroughs in New York City as our sample. A second purpose was to compare the variables of migration with those of ethnic membership in terms of their relative ability to predict the current distribution of social health and mental health problems over the health areas of the two boroughs. A third purpose, similar to the second, was to compare their relative efficacy in predicting the future distribution of social and health problems five years hence. The latter purpose has obvious implications for planning the extent and type of intervention programs needed in the future.

The observations of this study do not refer to individuals but to residents within defined geographical or social areas. Demarcated social areas, like objects or people, may be described along many dimensions and at different points in time to yield a meaningful profile of descriptive data. These observations may then be manipulated with standard statistical procedures to produce a set of interpretable results. It must be noted that generalizations regarding, for example, relationships among variables are limited to the units of observation and may not be extrapolated to relationships among characteristics of individuals. This would constitute the "ecological fallacy" (Cartwright, 1968).

The above procedure, usually referred to as social area analysis, is described in terms of its history and development in a comprehensive article by Cartwright (1968). It has been developed to a sophisticated level by Tryon and his colleagues (1955, 1967, 1968) over a fifteen-year period. In his most recent work, Tryon (1967, 1968) has applied multivariate statistical procedures, computed with the comprehensive BC TRY computer system (Tryon and Bailey, 1966), to census data of the San Francisco Bay region in order to identify basic descriptive dimensions, social areas (census tracts) and, with the above dimensions as variables, a meaningful typology of neighborhoods. Tryon (1967) has demonstrated the remarkable stability of such neighborhood types over time as well as the powerful linkage of his three basic dimensions (socioeconomic independence, family life, assimilation) to voting preference and other social behaviors.

Population and Sample

The unit of observation in this study is the health area. Some years ago, the five boroughs of New York City were divided into 347 health areas by the Department of Public Health. The 176 health areas of two of these boroughs, the Bronx and Brooklyn, form the total sample of this study. For comparative purposes this total sample will be subdivided into the 63 health areas of the Bronx and the 113 health areas of Brooklyn. In the sense that all health areas of the Bronx and Brooklyn were observed, the sample is equal to the population to which we wish to generalize. However, the results on the Bronx and Brooklyn may hold for the remaining boroughs, or results for 1960 may generalize over time to 1965.

According to the 1960 census, the 63 health areas of the Bronx contain an average of 22,565 residents with a standard deviation of 6,910. The figures for the 113 health areas of the Brooklyn census are 23,207 and 6,901 respectively.

Variables

As previously indicated, health areas are composed of census tracts and therefore may be described on all census variables. The Department of Health reports an extensive fund of information on each health area. A number of private and public agencies also report frequencies and rates by health area. In this study we focused primarily on population characteristics of health areas as related to a series of illness, mental illness and behavioral attributes.

The following population characteristics for each of the 176 health areas of the Bronx and Brooklyn were derived from 1960 census data:

(1) The number of migrants from the South, defined as living in the southern states in 1955.
(2) The number of migrants from abroad, defined as living abroad in 1955.
(3) The number of relatively permanent Negro residents, defined, in census terminology, as the nonwhite residents minus the migrants from the South, arriving after 1955.
(4) The number of relatively permanent Puerto Rican residents, defined, in census terminology, as Puerto Rican residents minus migrants from abroad who arrived after 1955.

Since, in (3) and (4) above, the number of migrants is small compared to the number of residents, the correlation between, for example, the number of Puerto Rican residents minus the migrants from abroad is .99. Therefore, for correlation purposes, the two indicators would provide the same results. However, in this study, we are interested in defining the number of migrants and the number of relatively permanent residents for comparative purposes.

The following variables were selected as indicators of the health, mental health and social conditions of each of the 176 health areas of the Bronx and Brooklyn:

(1) The number of persons fourteen years and older who were divorced or separated in 1960. The number of persons on welfare in 1962.
(2) The number of people admitted to state and city mental hospitals in 1961 and 1965.
(3) The number of infant deaths under one year in 1960 and 1965.
(4) The number of premature births (less than 2501 grams) in 1960 and 1965.
(5) The number of births with prenatal care beginning in the seventh or eighth month of pregnancy in 1960 and 1965.
(6) The number of reported homicides in 1960 and 1965.
(7) The number of arrests of persons, seven to twenty years of age in 1960 and 1965.

Data Analysis

Each health area is described in terms of residents migrating from the South (MS, an estimate of the number of migrating Negroes or "Southerners") and abroad (MA, an estimate of the number of migrating Puerto Ricans or those from "abroad") as well as the estimated number of Negroes (PN) and Puerto Ricans (PPR) with a relatively longer period of residence in New York. Thus PN and PPR indicate the number of relatively permanent Negroes and Puerto Ricans respectively: MS and MA represent Negro and Puerto Rican migrants respectively.

In the tables which follow the four population variables are related singly and in various combinations to a series of indicators of health, mental health and social conditions in the Bronx and Brooklyn, usually at two points in time.

Specifically the following statistics were computed:

(1) As descriptive data, the mean and standard deviation of each social and health variable for the Bronx, Brooklyn and the two boroughs combined were computed. With four exceptions, all variables were described at two points in time, usually 1960 and 1965.

(2) Linear relationships were computed between the four population variables (MS, MA, PN, PPR) and each social or health variable for the Bronx, Brooklyn and the two boroughs combined. In most cases the correlations were computed with the social or health variable observed in 1960 and 1965.

The same pattern of relationships was determined for the weighted linear combination of five migration variables and the two estimates of the number of permanent residents. The five migration variables combined were MS, MS squared, MA, MA squared and MA times MS. This combination considers the possibility of curvilinear relationships and the predictive potential of the interaction effect of MS and MA. Subsequent work has shown that curvilinear relationships and the interaction effect contribute little or nothing to the prediction of social or health problems. However, it will be seen that the linear combination of MS and MA increases the predictive power considerably in most cases. PN and PPR were also combined in a weighted linear combination to determine the extent to which the frequency of social and health problems could be predicted from these population variables.

In almost all cases correlations are between two frequencies. In general the frequencies refer to the number of people in a particular population category and the number of people experiencing a certain health or social problem. If the frequencies are highly correlated over inferences regarding a cause and effect sequence. We can compare the degree of relationship among population variables and indicators of social problems as we shift across boroughs and we can ascertain the ability of population variables to predict social and health problems five years later. It should be kept in mind that our values are based on

[230] BEHAVIOR IN NEW ENVIRONMENTS

populations rather than samples and therefore are not subject to the usual sample fluctuations. Therefore, some comparisons may be made without tests of significance. For the purpose of this presentation we shall only comment on the more obvious patterns of similarity and difference.

Results

The Migration Variables

As previously indicated, migrants from the South and migrants from abroad were identified as being primarily Negroes and Puerto Ricans. This is an inference rather than a direct count of, for example, the number of Puerto Ricans among migrants from abroad. This inference is based on the high relationships between the number of migrants and the number of persons in the ethnic group of which the migrants are assumed to be members.

Table 1 presents the pattern of relationships which indicates the characteristics of health areas into which migrants move. The number of

TABLE 1
RELATIONSHIPS AMONG MIGRATION VARIABLES AND POPULATION CHARACTERISTICS OF THE HEALTH AREAS OF THE BRONX AND BROOKLYN

	Bronx N=63 H.A.		Brooklyn N=113 H.A.		Both N=176 H.A.	
	MS[a]	MA[a]	MS	MA	MS	MA
NUMBER OF RESIDENTS						
White	-39[b]	-40	-49	-20	-41	-28
Nonwhite	79	27	92	19	89	19
Puerto Rican	48	88	21	80	18	82
NUMBER OF BIRTHS						
Nonwhite	82	36	93	24	92	24
Puerto Rican	48	91	21	84	20	87
Number of Persons on Welfare	64	64	64	51	58	56
Median Education	-21	-62	-09	-38	-10	-46
Median Family Income	-36	-53	-43	-52	-36	-52
Number of Nonwhites and Puerto Ricans	74	72	86	45	78	54

[a] See note, table 2b, for definition of terms.

[b] All correlations are in .XX format.

migrants from the South who take up residence in the Bronx varies directly with the number of nonwhite (R = .79) and the number of nonwhite births (R = .82). In other words (across the 63 health areas of the Bronx) as the number of migrants from the South increases, so does the number of nonwhite persons and nonwhite births. This same pattern is observed across the Brooklyn health areas where the number of migrants from the South is highly correlated with the number of nonwhites (R = .92) and the number of nonwhite births (R = .93). When the health areas of the two boroughs are combined, the relationships between the number of migrants and the number of nonwhites and nonwhite births are .89 and .92 respectively, indicating that, generally, migrants from the South are moving into areas as a direct function of the number of nonwhites already in the area.

The same is true for migrants from abroad. Their number is correlated .88 with the number of Puerto Ricans and .91 with the number of Puerto Rican births in the Bronx; for Brooklyn the values are .80 and .84, while for the two boroughs combined the figures are .82 and .87. Thus, across health areas, the number of migrants from abroad is strongly linked to the number of Puerto Ricans.

It is also apparent that as the number of migrants increases, so does the number of welfare recipients, while educational level and median family income decreases. Thus migrants tend to move into health areas with a relatively large number of persons belonging to ethnic minority groups, and with low educational levels, low median family incomes and high welfare rates. In view of the above realtionships it appears likely that a very large proportion of migrants are drawn from minority groups with low economic status. Although this interpretation is not based on direct observation, and hence must be considered in subsequent interpretations, it seems highly probable that migrants from the South are primarily Negroes and migrants from abroad are predominantly Puerto Ricans. In later work it is hoped that cross tabulated census data will be available so that direct counts can be made.

Family Disruption and Socioeconomic Level

Table 2a presents the extent of family disruption and an indication of socioeconomic conditions in the health areas of the Bronx, Brooklyn and the two boroughs combined. Family disruption is defined as the number of persons fourteen years and older who are divorced or separated while socioeconomic conditions are defined by the total number of people on welfare. These indicators of social and economic conditions are, as one would expect, highly correlated: 80 for the Bronx, .70 for Brooklyn and .73 for both. The average amount of divorce/separation per health area is slightly higher in the Bronx; the average number of persons on welfare is also higher in the Bronx even though there are 600 less people per health area in that borough.

TABLE 2a

FREQUENCY OF DIVORCED/SEPARATED AND OF PERSONS ON WELFARE IN THE HEALTH AREAS OF THE BRONX AND BROOKLYN

	Bronx D/S 1960	Bronx Welfare 1962	Brooklyn D/S 1960	Brooklyn Welfare 1962	Both D/S 1960	Both Welfare 1962
MEAN	745	1067	728	1022	734	1038
S.D.	437	1159	526	1146	495	1148
N.	63	63	113	113	176	176

TABLE 2b

POPULATION CHARACTERISTICS OF THE HEALTH AREAS OF THE BRONX AND BROOKLYN AS RELATED TO THE FREQUENCY OF DIVORCE/SEPARATION AND NUMBER OF PERSONS ON WELFARE

	Bronx D/S 1960	Bronx Welfare 1962	Brooklyn D/S 1960	Brooklyn Welfare 1962	Both D/S 1960	Both Welfare 1962
M. STATUS						
MS	79	64	91	64	85	58
MA	71	64	41	51	50	56
5MPV	89	76	93	78	90	76
R. STATUS						
PN	76	71	89	70	85	68
PPR	74	77	29	68	45	69
PN, PPR	89	88	89	88	89	87

NOTE: M. Status equals migration status: MS equals migrants from the South, assumed to be primarily Negroes; MA equals migrants from abroad, assumed to be mostly Puerto Ricans; 5MPV equals five migration predictor variables (see data analysis). R. Status equals residence status; PN equals permanent Negroes. PPR equals permanent Puerto Ricans; PN, PPR indicates the two above population values in combination (see analysis). All relationships on line with 5MPV and PN, PPR are multiple correlations: other values are product moment correlation coefficients; all values are in .XX format.

Divorce and Separation

Table 2b makes it apparent that migration (MS and MA) is strongly related to divorce/separation in both the Bronx and Brooklyn, although in Brooklyn MS is a much more powerful predictor variable than is MA. When MS, MS squared, MA, MA squared and MS×MA are combined, the multiple correlation with the number of divorce/separations is very high for both the Bronx (.89) and Brooklyn (.93).

The pattern of correlations among those variables indicating relatively permanent residence (PN and PPR) and family disruption is extremely similar to the pattern of relationships among migration variables and family disruption. When PN and PPR are combined to predict the frequency of divorce/separation, the multiple R is .89, exactly that obtained in predicting the same variable from migration variables (5 MPV) for the Bronx. For Brooklyn we find that the number of migrants from the South and the number of relatively permanent Negroes play a dominant role in predicting frequency of divorce/separation while both the number of migrants from abroad and the number of relatively permanent Puerto Ricans play minor roles in predicting divorce/separation. This is contrary to the market similarity of relationships in the Bronx and suggests that migrants from abroad who go to Brooklyn are different in this respect from those who go to the Bronx while the converse is true for migrants from the South. Again this same pattern holds true for relatively permanent Negroes and Puerto Ricans when we compare the Bronx and Brooklyn, suggesting that family disruption plays different roles in the two ethnic minorities as we move from borough to borough. In Brooklyn we find that separation/divorce is quite accurately predicted from either a weighted combination of the migration variables ($R = .93$) or the two permanent residence variables combined ($R = .89$).

The pattern of correlations for the boroughs combined (BOTH) reflects to a greater degree the pattern for Brooklyn simply because Brooklyn has almost twice as many health areas (113) as the Bronx (63). These six correlations are based on observations of 176 health areas which contain four million people and must therefore represent reasonably stable relationships. We note extremely similar correlation profiles for migrants and permanent residents. The striking differences are between Negroes and Puerto Ricans where family disruption, as measured from census data, plays a much more powerful role in Negro families. Later we will see if similar patterns emerge for other indications of social and family conditions.

Welfare

In the Bronx, relationships among migration variables (MS and MA) and the number of persons on welfare are modestly lower than for the variables indicating more permanent residence (PN and PPR). As expected, the overall prediction from migration variables is lower (.88). For Brooklyn the pattern is remarkably similar to that for the Bronx, although MS is modestly higher in relation to persons on welfare than is MA. This modest difference does not hold up when the predictive power of PN is compared with that of PPR and is certainly not similar to the relationships among population variables and family disruption.

When the boroughs are combined there is no difference in the predictive power of MS or MA and none between that of PN and PPR. However, migration status is less strongly linked to welfare than is more permanent residence. The multiple R of .76 for the combined migration variables accounts for 58 percent of the variability in the number of persons on welfare while the combined permanent residence variables accounts for 76 percent of this variability. It is clearly apparent that recent arrivals to the city do not end up on welfare roles more quickly than more permanent residents, as is one of the common stereotypes.

We turn now to variables which are more direct indicators of physical and mental illness and social behavior.

Admission to Hospitals

As indicated in Table 3a there has been a marked increase in hospitalization for mental illness from 1961 to 1965, particularly in the Bronx. In the boroughs combined 2,316 more people were admitted in 1965, representing a 35 percent increase in admissions.

Relationships

The number of migrants from the South (MS) and abroad (MA) play identical roles in predicting the number of mental hospital admissions for Bronx health areas in 1961. In Brooklyn for 1961 the number of migrants from the South accounts for 58 percent of the variability in admissions over the 113 health areas while the number of migrants from abroad accounts for only 23 percent of this variability. Clearly the two migration variables have equal predictive power in the Bronx while MS dominates MA by a considerable margin in Brooklyn. When the 113 health areas of Brooklyn are combined with the 63 health areas of the Bronx, MS continues to dominate as a predictor variable, accounting for 56 percent as compared to 20 percent of the variability in number of admissions.

TABLE 3a
FREQUENCY OF ADMISSION TO STATE AND CITY MENTAL HOSPITALS IN THE HEALTH AREAS OF THE BRONX AND BROOKLYN IN 1961 AND 1965

	Bronx 1961	Bronx 1965	Brooklyn 1961	Brooklyn 1965	Both 1961	Both 1965
MEAN	30.1	48.2	41.4	51.6	37.3	50.4
S.D.	11.9	16.8	21.5	25.5	19.3	22.8
N	63	63	113	113	176	176

TABLE 3b
POPULATION CHARACTERISTICS OF THE HEALTH AREAS OF THE BRONX AND BROOKLYN AS RELATED TO THE FREQUENCY OF ADMISSION TO STATE AND CITY MENTAL HOSPITALS FOR 1961 AND 1965

	Bronx 1961	Bronx 1965	Brooklyn 1961	Brooklyn 1965	Both 1961	Both 1965
M. STATUS						
MS	66	40	76	68	75	64
MA	66	42	48	44	45	41
5MPV	77	53	82	78	80	72
R. STATUS						
PN	56	46	70	63	74	66
PPR	60	48	30	33	36	33
PN, PPR	68	56	72	66	77	69

NOTE: See table 2b for definition of terms.

The patterns of relationship among the more permanent residents, PN and PPR and frequency of hospital admission are very similar to those for the migration variables, MS and MA, although the magnitude of the relationships is somewhat lower. As expected, the prediction of admissions from the combination of migration variables (5MPV) is also higher than for the weighted combination of PN and PPR. When both boroughs are considered, however, the magnitude of the multiple R for the migration variables (.80) is almost equal to that for the permanent residence variables (.77). In general, over both boroughs the frequency of Negroes plays the dominant role in predicting hospital admissions although there are important differences between the Bronx and Brooklyn. Population characteristics in combination play an important part in predicting hospital admissions, accounting for a minimum of 46 percent of the variability and a maximum of 68 percent.

The profiles of relationships for 1965 are similar, but lower, when compared to those for 1961. Inspection of the multiple correlations for the boroughs combined indicates that population values account for half of the variability in hospital admissions in 1965. (R = .72 for 5MPV, .69 for PN, PPR). This is impressive when the marked changes in admission rates and all of the other possible influences are considered.

Infant Death

The number of infant deaths of children under one year has increased slightly over the five-year period 1960 to 1965, as indicated in Table 4a. The distribution of infant deaths as related to the four population variables and selected combinations of these variables follows.

TABLE 4a

FREQUENCY OF INFANT DEATH IN THE HEALTH AREAS OF THE BRONX AND BROOKLYN IN 1960 AND 1965

	Bronx 1960	Bronx 1965	Brooklyn 1960	Brooklyn 1965	Both 1960	Both 1965
MEAN	10.6	11.7	13.2	13.8	12.3	13.1
S.D.	6.0	7.8	9.3	10.7	8.4	9.8
N	63	63	113	113	176	176

TABLE 4b

POPULATION CHARACTERISTICS OF THE HEALTH AREAS OF THE BRONX AND BROOKLYN AS RELATED TO FREQUENCY OF INFANT DEATH FOR 1960 AND 1965

	Bronx 1960	Bronx 1965	Brooklyn 1960	Brooklyn 1965	Both 1960	Both 1965
M. STATUS						
MS	49	55	81	64	77	62
MA	68	75	44	56	47	59
5MPV	72	79	86	77	83	77
R. STATUS						
PN	41	44	84	65	77	61
PPR	74	73	47	42	46	46
PN, PPR	75	74	89	71	82	69

NOTE: See table 2b for definition of terms.

Relationships

Over the 63 health areas of the Bronx, the number of infant deaths is strongly linked to the number of migrants from abroad (.68) and somewhat less strongly related to the number of migrants from the South (.49). This pattern is reversed for Brooklyn, where MS is very strongly related to infant death (.81) and MA is related to infant death at a much lower magnitude (.44).

The relationships of the more permanent residence variables (PN and PPR) to frequency of infant death follows the above pattern with a high degree of regularity. The PN·ID relationship is .41; for PPR·ID it is .74. Again, for Brooklyn, the pattern is reversed with PN·ID equal to .84 and PPR·ID at .47.

For the two boroughs combined the Brooklyn pattern dominates, with migrants from the South the most powerful predictor at .77 and migrants from abroad at .47. The correlations are remarkably similar for relatively permanent Negroes (.77) and Puerto Ricans (.46). The weighted combination of the two migration variables accounts for 69 percent of the variability in infant deaths while the two permanent residence variables explain 67 percent of the above variability. Not only are these population variables powerful predictors of the distribution of infant death, but they play remarkably similar roles among migrants and more permanent residents, with the exception of extreme role differences in the two boroughs.

Study of the relationships for 1965 indicates a reasonably similar pattern as that for 1961, although in Brooklyn the ethnic differences tend to converge. As a result the overall (N = 176) differences between the predictive power of the two migration variables and the two permanent residence variables is considerably less for 1965. As expected, the predictive power for 1965 is less than that for 1961. However, for the two boroughs combined the population variables remain quite powerful for a five year prediction with multiple R's of .77 and .69 for the migration and permanent residence variables respectively.

Premature Birth

Table 5a indicates an increase in premature births in both the Bronx and Brooklyn. The average increase over the 176 health areas of the Bronx and Brooklyn is 4.3 premature births. The following section presents the relationships among the four population variables and the frequency of premature birth in 1960 and 1965.

[238] BEHAVIOR IN NEW ENVIRONMENTS

TABLE 5a
FREQUENCY OF PREMATURE BIRTH IN THE HEALTH AREAS OF THE BRONX AND BROOKLYN

	Bronx 1960	Bronx 1965	Brooklyn 1960	Brooklyn 1965	Both 1960	Both 1965
MEAN	40.7	46.8	49.5	52.9	46.4	50.7
S.D.	22.4	26.8	32.1	34.6	29.3	32.1
N	63	63	113	113	176	176

TABLE 5b
POPULATION CHARACTERISTICS OF THE HEALTH AREAS OF THE BRONX AND BROOKLYN AS RELATED TO THE FREQUENCY OF PREMATURE BIRTH FOR 1960 AND 1965

	Bronx 1960	Bronx 1965	Brooklyn 1960	Brooklyn 1965	Both 1960	Both 1965
M. STATUS						
MS	77	66	80	70	78	68
MA	75	78	40	48	47	55
SMPV	90	86	83	77	84	79
R. STATUS						
PN	69	58	82	73	79	69
PPR	78	73	45	36	48	44
PN, PPR	87	79	87	76	85	75

NOTE: See table 2b for definition of terms.

Relationships

The migration variables (MS and MA) play identical and prominent roles in predicting premature birth for 1961 in the Bronx. The situation changes for Brooklyn where the number of migrants from the South is highly (R = .80) related to the number of premature births while the number of migrants from abroad is moderately (R = .40) related to this indicator of reproductive problems. The relationships for the Bronx and Brooklyn combined are most similar to those for Brooklyn with MS explaining 61 percent of the variability in premature births and MA accounting for 22 percent.

The pattern of relationship is very similar for the variables representing permanent residence. The exception is in the Bronx where, for 1961, PPR accounts for a larger proportion (61 as compared to 41 percent) of the variability in premature births than does PN.

When the migration and permanent residence variables are combined to predict premature birth, all multiple correlations exceed .80, indicating that population variables are strong predictors of the distribution of premature birth over health areas. Migration variables in combination are no more powerful than permanent residence variables in this role. For the 176 health areas of the Bronx and Brooklyn the multiple R's are virtually identical (.84 and .85).

The pattern of relationships for the 1965 data is very similar to that for 1961 with most correlations somewhat lower in magnitude. All multiple correlations equal or exceed .75, indicating that combinations of migration or permanent residence variables play an equal but prominent role in predicting the distribution of premature births five years after the population variables were observed.

TABLE 6a

FREQUENCY OF BIRTH WITH PRENATAL CARE BEGINNING IN THE SEVENTH OR EIGHTH MONTH IN THE HEALTH AREAS OF THE BRONX AND BROOKLYN

	Bronx		*Brooklyn*		*Both*	
	1960	*1965*	*1960*	*1965*	*1960*	*1965*
MEAN	60.6	81.7	69.9	88.5	66.6	86.0
S.D.	51.4	64.3	68.6	82.8	63.0	76.6
N	63	63	113	113	176	176

TABLE 6b

POPULATION CHARACTERISTICS OF THE HEALTH AREAS OF THE BRONX AND BROOKLYN AS RELATED TO THE FREQUENCY OF BIRTH WITH PRENATAL CARE BEGINNING IN THE SEVENTH OR EIGHTH MONTH FOR 1960 AND 1965

	Bronx		*Brooklyn*		*Both*	
	1960	*1965*	*1960*	*1965*	*1960*	*1965*
M. STATUS						
MS	72	58	79	68	76	64
MA	83	82	61	63	66	68
5MPV	92	87	90	83	90	84
R. STATUS						
PN	69	53	83	70	80	66
PPR	89	80	63	53	65	58
PN, PPR	96	83	95	80	93	79

NOTE: See table 2b for definition of terms.

Prenatal Care Beginning in the Seventh or Eighth Month

Between 1960 and 1965 the number of expectant mothers who did not obtain medical care until the seventh or eighth month of pregnancy increased by approximately 3,500 in both boroughs. The relationships of late prenatal care to the four population variables are presented in the following section.

Relationships

In the Bronx (1960) the number of migrants from abroad (MA) and the number of Puerto Ricans who are relatively permanent residents (PPR) are excellent predictors of the number of expectant mothers receiving late prenatal care. The number of migrants from the South (MS) and more permanent Negroes (PN) play less powerful but still prominent roles in predicting this later prenatal care. The pattern is inverted for Brooklyn where MS and PN exceed MA and PPR in the magnitude of relationships. The Brooklyn pattern holds for Brooklyn and the Bronx combined, with minor exceptions.

In all of the six cases the multiple correlations for 1960 data are equal to or exceed .90. Permanent residents are slightly higher on all of the three possible comparisons. Late prenatal care is thus strongly linked to population characteristics.

The 1965 data shows an interesting trend with MS and PN decreasing in strength and MA and PPR retaining essentially the same ability to predict. Thus previous patterns of differences (between MS and MA, PN and PPR) are increased while the multiple correlations for 1965 data are generally decreased. However, the population variables retain a prominent role as predictors of significant behavior five years later even though many other factors obviously influence the use of prenatal facilities.

Homicide

Almost twice as many homicides were reported for the Bronx and Brooklyn in 1965 as in 1960. Approximately 167 more people were reported killed in 1965 than in 1961. As indicated in Table 7a, the rate of increase was most pronounced in the Bronx. Relationships of frequency of homicide and population variables are presented in Table 7b.

Relationships

In 1960 the number of migrants from abroad was moderately related to the frequency of homicide in the Bronx while the number of migrants from the South plays a less important role in this respect. The roles of the migration variables are reversed for Brooklyn and in the two boroughs combined the correlation of MS with frequency of homicide is .71 while for MA it is .34.

The pattern and magnitude of correlations between the permanent residence variables and homicide is a close duplication of that for the migration variables. The multiple correlations of the migration variables with homicide are very similar to those for the permanent residence measures; over the 176 health areas of the Bronx and Brooklyn, the multiple R's are .73 and .74 respectively.

For the 1965 data the relationships of MS and MA to homicide tend to converge, especially for the Bronx. This holds true for PN and PPR as they relate to homicide. For the two boroughs combined there is almost an exact duplication of results when the relationships of migration variables to homicide are compared to those of permanent residence variables to homicide.

TABLE 7a

FREQUENCY OF REPORTED HOMICIDE IN THE HEALTH AREAS OF THE BRONX AND BROOKLYN IN 1960 AND 1965

	Bronx 1960	Bronx 1965	Brooklyn 1960	Brooklyn 1965	Both 1960	Both 1965
MEAN	.65	2.22	1.22	1.82	1.02	1.97
S.D.	.99	2.20	1.75	2.43	1.54	2.36
N	63	63	113	113	176	176

TABLE 7b

POPULATION CHARACTERISTICS OF THE HEALTH AREAS OF THE BRONX AND BROOKLYN AS RELATED TO THE FREQUENCY OF REPORTED HOMICIDE FOR 1960 AND 1965

	Bronx 1960	Bronx 1965	Brooklyn 1960	Brooklyn 1965	Both 1960	Both 1965
M. STATUS						
MS	44	46	73	72	71	62
MA	64	51	29	33	34	39
5MPV	69	60	75	74	73	68
R. STATUS						
PN	31	51	80	71	72	64
PPR	62	52	35	29	33	39
PN, PPR	62	61	82	72	74	69

NOTE: See table 2b for definition of terms.

Approximately 50 percent of the variability in the frequency of homicide over health areas is explained by combinations of population variables.

Arrests of Persons Seven to Twenty Years Old

Frequency of arrests as indicated in Table 8a has increased substantially in the Bronx and Brooklyn over the period 1961 to 1965. The increase has been somewhat greater in the Bronx. In the two boroughs combined there were approximately 3,500 more arrests in 1965 than in 1961. The relationships of the number of arrests by health area to the four population variables are presented in Table 8b and will be described in the following section.

TABLE 8a

FREQUENCY OF ARRESTS IN THE HEALTH AREAS OF THE BRONX AND BROOKLYN FOR 1961 AND 1965

	Bronx 1961	Bronx 1965	Brooklyn 1961	Brooklyn 1965	Both 1961	Both 1965
MEAN	114.9	137.3	93.0	109.4	107.1	127.3
S.D.	101.0	99.5	74.8	73.5	92.9	91.8
N	63	63	113	113	176	176

TABLE 8b

POPULATION CHARACTERISTICS OF THE HEALTH AREAS OF THE BRONX AND BROOKLYN AS RELATED TO FREQUENCY OF ARRESTS FOR 1961 AND 1965

	Bronx 1961	Bronx 1965	Brooklyn 1961	Brooklyn 1965	Both 1961	Both 1965
M. STATUS						
MS	64	70	67	64	66	63
MA	75	77	51	56	55	59
5MPV	86	85	81	79	81	78
R. STATUS						
PN	71	60	74	69	74	67
PPR	85	79	66	59	63	57
PN, PPR	94	84	90	82	88	79

NOTE: See table 2b for definition of terms.

Relationships

In the Bronx the two migration variables are strongly related to the number of arrests for 1961, with MA assuming the more prominent role, while in Brooklyn this order is inverted with MA assuming the lesser role. The combination of the two boroughs results in a moderately high relationship of .66 for MS and a somewhat lower correlation of .55 for MA. For the migration variables combined the three multiple correlations exceed .80. They are .86 and .81 for the Bronx and Brooklyn respectively.

When permanent residence variables are combined and related to frequency of arrest a pattern of relationships is formed which is very similar to that of the combined migration variables, except that the magnitude of the multiple correlations is generally higher for the permanent residence variables. For the permanent residence variables the multiple correlations are .94, .90 and .88 for the Bronx, Brooklyn and the two boroughs combined respectively. For the migration variables the same relationships are .86, .81 and .81. Thus the distribution of arrests of youth among health areas is accurately predicted from a knowledge of their population compositions.

The relationships for the 1965 data are similar to those for 1961, although the magnitude of the correlations is somewhat lower for the PN and PPR variables. When the two boroughs are combined the multiple R of .78 for the migration variables is only slightly less than that obtained for the 1961 data, where R = .81. For the permanent residence variables the multiple R's are somewhat lower for the 1965 data. Population variables, however, are important predictors of the distribution of arrests, even five years after the population values are observed.

Summary of Results

(1) Population characteristics of the 176 health areas of Brooklyn and the Bronx, representing four million people, played important roles in predicting the distribution of physical illness, mental illness, family disruption, socioeconomic conditions, deviant behavior, premature birth and infant mortality among these health areas.

(2) Among the population variables the number of migrants and the number of relatively permanent residents of the same ethnic group played virtually identical roles in predicting the above indicators of social, health and mental health problems. The one possible exception was in predicting hospitalization for mental illness, where the number of migrants was a more effective variable than the number of permanent

residents of the same ethnic group. It appears that migrants display patterns of behavior which are very similar to those of permanent residents of the same ethnic group within a given health area.

(3) Ethnic membership, however, in both the migrant and permanent resident groups, played very different roles in predicting the distribution of social and health problems, both within and across boroughs. Thus the roles of ethnic membership, while always significant and frequently very powerful as predictor variables, were far too inconsistent to allow generalization from borough to borough or, by inference, from city to city.

(4) Even though individual ethnic variables played radically different predictor roles, a combination of the two ethnic variables for migrants displayed virtually identical multiple correlation patterns as did a combination of the same ethnic variables for permanent residents. Thus a weighted linear combination of the number of citizens in a health area belonging to the two ethnic groups was a powerful predictor battery. This two variable battery displayed very similar predictor roles for both Brooklyn and the Bronx.

(5) Population variables are powerful predictors of the future distribution of health and social problems. Therefore current and anticipated population patterns provide one important basis for predicting the distribution and changes in the distribution of social problems throughout the sectors of cities.

Implications for Further Research

The ability to predict the future distribution of social problems among the sectors of our great cities would seem to be of obvious importance if we are to adequately plan and subsequently evaluate programs of intervention and reconstruction. The results of this study indicate the ability to predict, with considerable accuracy, the distribution of social problems among health areas five years beyond the observation of only two population variables. Based on the work of Tryon (1968) and unpublished work of the writers, it seems highly probable that a comprehensive and carefully selected battery of predictor variables could predict, with great accuracy, the distribution of social problems at least ten and very likely twenty years into the future. The use of smaller units of observation and the consideration of nonlinear relationships could only enhance the accuracy of prediction. The role of migration, particularly if it were measured in a more comprehensive and meaningful manner by the census, could be accurately analyzed within a data analytic framework which compared the relative efficiency of hypothetically selected set of predictor variables, including their nonlinear and interaction components.

This study supports the conclusion that, within health areas, migrants and more permanent residents of the same ethnic group become involved in similar patterns of social problems. An explanation of this finding is not readily apparent and may be a function of several factors. It may derive from the influence of the more permanent residents who communicate and display patterns of behavior which they approve and recommend to the newcomers. Or, it may be that migrants are somehow distributed into health areas in which their established behavior patterns are very similar to those of their ethnic counterparts. This may come about because of established communication networks between the migrant's place of origin and potential areas of new residence. Other determinants, such as seeking the same socioeconomic level, the proximity of other ethnic groups, the nature of housing conditions or the influence of religious values, may play important roles in explaining the distribution of problems in adaptation. A comparison of areas where migrants differ markedly from their neighbors with areas where migrants and their neighbors are very similar, may contribute to our understanding of the distribution of social and health problems as related to migration.

The extent of ethnic group membership is an effective predictor of the distribution of social and health problems among health areas. However, the role of measures of ethnic membership varies greatly within and across boroughs, indicating that different ethnic groups exhibit very different adaptation patterns both within and across boroughs and that similar ethnic groups display very different adaptation patterns in the two boroughs.

Studies designed to explain these marked differences in adaptation patterns may provide an understanding of their determinants. For example, a comprehensive comparison of sectors of cities matched on population and economic characteristics, but very different in terms of rates of health and social problems, may provide at least a partial understanding of the causes and development of these differences.

The results of this study suggest that a study of one city is not an adequate basis for making generalizations about the behavior patterns of important segments of our urban populations, including, of course, the rich variety of migrants who enter and leave our cities. It would appear that replication with refinements of some of the studies which now provide our fragmentary understanding of migration patterns, the process of migration and the problems of adaptation are very much in order if we are to comprehend more accurately the cause and effect of migration, an ever increasing phenomenon.

NOTE

1. Health Areas are units of observation defined by the New York City Department of Public Health. Health Areas are composed of census tracts and contain, on the average, 23,000 residents with a standard deviation of 7,000 residents.

REFERENCES

FARIS, R and H. W. DUNHAM (1939) Mental Disorders in Urban Areas. Chicago: Univ. of Chicago Press.
FREEDMAN, R. (1950) Recent Migration to Chicago. Chicago: Univ. of Chicago Press.
FRIED, M. (1966) "The role of work in a mobile society." Pp. 81-104 in S. B. Warner (ed.) Planning for a Nation of Cities. Cambridge, Mass.: MIT Press.
--- (1964) "Effects of social change on mental health." American J. of Orthopsychiatry 34.
GERARD, D. L. and L. G. HOUSTON (1953) "Family setting and the social ecology of schizophrenia." Psychiatry Q. 27: 90-101.
GOLDHAMER, H. and A. W. MARSHALL (1949) Psychosis and Civilization: Two Studies in the Frequency of Mental Disease. Glencoe, Ill.: The Free Press.
GUTTENTAG, M. and F. DENMARK (1965) "Psychiatric labelling: role assignment based on the projective test performance of in-migrants." International J. of Social Psychiatry 11: 131-137.
HADDEN, J. K. and E. F. BORGATTA (1965) American Cities: Their Social Characteristics. Chicago: Rand McNally.
HAMMER, M. and E. LEACOCK (1961) "Source material in the epidemiology of mental illness." Pp. 418-486 in J. Zubin (ed.) Field Studies in the Mental Disorders. New York: Gruene & Stratton.
KEELER, M. H. and M. M. VITOLS (1963) "Migration and schizophrenia in North Carolina Negroes." Amer. J. of Orthopsychiatry 33: 554-557.
KLEINER, R. J. and S. PARKER (1965) "Goal-striving and psychosomatic symptoms in a migrant and non-migrant population." Pp. 78-85 in M. Kantor (ed.) Mobility and Mental Health. Springfield, Ill.: Charles C. Thomas.
--- and S. PARKER (1959) "Migration and mental illness: a new look." American Sociological R. 24: 687-690.
KLINEBURG, O. [ed.] (1944) Characteristics of the American Negro. New York: McGraw-Hill.
LAZARUS, J., B. Z. LOCKE and D. S. THOMAS (1963) "Migration differentials in mental disease." Milbank Memorial Fund Q. 41: 25-42.
LEE, E. S. (1963) "Socio-economic and migration differentials in mental disease, New York State, 1949-1951." Milbank Memorial Fund Q. 41: 249-268.
LOCKE, B. Z. and H. J. DUVALL (1964) "Migration and mental illness." Eugenics Q. 11: 216-221.

MALZBERG, B. (1962) "Migration and mental disease among the white population of New York State, 1949-1951." Human Biology 34: 89-98.

――― and E. S. LEE (1956) Migration and Mental Disease. New York: Social Science Research Council.

MURPHY, H. B. M. (1965) "Migration and the major mental disorders." Pp. 5-29 in M. Kantor (ed.) Mobility and Mental Health. Springfield, Ill.: Charles C Thomas.

――― (1961) "Social change and mental health." Milbank Memorial Fund Q. 39: 385-446.

SHRYOCK, H. S. (1964) Population Mobility within the United States. Chicago: Univ. of Chicago Press.

SROLE, L., T. S. LANGNER, S. T. MICHAEL, M. K. OPLER and T. A. RENNIE (1962) The Midtown Manhattan Study. New York: McGraw-Hill.

THOMLINSON, R. (1962) "The determination of a base population for computing migration rates." Milbank Memorial Fund Q. 40: 356-366.

TIETZE, C., P. LEMKAU and M. COOPER (1942) "Personality disorder and spatial mobility." American J. of Sociology 48: 29-39.

TRYON, R. C. (1968) "Comparative cluster analysis of social areas." Multivariable Behavioral Research 3: 213-232.

――― (1967) "Predicting group differences in cluster analysis: the social area problem." Multivariable Behavioral Research 2: 453-475.

――― (1955) Identification of Social Areas by Cluster Analysis. Univ. of Calif. Publ. Psych. 8, 1-100. Berkeley: Univ. of Calif. Press.

――― and D. E. BAILEY (1966) "The BC TRY computer system of cluster and factor analysis." Multivariable Behavioral Research 1: 91-111.

WILSON A. W., G. SAVER and P. A. LACHENBRUCH (1965) Residential mobility and psychiatric help-seeking." American J. of Psychiatry 121: 1108-1109.

Chapter 11

Mexican Americans of Texas: *Some Social Psychiatric Features*

HORACIO FABREGA, JR.

This report will review recent studies we have conducted dealing with psychiatric problems of Mexican American subjects of Texas. The early portion of the report will provide a description of some of the social and cultural characteristics of this ethnic group and emphasis will be given to information which pertains to Mexican American residents of Texas.[1] Psychiatric problems will be examined in relation to these social and cultural characteristics. Central to the orientation of the psychiatric studies to be reviewed is the knowledge that Mexican Americans have participated and are participating in the migration-acculturation process and experience.

Migration and acculturation traditionally have been regarded as separate processes. In addition, the effects of these processes on human adaption usually have been investigated using distinct procedures and methods. It is important to emphasize that each of these processes is closely related to the other when focus is placed on human behavior. This is so because, first of all, both processes actually involve people in a relationship to an environment that is changing or changed. Secondly, when the sociocultural unit used in the analysis of a situation involving international migration is

---

Author's Note: *The preparation of this manuscript was made possible through the support of the Hogg Foundation for Mental Health. The help of Miss Carole Ann Wallace is gratefully acknowledged.*

the individual family or a particular social group, a context that approximates an acculturative process is also set up. This is the case because the particular group involved in the migration is likely to come in new "contact" with another group that will probably have different sociocultural patterns. Regardless of how well insulated this new unit is, some change in its cultural patterns is likely to take place. On the other hand, if a historical emphasis is included in the approach to a study of acculturation, a process involving the influx of people with new values, patterns, and so on (an earlier migration?) will be found to have antedated and contributed significantly to the current situation. For these reasons, when a study is planned dealing with the behavior implication of either acculturation or migration, it will be found useful to rely on a perspective that encompasses both of the processes.

In view of this relationship between acculturation and migration, it may appear somewhat surprising that the concepts, approaches and methods used to investigate the psychiatric facets of these related processes usually differ. Studies dealing with the behavioral consequences of acculturation usually involve direct participation and observation of behavior and proceed to the specification of tendencies and patterns. Very often, questionnaires or tests dealing with psychological parameters are used. On the other hand, investigations dealing with the effects of migration follow the demographic or ecological model. There is often little focus on psychological data, and variables that intervene between the displacement and hospitalization (a frequent dependent variable) are not emphasized. It should be kept in mind, however, that there are some real differences in the contexts or situations wherein the processes have been investigated (Spiro, 1955; Beals, 1951). Studies of migration, for example, usually involve individuals who have some formal education and who share the Western tradition. Quite frequently, these individuals have voluntarily decided to move to a new urban industrial setting with the intention of adopting the values and emphases of the receiving nation. Consequently, strictly sociodemographic variables appear relevant. Acculturation, as initially defined and studied, on the other hand, usually dealt with individuals of preliterate settings (sometimes in a colonial status) whose life style and background was radically different from that of the dominant or contract culture. In addition, the individuals had not voluntarily chosen the context, and assimilation or value adoption appeared not to be the goal. Rather, a part of the Western system was spreading towards them threatening the values to which they were attached. Purely sociodemographic variables were often not applicable or relevant in the analysis of these situations.

This report deals with the Mexican American ethnic group. The focus is on some of the psychiatric problems that are associated with this group. Characteristics of the Mexican American's position in the dominant Anglo American society make it desirable that any investigation focused on psychiatric problems employ a perspective which encompasses parameters associated with both the migration and acculturation processes. As will be touched upon subsequently, Mexican Americans have a distinct and traditional set of cultural values and premises that differ from those of the Anglo Americans. Consequences of the Mexican American's attempts to cope with and adapt to the Anglo American system approximate those that have been classically identified with situations of culture contact or acculturation. At the same time, Mexican Americans, whether born in Mexico or in this country, share with the Anglo Americans characteristics of the Western tradition. In addition, individuals who have come to the United States from Mexico have moved relatively recently, quite often with the intention of settling here permanently. Studies focused on the Mexican American's adaption in Anglo society, thus, need also to follow a rationale and employ procedures similar to those of migration studies.

BACKGROUND FEATURES OF MEXICAN AMERICANS

GENERAL SOCIOHISTORICAL DIMENSIONS

Mexican Americans are mainly concentrated in the five Southwestern states: Arizona, California, Colorado, New Mexico and Texas.[2] Inspection of population census data shows that it is a rapidly growing minority group: in the five states, the average increase in population between 1950 and 1960 was fifty-one percent. As is generally known, the five-state area where the Mexican Americans are now concentrated originally belonged to Mexico. Some of the "Mexican Americans" of today are descendants of the people who lived in these territories before formal incorporation into the continental United States. This segment of the group, however, is rather small, and in many places is not referred to as Mexican American but rather as Spanish American. Information characterizing the population prior and just subsequent to the annexation of these territories is not accurate or reliable. "In the late years of the nineteenth century, the migrations were still highly informal and intermittent, though probably significant in relation to the sparse total population in the areas both north and south of the border" (Grebler, 1966: 19). In the first decade of the twentieth century, migration became easier, less costly, but was still only of modest proportions.

Beginning with the Mexican revolutionary period of 1910, migration to the United States became more significant. Besides the "push" which yielded immigrants deprived and threatened by the revolution, in the latter part of the decade, this was reinforced by the "pull" of American farm labor requirements. Immigration from Mexico to the United States was highest in the decade of the 1920s with close to half a million reported entering during this decade. The economic deprivations resulting from the Civil War in Mexico and the existing prosperity in the United States were possible factors associated with this peak immigration phase. The Mexican immigrants provided a low-wage labor supply for agriculture and manufacturing and the growing dependence on this supply was reflected in the opposition of business to an extension of the new immigration quotas to the Western Hemisphere.

Immigration drastically declined and even reversed during the 1930s. The general slump in business decreased the demand for farm labor, and the Mexican economy was experiencing a recovery from the revolution years. During this decade, only 27,900 Mexicans entered on permanent visas in contrast to almost half a million in the 1920s. In terms of the total immigration to the United States during the 1930s, Mexico constituted only four percent versus eleven percent in the 1920s as country of origin. During this decade, displaced farmers from the Dust Bowl and urban unemployed filled the need for low-wage laborers. Repatriation of persons of Mexican descent took place on a large scale and legal and illegal immigrants, temporary workers and permanent residents, U.S. citizens and aliens were affected by mass shipments of persons to Mexico to avoid the expenses of welfare cases. Much of this effort was organized by local authorities and private welfare agencies with assistance from the Mexican government. During the decade, 89,000 Mexican aliens departed with experiences that strengthened mistrust of American society.

The Bracero program, instituted in 1942 as a bilateral government agreement, provided a regulated recruitment procedure to fill the labor shortage created during World War II. The program provided many advantages and guarantees for the temporary Mexican worker and during its life became an increasingly important source of labor movement. From a modest beginning of 4,203 workers, the program peaked in 1956 with the movement of some 445,000 laborers. In 1951, a public law was enacted which replaced the earlier executive agreements for the temporary workers. In addition to the Bracero, an important source of labor during this period was the illegal movement of Mexican workers into the area to avoid the expenses and red tape of official programs.

During the 1950s, a strong effort was made to control the illegal movement. Willful importation, transportation and harboring of illegal aliens was made a felony in 1952. Many injustices were done, for when confronted, a person had to present immediate documentary proof that he had legal status in the country or he could be sent back to Mexico. In the first half of the decade, 3.8 million (including double-counts) were expelled. Feelings of antagonism toward American society were strengthened. The expulsions increased permanent immigration and the Bracero movement, both of which peaked in 1956. Approximately 293,500 permanent visa immigrants from Mexico entered the United States in the 1950s (over fifteen percent of total immigration in the last half of the decade).

In the first half of the 1960s, the fifteen percent of total immigration to the United States was maintained by permanent visa immigrants from Mexico. In 1964, the Bracero program was terminated and reliance on foreign labor supplies in the Southwest was ended. In 1965, Congress extended the immigration quotas to the Western Hemisphere to take effect within three years, with a ceiling of 120,000 per year on immigrants. About 1.3 million Mexicans were reported to have entered the United States for permanent residence between 1900 and 1964. In 1960, 1.7 million (forty-five percent of the 3.8 million Mexican American population of the United States) were born in Mexico or were of Mexican or mixed parentage.

SOCIODEMOGRAPHIC FEATURES OF THE MEXICAN AMERICANS OF TEXAS

Enumerations from the 1960 census show that the Spanish surname group in Texas was approximately fifteen percent of the total population for the state.[3] California's Spanish surname population exceeded that of Texas in number, but the group constituted only nine percent of the population of that state. Eighty-two percent of the Spanish surname population of the Southwest was in California and Texas. The Spanish surname subpopulation of Texas was concentrated south and west. Along the Mexican border, none of the counties had less than twenty-five percent of their population in this ethnic group. The fifteen counties with a proportion of sixty percent or more Spanish surname persons were all in the southern Rio Grande Valley region where over one-fourth of this subpopulation in Texas were counted. The other minority group in Texas (nonwhite) was primarily in the eastern portion of the state, and no county had a twenty-five percent representation from each of the two minority ethnic groups. The percentage growth from 1950 to 1960 was greatest for the Spanish surname population which was seventy percent greater than either the Anglo or nonwhite groups. This growth was attributed not only to natural increase but to the large net migration from Mexico which was approximately eighty percent of the total immigration to Texas from 1955 to 1960. There was a noted movement of the Spanish surname group from rural to urban areas and over seventy-eight percent of the group was classified as urban in 1960. In spite of this "urbanization" of the group, however, a substantial proportion of the subpopulation was still identified with agricultural work.

Delineating three age categories (young, 0-14 years; working age, 15-64 years; and old, 65 years and over) revealed that the Spanish surname population was younger than the Anglo with approximately fifty percent more persons in the "young" group. The nonwhite population was intermediate between the other two. A dependency ratio (number of "young and old" per 100 "working age") showed the Spanish surname with 93 as opposed to 63 per 100 for Anglo. The sex ratio (number of males divided by number of females) gave the Spanish surname and nonwhite populations lower ratios than the Anglos in the 15-64 years category and higher ratios in the 65 years and over group. A child-woman ratio (number of children under 5 years of age divided by number of women 15-49 years of age) as a measure of fertility showed the Spanish surname population higher than the Anglo and nonwhite. Percent families with four or more children under 18 years of age was twenty-seven percent for the Spanish surname group, seventeen percent for the nonwhite and eight percent for the Anglo. Proportion of families with seven or more persons was approximately twenty-five percent for the Spanish surname versus four percent for the Anglo.

Educational attainment was perhaps one of the most striking differentials in the Texas population with the Spanish surname group occupying the lowest position. Twenty-three percent of that group were 25 years of age or over with no education, opposed to 5.4 percent for nonwhites and 1.1 percent for Anglos. In 1960, the median school years completed was 11.5, 8.1, and 6.1 for Anglo, nonwhite and Spanish surname, respectively. Similar to the Anglo and nonwhite groups, 80.2 percent of the Spanish surname group 5-15 years of age had been enrolled in school. However, for the age group 16-19 years, Anglos had 64.3 percent, nonwhites 57.6 percent and Spanish surname only 46.2 percent enrolled, showing a tendency for the Spanish surname students to drop out of school before age 19. For the decade 1950 to 1960, the increase in those persons 25 years of age and older who had completed 9-12 years of education was greatest for the Spanish surname population at sixty-two percent. The nonwhites had an increase of forty-two percent and the Anglos had only a slight gain. Still, the percentage of individuals having this degree of education in 1960 was only sixteen percent for the Spanish surname group as opposed to thirty-one percent and 46.5 percent for the nonwhite and Anglos, respectively. In general, those Spanish surname persons living in metropolitan areas had higher levels of education.

The percent of the labor force 14 years of age and over unemployed in 1950 and 1960 was highest for the Spanish surname population though the nonwhite group was not far below it. Thirty percent of the Spanish surname heads of families worked sporadically if at all in the year preceding the census, a figure substantially above the twenty-two percent of the other groups combined. The Anglo males had 6.4 percent employed in farm and other labor categories, but the Spanish surname and nonwhite groups had approximately one-third each. Spanish surname males other than laborers were concentrated in the craftsmen and operatives categories. For females, occupations in clerical and sales positions seemed fairly open to the Spanish surname group and a good many of them were employed as operatives. Private household and service workers were also well represented in this group. Both Anglos and nonwhites had approximately eight percent of their employed population in the agricultural industry, but the Spanish surname group had seventeen percent, pointing out the nature of the work that a good part of this group is engaged in.

The Spanish surname group median income is currently approximately half of the Anglo median. For employed males 14 years of age and older, the median dropped below one-half the Anglo median for the Spanish surname group. Family income below $3,000 included one-fifth of the Anglos and over one-half of the other subpopulations. The median family income was $5,636, $2,913 and $2,614 for the Anglos, Spanish surname group, and nonwhites, respectively. The income of the Spanish surname population was lowest near the Mexican border. Nonmetropolitan counties in this area had some of the lowest living standards in the nation.

For the first time in 1960, a sizeable proportion (i.e., over fifty percent) of Texas' Spanish surname population were natives of native parentage. The foreign born made up only fourteen percent of this group and natives of foreign or mixed parentage were thirty-one percent. The natives of native parentage were found to have more education, but the natives of foreign or mixed parentage were very similar to them. The lowest median income for employed males 14 years of age and over was in the foreign born group and the highest was in the natives of foreign or mixed parentage group rather than the natives of native parentage. This would suggest that longer residence in the United States does not necessarily mean greater assimilation for the Spanish surname population. For males, farm labor was higher in the foreign born group and the number of private household workers was greatest for foreign born females.

SOME CULTURAL FEATURES

The cultural features of the Mexican Americans of Texas have been studied in the lower Rio Grande region and have been reviewed in previous publications (Madsen, 1964; Heller, 1966; Rubel, 1966; Samora, 1966). It has been stated that general suspicion and mistrust often lie beneath the outward friendliness that sometimes characterizes Anglo and Mexican American relations. To a large extent, the general mistrust and/or discrimination that seems to punctuate interethnic relations is a matter of class rather than culture or race. The fact that full participation in the Anglo American school system often is not possible seems to accentuate the plight of the Mexican American. In South Texas, the two groups live separated in a social and spatial or geographic sense, each with a cultural image of the other that suggests resentment and antagonism. In the larger cities of the state, many Mexican Americans are found living in concentrated, ghetto-like circumstances.

Spiritual and cultural bonds are felt to unite the Mexican American group. A key to understanding the Mexican American is appreciation of the importance of his commitment to his family. "The most important role of the individual is his familial role and the family is the most valued institution in Mexican American society. The individual owes his primary loyalties to the family, which is also the source of most affective relations" (Madsen, 1964: 17). Extended family-of-origin ties, family allegiance and closeness to relatives are felt to be sources of psychological support to the individual and he is expected to value these relationships. Great respect is shown to older relatives and separate interest of family members are subordinated to family considerations.

With regard to the behavior expectation of the sexes, it can be said that general acknowledgment is made of the male's superiority and authority. Respect as well as protection is shown females because it is felt they are more vulnerable. Strict authority of the father is unquestioned and the mother is viewed as the supportive and self-sacrificing family member. Men are accepted as hard, unyielding and strong while women are expected to be soft and destined for much suffering and pain. These sex "roles" permeate the gender identity described for this ethnic group. Much has been written regarding the Mexican American's regard for dignity. Jobs are positively valued only if the individual employee is treated in a way that makes him feel he is personally respected by the employer; considerations of salary, promotions and impersonal "work conditions" being secondary. The "folk" ideas and beliefs of the Mexican Americans regarding medical phenomena have been carefully studied. Western scientific premises about the nature of disease, causation and cure, judging from interviews and observed medical practices, are not accepted, and instead, a rich and elaborate native sytem prevails (Rubel, 1960 and 1964). The interpersonal relations of Mexicans, as related by some investigators, exhibit a "passive" coping style. Life situations are viewed from the standpoint of endurance as opposed to an "active" coping style of Anglos who emphasize mastery of their fate. Mexicans see the environment as "uncontrollable" and orient to the present rather than the future (Diaz-Guerrero, 1964, 1963, 1955).

PSYCHIATRIC ASPECTS OF THE MEXICAN AMERICAN GROUP

This section will present information dealing with the psychiatric problems of Mexican Americans. Initially, the results of Professor Jaco's epidemiologic survey (1960) of treated psychiatric disease in Texas, which dealt with the ethnicity variable, will be briefly reviewed. Following this, emphasis will be given to three recent studies conducted in Texas which have investigated the ways in which social and cultural factors relate to psychiatric symptomatology.

For purposes of introduction, it should be emphasized that research involving the relationship between the social environment and psychiatric problems has usually employed an etiological perspective. The literature dealing with the psychiatric implications of both acculturation and migration reveals this primary emphasis on considerations of etiology or development of symptoms and/or disease. The main focus seems to be "Do these sociocultural processes lead to or produce disability?" or "Are the individuals involved prone to develop more psychopathology?" The first study to be reviewed subsequently employs a related perspective. The social characteristics and value indentification of Mexican American patients and nonpatients are compared. The aim was to clarify in what ways psychiatric symptoms relate to social productivity. An equally legitimate and important emphasis in social psychiatric studies of this type involves treating the social and cultural variables as situational ones that affect the motivation to request psychiatric treatment. From this perspective an important question becomes "How do these processes

[258] BEHAVIOR IN NEW ENVIRONMENTS

affect those individuals who already manifest symptoms, regardless of how or why they developed these symptoms?" An understanding of the relationship between a group's psychiatric problems and the request for psychiatric treatment requires dealing with the issues of (a) family values that affect the degree of tolerance shown the symptomatic person in the home, (b) the way psychological and physical symptoms are traditionally handled in the group, (c) nature of the beliefs held about "illness" and its etiology, (d) meanings attached to the scientific medical system or formalized methods of treatment, and (e) biases or other factors affecting the type and length of treatment that is administered to members of this social group. In other words, many issues affecting the persons already showing behavioral decrement need to be considered when studying the role of the social environment in psychiatric disability. Investigators often overlook these social psychiatric issues which are also importantly linked to the processes of acculturation and migration. To the extent, of course, that immigrants or individuals involved in acculturation belong to differing social classes, then data showing that this variable has a bearing on access, utilization, type and length of treatment is relevant. Work is needed which will deal precisely with the role of social class and ethnicity in psychiatric treatment. The last two studies that will be reviewed in this report deal with these related problems.

JACO'S EPIDEMIOLOGIC SURVEY

Professor E. Gartly Jaco (1960) conducted an epidemiologic survey of treated psychiatric illness in the state of Texas which included the years 1951-1952. "All bona fide residents of the state of Texas who were diagnosed as having a psychosis, in accordance with the current classification of the American Psychiatric Association, and who sought psychiatric treatment for the first time in their lives... were counted as 'cases' in this survey" (Jaco, 1960: 22). Data was used which was furnished by psychiatrists in private practice, record rooms of all private hospitals with psychiatric facilities, teaching hospitals and the public supported institutions, including state hospitals situated within Texas and the surrounding states. Information on demographic features and clinical diagnosis was collected. Significant differences in incidence rates of the psychoses were found between the Anglo, Mexican and Negro American ethnic groups. The dominant Anglo American group exhibited a far higher total psychotic incidence than the two subdominant groups, the Mexican American group showing the lowest rate. The rates of psychosis were analyzed not only in terms of ethnicity, but also in terms of a host of sociological variables (age, sex, marital status, etc.). In addition, incidence rates were examined with respect to place of birth (native-born Texans as opposed to those born outside the state). Results were too numerous to summarize here. In general, migration was not a factor in the psychotic rates. Neither type of disorder nor treatment source varied significantly between migrants and nonmigrants. The outstanding exception was the very high rate observed among migrant Mexican American females who showed a much higher rate than native-born Mexican American females. The migrant Mexican American group tended to show a higher rate than the native group, but migration was of less consequence to the other ethnic group.

Based on the results of this survey, the conclusion was tentatively drawn that the Mexican American group had an actual lower incidence of psychotic illness. The group was depicted as "a warm, supportive and reasonably secure subcultural group." Psychologically protective factors were felt to inhere in the native cultural patterns generally and the stability of the family specifically. The possibility that the lower incidence figure was the result of the family's greater tolerance for deviant psychotic behavior in the home (leading to the avoidance or delay of seeking hospitalization or medical care) was also considered. The latter, however, was felt to be insufficient as an explanation for the lower incidence of psychotic illness in this ethnic group.

VALUE IDENTIFICATION IN PATIENTS AND NONPATIENTS

This study (Fabrega and Wallace, 1968) involved comparing Mexican American patients and nonpatients of South Texas in terms of social background variables and value identification. Seventy-six Mexican Americans accepted for outpatient psychiatric care during a six-month period at an adult mental health clinic in South Texas were used as the patient sample. Data was collected in interviews which were conducted in Spanish by an American of Mexican descent who was intimately familiar with the background culture and with the communicational style. Forty-eight persons comprised the nonpatient group and were selected using a probability sampling plan from a representative small town served by the clinic. The same individual that had interviewed the patients in the clinic setting also visited the selected homes. This person functioned as a lay interviewer and was unaware of the design, aim and implication of the study. Only adult members of households were interviewed.

In both settings, data was collected on a variety of social background characteristics (e.g., age, sex, marital status, place of birth, family size, religion, education, occupation, income, public assistance, number of household moves in previous two years, etc.). In addition, a questionnaire was used to assess the degree of traditional value identification of the subjects. Each questionnaire item consisted of a brief story which involved individuals in a social situation. Themes such as family solidarity, behavior toward parents and relatives, attitude and orientation toward work, interpersonal relations, and folk medical emphases were presented. A conflict was posed in the story and a choice between two alternatives was required of the respondent. His choice resolved the conflict and at the same time reflected his underlying value orientation. Responses coded traditional were those emphasizing (a) lack of individual autonomy vis-a-vis family and elders, (b) particularistic as opposed to universalistic value choices (in work settings or in personal relations with institutions), (c) acknowledgment of folk as opposed to scientific medical constructs, (d) disavowal of the English language at home, and (e) stereotyped conceptions regarding gender differences in role prescriptions and behavior. "Traditionalistic" options, thus, reflected commitment to and agreement with the values of the native Mexican culture. If the respondent chose the alternate option, this was taken to mean that he did not follow the traditional emphases of the Mexican group in those behavioral situations depicted by that item. Conflict situations were selected which focused sharply on issues, social relations, and strategies involving means-ends choices that were known to saliently distinguish the Anglo from the Mexican American sociocultural premises. Options resolving the conflicts were structured to exemplify Mexican as opposed to Anglo values. As a result of this coding rationale, the manner in which an individual answered or resolved the social situations presented in the questionnaire was used to measure his value identification.

Analysis of social background data showed that a higher proportion of marital instability was found to exist in the patient sample. Significant differences were found when the two groups were compared on variables related to unemployment and "outside" financial support. In the patient group, more families were unemployed and more instances of support from sources outside the household were reported. In addition, a median test was used to compare the family income of the groups. Again, the patient group was found to fall significantly below the median. In addition, there was a trend for nonpatients to have higher levels of education and occupational ratings (see Table 1).

TABLE 1
BACKGROUND CHARACTERISTICS OF THE PATIENT AND NONPATIENT SAMPLES

Characteristics of Individuals	Patients N=76	Nonpatients N=48
Age:		
Mean	38.1	36.7
Standard Deviation	14.3	12.1
Median	34.5	32.8
Range	18-76	18-66
% Males	43.4	29.2
% Married	46.1	85.4
% Separated of divorced	13.1	6.3
% Widowed	7.9	8.3
Mean number of children[a]	4.4	4.1
Standard Deviation	3.2	2.5
Median number of years' education	5.4	7.0
% Catholic	81.6	91.7
% Protestant	15.8	8.3
% Noncitizens	19.7	20.8
% Born in Texas	65.8	77.1
% Born in Mexico	27.6	22.9
% of Native-born parents	31.6	25.0
% of Foreign-born parents	39.5	50.0
Mean number household moves in preceding 3 years	.8	.8
Standard Deviation	1.2	1.1
% Non-English speaking	36.8	22.9
Mean numbers of years' residing in Texas	30.0	32.1
Standard Deviation	16.9	11.3
Characteristics of Families		
Median family income[b]	$1850	$3050
% Families unemployed[b]	52.6	18.8
% Families nonself-supporting[b]	34.2	2.1
Mean number adults living with person	1.8	1.7
Standard Deviation	1.3	1.2

[a] Single individuals were excluded for this computation.
[b] Using the X^2 test, differences between the two groups were found to be significant beyond the .001 probability level.

The questionnaire items were subjected to scalogram analysis, and a scale reflecting degree of traditionalism was obtained. The two groups were compared to detect a possible association between group status and degree of traditionalism. The two groups did not show chi-square differences which would indicate that either group had a significantly higher proportion of traditionalistic *or* nontraditionalistic individuals. The mean traditional scores of both groups were compared and no significant differences resulted when the t-test was applied. However, when the three scale regions (nontraditionalistic, mixed or intermediate, and very traditionalistic) were compared, results showed that each group was distributed differently along the scale (see Table 2). It was found that each of the four extreme scale positions had proportionately more nonpatients than patients. In other words, although nonpatients were equally divided between extreme and intermediate regions, more than three-quarters (seventy-eight percent) of the patients were found in the central part of the scale. Nonpatients thus took either a very traditional posture or one that was its converse, namely very nontraditional. Patients, on the other hand, seemed to follow both traditional and nontraditional patterns to a moderate extent, having proportionately fewer individuals with identifications manifesting commitment to either pole of the value continuum.

TABLE 2

VALUE SCALE POSITIONS AND DISTRIBUTION OF
SUBJECTS ON SCALE CONTINUUM[a]

A. SCALE POSITIONS

Nontraditionalistic	Mixed or Intermediate	Very Traditionalistic
0-1	2-6	7-8

B. NUMBER OF SUBJECTS BY SCALE POSITION

	0-1	2-6	7-8		Intermediate	Extremes
Patients	7	59	10	Patients	59	17
Nonpatients	11	24	13	Nonpatients	24	24

[a.] The first part of the table indicates the scale positions (or value scores) in the different scale regions used in the analysis. The second part shows the number of individuals of each group in the different scale regions.

In summary, results indicated that in the setting studied, the nonpatient group was more successful (in terms of social functioning) and also manifested values and attitudes that indicated "exclusive or total" value orientations. Viewing the results descriptively, if nonpatients (who indicated greater social productivity) reflected traditional (Mexican) values in relation with family, for instance, they were also more likely to espouse folk conceptions, acknowledge Mexican masculinity-femininity notions, and prefer social relationships in work situations that could be said to be particularistic. Or vice versa, if nontraditional (Anglo American) options were chosen in one type of social activity or domain, these same types of options were more likely to be followed in others. For the nonpatients, thus, there was greater similarity between the type of underlying value orientation that influenced choices in differing social circumstances. It was thought possible that orientations of that type might enable greater psychological organization and hence improve social performance by allowing individuals to function with fewer (more general) premises. Thus, by relying exclusively on either Mexican or Anglo choices in many differing behavioral contexts, psychological consistency could be maintained and productivity increased.

ETHNIC DIFFERENCES IN PSYCHOPATHOLOGY AMONG HOSPITALIZED SCHIZOPHRENICS

The purpose of this study (Fabrega, Swartz and Wallace, 1968a; Fabrega, Swartz and Wallace, 1968b) was to analyze and interpret the clinical differences that resulted when Mexican American schizophrenics were matched and compared with Negro and Anglo schizophrenics. The study dealt with hospitalized patients and involved the use of structured psychiatric rating scales. The results of the comparisons were analyzed using (a) hypotheses derived from previous psychological and anthropological formulations dealing with the Mexican cultural patterns, and (b) assumptions based on previous psychiatric epidemiological work dealing with the Mexican American. Specifically with regard to Jaco's study (1960), it was assumed that the lower incidence figure he reported for the Mexican Americans was the result of the family's unwillingness to seek psychiatric care rather than a result of an actual lower amount of psychotic illness. It was hypothesized that Mexican American schizophrenics would show higher measures of disorganization, regression and grossly psychotic behavior because hospitalization was more likely to have occured later in the course of their illness when greater impairment was manifest.

Using background psychocultural information, then, and also the results of a previous epidemiological study (both of which concerned the Mexican American ethnic group), a series of predictions were made regarding how the Negro and Anglo groups would differ from the Mexican American group on the particular clinical indices used in the study. The probability sample of patients used in the Administrative Survey of the Texas State Mental Hospitals (Pokorny and Frazier, 1966) comprised the study's population and the patients investigated were drawn from that group. All patients were between the ages of 18 and 60, had been hospitalized less than two years, and were classified as schizophrenics. The sample included nineteen Mexican American schizophrenics who were carefully matched with schizophrenics from each of the other two ethnic groups represented (Anglo and Negro) on the variables of sex, age, IQ estimate, number of previous psychiatric hospitalizations and when possible education. The following information was used in the analysis: (a) Psychiatrist's scaled evaluations of patients' psychopathology, classified in the manner of Overall and Gorham (1962) [This expanded version of the Brief Psychiatric Rating Scale (PRS) consists of thirty-six items that refer to prominent psychological symptoms, pathological defenses and inappropriate interpersonal behavior.]; (b) Results of having ward nurses, who had repeatedly observed the patients' behaviors, complete the Nurses Observation Scale for Inpatient Evaluation (NOSIE). This questionnaire (Honigfeld, Gillis and Klett, 1966) contains thirty items that deal with features of patients' ward behavior judged from the social appropriateness and organization standpoint. Previous factor analytical studies yielded a method for computing scores on three positive (social competence, social interest and personal neatness) and three negative (irritability, manifest psychosis and retardation) factors as well as a composite score, total patient assets, reflecting overall ward adjustment.

Tables 3 and 4 list and describe the set of items that were used in the study. Items on which specific predictions were offered (Table 3) are marked in such a way as to indicate whether culture-psychodynamic reasons or considerations of psychoticism-regression served as the basis of the prediction. Many of the PRS items suggest chronicity, psychoticism and regression. Items 3 (emotional withdrawal), 4 (conceptual disorganization) and 7 (mannerisms and posturing) are a few examples of this, and, consequently, were predicted to be more pronounced in the Mexican American group. The reported tendency of the Mexican culture to sanction passive coping and psychological "endurance" might lead to greater depression, acknowledgment and introjection, but less projection

TABLE 3
COMPARISONS OF MEAN SCORES FOR THE THREE ETHNIC GROUPS ON THE PSYCHIATRIC RATING SCALE[a]

PRS Item	Basis of Prediction – Clinical Severity	Basis of Prediction – Culture Psychodynamics	Prediction	Mexican American versus Anglo Group	Mexican American versus Negro Group
1. Somatic Concern	x		+	n.d.	−
2. Anxiety				−	+[b]
3. Emotional Withdrawal	x		+	+	+
4. Conceptual Disorganization	x		+	+	+
5. Guilt Feelings				−	+
6. Tension				+	+
7. Mannerisms and Posturing	x		+	+[b]	+
8. Grandiosity				+	−
9. Depressive Mood		x	+	−	+
10. Hostility				+	−
11. Suspiciousness				n.d.	+[b]
12. Hallucinatory Behavior	x		+	+	+
13. Motor Retardation	x		+	+	+
14. Uncooperativeness				+	−
15. Unusual Thought Content	x		+	+	+
16. Blunted Affect	x		+	+	+
17. Excitement	x		−	+	+
18. Disorientation	x		+	+	+
19. Memory Impairment	x		+	+	+
20. Elevated Mood				+	−
21. Manipulation		x	−	−	−[b]
22. Phobias				−	−
23. Passive Dependence		x	+	−	n.d.
24. Compulsive Acts				−	−
25. Rigid Perseverence	x		+	+	+
26. Obsessive Thoughts	x		+	+	n.d.
27. Projection of Blame		x	−	−	−
28. Introjection of Blame		x	+	+	+
29. Dramatization		x	−	n.d.	−
30. Impulsiveness		x	−	+	+
31. Denial		x	−	−	−
32. Inadequacy		x	−	−	−[b]
33. Emotional Lability		x	+	+	+
34. Egocentricity				+	+
35. Sexual Identification		x	−	−	+
36. Intellectualization		x	−	−	+

[a] The '+' symbol indicates a higher mean for the Mexican American group, the '−' symbol indicates a lower mean, and 'n. d.' indicates no difference in means.
[b] Comparison between the Mexican American and Anglo or Negro groups showed significant between-group differences beyond the 0.05 level of confidence.

TABLE 4
MEAN FACTOR SCORES
FOR THREE ETHNIC GROUPS ON THE
NURSES OBSERVATION SCALE FOR INPATIENT EVALUATION

	Anglo (N=19)	Mexican American (N=19)	Negro (N=19)
Positive Factors			
Social Competence	27.68	27.05	33.58
Social Interest	17.26	12.63	16.63
Personal Neatness	21.68	21.05	24.11
Negative Factors			
Irritability	7.89	8.42	6.21
Manifest Psychosis	3.47	4.84	2.63
Retardation	8.53	10.74	5.58[a]
Total Patient Assets	142.74	132.74	155.89[a]

[a] Comparison between Mexican American and Negro groups showed significant between-group difference.

of blame. This reasoning (based on culture-psychodynamic considerations) led to predictions on those items measuring the particular symptomatic trend (items 9, 27, 28 and 31). Despite the fact that some items were predicted to show lower mean scores in the Mexican American group, on the whole, it was anticipated that in each comparison involving the PRS, the Mexican American group would show a greater proportion of higher (more serious) scores. This view was held because, by focusing on symptomatic trends, results on the PRS in a general way indicate gross impairment or pathology. With regard to the NOSIE, all three of the positive factors were predicted to be lower in the Mexican American group, whereas the Negro and Anglo groups were expected to show a lower mean score on each of the three negative factors. This was expected because the behavior of the Mexican American schizophrenics was thought to reflect greater regression or psychoticism.

Results of the psychopathology judgments made by psychiatrists indicated that the direction of the mean score difference was as predicted in a statistically significant proportion of instances. In each comparison, it was the set of predictions made on the basis of considerations of regression and/or psychoticism that was more consistently supported. The predictions that were based on culture-psychodynamic reasons were not significantly supported by the results, though they were accurate more often than not. The mean scores of the ward social behavior scales that the nurses had completed conformed to predictions. When compared to the Mexican American group, the Negro and Anglo groups both received higher mean scores on all three of the positive factors. In addition, the three negative factor scores were higher in the Mexican American group in both comparisons. In general then, results of analyzing nurses' judgments of the ward behavior of these schizophrenics tended to show that the Mexican American group was more socially disorganized and regressed, particularly when compared with the Negro subsample. Thus, psychiatrists' estimates of specific psychopathological trends and nurses' judgments made independently seemed to show greater psychoticism in the Mexican American patients (see tables 3 and 4).

These results could be interpreted to support the hypothesis that the Mexican Americans as compared to their Negro and Anglo counterparts, are hospitalized at a point in the course of their illness when more symptoms reflecting psychoticism and/or regression are manifest. This would suggest that the families of Mexican American patients may be more tolerant of deviant psychotic behavior than families of the other two groups, and, consequently, delay seeking help of hospitalization longer. Perhaps native curers are sought first and families are advised to cope with the disturbance in various ways, all of which entail that the symptomatic individual stay at home. In addition, among Mexican Americans, the mutual support that exists between members of a family and the sense of obligation toward a member who falls ill may be forces that retard hospitalization because of an unwillingness on the part of the family and the ill individual to separate. The result of these issues is that only when the behavior is grossly uncontrollable might hospitalization be resorted to. If this reasoning is correct, it suggests that the lower incidence rate of psychotic illness that was observed by Jaco in his survey might not accurately reflect the extent of psychotic illness in his group. In other words, many new psychotic behavioral disorders that would ordinarily lead to hospitalization in the case of the Negro or Anglo would be handled by the more tolerant relatives of the Mexican American at home or in the community by appeal to folk institutions. A survey of treated illness would then show only severe cases in the Mexican American group, whereas in the Anglo and Negro groups, both severe and less severe cases would be represented.

Another interpretation is possible regarding the greater psychoticism and social disorganization shown by the Mexican American group of schizophrenics. This is that the Mexican American patient may be more prone to regression once he is hospitalized. This could result because in the Mexican American patient the social environment of the hospital might produce considerable feelings of alienation. Similarly the meaning that hospitalization has for the Mexican American patient might be uniquely stressful to him. Both of these factors could promote regression following hospitalization. In general, then, this study did not allow determining if the greater psychoticism-regression of the Mexican American group reflected patient adjustment at a time preceding hospitalization or if it resulted from events which occured subsequent to admission into the hospital.

ANGLO AND MEXICAN AMERICAN PSYCHIATRIC OUTPATIENTS

The last set of findings to be reviewed concern individuals presenting themselves for outpatient care in a psychiatric clinic located in South Texas (Fabrega and Wallace, forthcoming). Analysis was carried out using two emphases: (a) comparing acculturated Mexican Americans with unacculturated Mexican Americans and (b) comparing each of these two subgroups with the Anglo Americans who presented for outpatient care in the same clinic. These comparisons were undertaken so as to deal with issues raised by the studies reviewed earlier in this presentation. The same psychiatric rating scale (PRS) was used in this study and the same rationale was adopted (predictions based on criteria of psychoticism-regression received principal emphasis) to analyze the Anglo-Mexican American comparisons and the acculturated Mexican American-unacculturated Mexican American comparisons.

All Mexican and Anglo American patients who requested assistance at this psychiatric clinic during a six-month period were used in this analysis. Both the method of collecting information and the variables used in the analysis were described earlier under the heading Value Identification in Patients and Nonpatients. Only patients who belonged to Hollinghead's class IV and V were used in this study. Degree of acculturation was estimated by means of a score derived from the following: (a) degree of education, (b) citizenship status, (c) extent to which the patient lived and participated with relatives or maintained an autonomous existence living only with his spouse and children, and (d) degree of competence with the English language. Acculturation scores, computed by scales involving these variables, estimated the extent to which a person of Mexican descent seemed to be assimilated into and identified with the Anglo American culture. Using this acculturation score, Mexican American psychiatric outpatients visiting the clinic were ranked, and two groups were formed which were termed the acculturated and unacculturated groups.

When compared to the Anglo outpatients, the Mexican American outpatient group as a whole had higher psychopathology ratings on a proportionately larger number of the items that reflect psychoticism and regression (the same items that were used to make predictions in the previous study). The probability value (obtained by the binomial expansion equation) reached the level required for significance. Interestingly, the trend for greater psychoticism in the Mexican American group was not evident when only acculturated subjects were compared to Anglo American outpatients. When the unacculturated subgroup of Mexican Americans was compared to the Anglo Americans, the evidence for greater indices of psychoticism-regression in the Mexican American group was highly significant. Similarly, results were significant when the unacculturated group was compared to the acculturated group of Mexican Americans (see Table 5). Thus, it appears that it is the unacculturated subset of Mexican Americans who present for treatment with clinical features that suggest greater psychoticism and regression.

The results of this latter study add some support for the belief that the factors and/or processes which lead to psychiatric consultation are related to sociocultural issues such as ethnicity and acculturation. More precisely, the personal and/or social issues that together lead to a request for psychiatric care differ in these variously differentiated social groups. One feature of this difference involves degree of acculturation and extent of clinical disability or regression. Compared to Anglo Americans and acculturated Mexican Americans, unacculturated Mexican Americans present themselves for psychiatric treatment with evidence of greater psychoticism and regression; acculturated Mexican Americans do not differ significantly from the Anglo American subjects when psychoticism and/or regression is used as an analytic variable. As was touched on earlier, this could mean that in the unacculturated subgroup, the patients and their immediate family members show differences in what can be termed "tolerance of symptoms," since they request consultation when symptoms of disorganization are more prominent. Acculturated Mexican Americans appear to be similar to Anglo Americans in terms of readiness to seek formal psychiatric help.

TABLE 5
COMPARISONS ON PRS ITEMS
INDICATING PSYCHOTICISM–REGRESSION IN PSYCHIATRIC PATIENT SUBGROUPS

		RESULTS OF COMPARISONS OF GROUP MEANS[c]			
		Anglo (N=28) versus			Acculturated (N=27) versus Unacculturated (N=32) Mexican Americans
Psychoticism-Regression Indices	Prediction[b]	All Mexican Americans (N=59)	Unacculturated Mexican Americans (N=32)	Acculturated Mexican Americans (N=27)	
1. Somatic Concern	+			x	x
3. Emotional Withdrawal	+	x	x		x
4. Conceptual Disorganization	+	x	x		x
7. Mannerisms and Posturing	+				x
12. Hallucinatory Behavior	+	x	x	n.d.	x
13. Motor Retardation	+	x	x	x	x
15. Unusual Thought Content	+	x	x		x
16. Blunted Affect	+	x	x		x
17. Excitement	–			x	
18. Disorientation	+	x	x	x	x
19. Memory Impairment	+	x	x	x	
25. Rigid Perseverence	+	x	x		x
26. Obsessive Thoughts	+				x
Significant Probability (two-tailed test) Values		.022	.022	—	.022

[a] Only patients in classes 4 and 5 of Hollingshead's "Index of Social Position" scheme were used in these comparisons.

[b] The '+' symbol indicates a prediction of a higher mean for the Mexican American or the unacculturated group, and the '–' symbol indicates a prediction of a lower mean.

[c] The 'x' indicates a conformance of the comparison results to the direction of the prediction made. The 'n.d.' indicates no difference.

SUMMARY

This report first reviewed some of the sociodemographic and cultural features of the Mexican Americans of Texas. The relatively disadvantaged social position of this ethnic group was apparent. It was emphasized that there is a need to regard the processes of acculturation and migration as equally relevant when the adaptation of the Mexican American is studied. Recent psychiatric studies that have been conducted on the Mexican American ethnic group of Texas were reviewed. Results suggest that compared with patients, Mexican American nonpatients are more socially productive and economically successful, implying greater assimilation into the dominant "Anglo society." They did not, however, manifest a less traditional or an Anglo "type" of value identification. Instead, they were found to be overrepresented at both ends of a value scale which measured traditionalism. Mexican American schizophrenics when matched and compared with Negro and Anglo American schizophrenics, showed greater indices of psychoticism, regression and social disorganization in the hospital. When outpatients of the Anglo and Mexican American ethnic groups were compared, it was shown that the Mexican American group again tended toward greater regression and psychoticism. In this regard, it was the group of unacculturated Mexican American patients which showed the most prominent clinical differences. These results point to underlying differences in such issues as the definition of "illness," the "need for treatment," and "tolerance of psychiatric symptoms."

NOTES

1. For a general review of descriptive characteristics of the Mexican American population group, the reports of the Mexican American Study Project of the University of California, Advance Reports 1 (Fogel, 1965), 2 (Grebler, 1966), 4 (Moore and Mittelbach, 1966), 5 (Mittelbach and Marshall, 1966), 6 (Mittelbach, Moore and McDaniel, 1966), 7 (Grebler, 1967), 10 (Fogel, 1967) and 11 (Moustafa and Weiss, 1968), the monograph by Browning and McLemore (1964) and the books by Madsen (1964), Heller (1966) and Rubel (1966) should be consulted.

2. This section relies on the material presented in Advance Report 2 of the Los Angeles Mexican American Study Project (Grebler, 1966).

3. The work of Browning and McLemore (1964) is a valuable exposition of the data collected during the 1960 U. S. Census on the Spanish surname population, and the details of this section are taken from their publication. Several of the University

of California Mexican American Study Project Advance Reports (see Note 1.) also present useful analyses. The approach used by the Bureau of the Census to identify the Mexican American is to use the classification "white persons of Spanish surname." Although there are sources both of overcount and undercount, the procedure adopted has proved to be effective. Issues related to this classification are elaborated upon by Browning and McLemore (1964).

REFERENCES

BEALS, R. L. (1951) "Urbanism, urbanization and acculturation." American Anthropologist 53: 1-10.
BROWNING, H. L. and S. D. MCLEMORE (1964) A Statistical Profile of the Spanish-Surname population of Texas. Austin, Texas: Bureau of Business Research, Univ. of Texas.
DIAZ-GUERRERO, R. (1964) "La dicotomia activo-pasivo en la investigacion transcultural." In Proceedings of the Ninth Congress of the Inter-American Society of Psychology. Miami, December 17-22.
——— (1963) "Respeto y posicion social en dos culturas." Pp. 116-137 in VII Congreso Inter-americano de Psicologia, Mexico, D. F.
——— (1955) "Neurosis and the Mexican family structures." American Journal of Psychiatry 112: 411-417.
FABREGA, H. and C. A. WALLACE (forthcoming) "Acculturation and psychiatric treatment: a study involving Mexican Americans."
——— (1968) "Value identification and psychiatric disability: an analysis involving Americans of Mexican descent." Behavioral Science 13 (September): 362-371.
FABREGA, H., J. D. SWARTZ and C. A. WALLACE (1968a) "Ethnic differences in psychopathology. II. Specific differences with emphasis on a Mexican American group." Journal of Psychiatric Research 6: 221-235.
——— (1968b) "Ethnic differences in psychopathology. I. Clinical correlates under varying conditions." Archives of General Psychiatry 19 (August): 218-226.
FOGEL, W. (1967) "Mexican Americans in Southwest labor markets." Mexican American Study Project Advance Report 10. Los Angeles: University of California (October).
——— (1965) "Education and income of Mexican Americans in the Southwest." Mexican American Study Project Advance Report 1. Los Angeles: University of California (November).
GREBLER, L. (1967) "The schooling gap: signs of progress." Mexican American Study Project Advance Report 7. Los Angeles: University of California (March).
——— (1966) "Mexican immigration to the United States: the record and its implications." Mexican American Study Project Advance Report 2. Los Angeles: University of California (January).
HELLER, C. S. (1966) Mexican American Youth: Forgotten Youth at the Crossroads. New York: Random House.

HONIGFELD, G., R. D. GILLIS and C. J. KLETT (1966) "NOSIE-30: a treatment-sensitive ward behavior scale." Perry Point, Md.: Central Research Laboratory Report No. 66.

JACO, E. G. (1960) The Social Epidemiology of Mental Disorders. New York: Russell Sage Foundation.

MADSEN, W. (1964) The Mexican Americans of South Texas. New York: Holt, Rinehart & Winston.

MITTELBACH, F. G. and G. MARSHALL (1966) "The burden of poverty." Mexican American Study Project Advance Report 5. Los Angeles: University of California (July).

MITTELBACH, F. G., J. W. MOORE and R. MCDANIEL (1966) "Intermarriage of Mexican Americans." Mexican American Study Project Advance Report 6. Los Angeles: University of California (November).

MOORE, J. W. and F. G. MITTELBACH (1966) "Residential segregation in the urban Southwest." Mexican American Study Project Advance Report 4. Los Angeles: University of California (June).

MOUSTAFA, A. T. and G. WEISS (1968) "Health status and practices of Mexican Americans." Mexican American Study Project Advance Report 11. Los Angeles: University of California (February).

OVERALL, J. E. and D. R. GORHAM (1962) "The brief psychiatric rating scale." Psychological Reports 10 (June): 799-812.

POKORNY, A. D. and S. H. FRAZIER (1966) Report of the Administrative Survey of Texas State Mental Hospitals, 1966. Austin, Texas: MH Project Grant No. 09235-01 of NIMH, sponsored by the Texas Foundation for Mental Health Research in collaboration with the Texas Department of Mental Health and Mental Retardation.

RUBEL, A. J. (1966) Across the Tracks: Mexican Americans in a Texas City. Austin, Texas: Published for the Hogg Foundation for Mental Health by the University of Texas Press.

——— (1964) "The epidemiology of a folk illness: Susto in Hispanic America." Ethnology 3: 268-283.

——— (1960) "Concepts of disease in Mexican American culture." American Anthropologist 62: 795-814.

SAMORA, J. [ed.] (1966) La Raza: Forgotten Americans. Notre Dame, Indiana: University of Notre Dame Press.

SPIRO, M. E. (1955) "The acculturation of American ethnic groups." American Anthropologist 57: 1240-1252.

Chapter **12**

Adaptation of Adolescent Mexican Americans to United States Society

ROBERT L. DERBYSHIRE

Culturally divergent ethnic categories and adolescents have much in common in the United States' society. Both are minorities excluded from the mainstream of American adult culture and each lacks adequate access to economic, political and social power. Each is struggling toward acceptance yet each has difficulty locating adequate and functional acculturative frames of reference for culturally integrated participation. Functional relations between simultaneous membership in culturally excluded minority worlds for different migratory generations and relationship of these forces to adolescent attitudes is the subject of this paper. Adolescent attitudes are based upon data gathered from Mexican American adolescents residing within the most economically depressed area of East Los Angeles. Identity crisis, as described by these attitudes, with its subsequent strain, resulting from role conflict, stimulated by the adolescent's desire to identify with family, peer and greater adult worlds, concomitantly influenced by the dominant culture's lack of acceptance of Mexican American cultural diversity and the role conflict it engenders, suggests that adolescence for Mexican American migrants in an urban setting is vulnerable to deviant behavior (Derbyshire, 1968).

Migration is a complex phenomena. The term does not imply a discrete, mutually exclusive category. However, as they do many other categories describing social phenomena, researchers and laymen frequently view the concept as implying that all persons fitting into the

---

Author's Note: *This paper was prepared for the conference on "Migration and Behavioral Deviance," November 4-8, 1968, Dorado Beach, Puerto Rico.*

category maintain similar social, cultural, and psychological characteristics.

The psychosocial dynamics of those who migrate are myriad. Therefore, one's socially visible pattern of moving from one nation to another, one social configuration to another, or one culture complex to another is viewed as the apparent unifying criterion of commonality. Behavioral similarities are often expected after placing into the same category all persons who have migrated. For example, even when an external force such as a revolution appears influential in a migratory pattern, there is every reason to believe that persons who migrate are not necessarily basically motivated to migrate because of the revolution, but the revolution provides them with a stimulus for responding to already unmet migratory needs. In other words, migration of a large number of Cubans, during and after the Cuban Revolution, may appear on the surface as a large segment of people being pushed out to "seek freedom" when more accurately the revolution may be the mechanism or the stimulus for acting upon and carrying out migratory wishes encouraged by the drawing forces of United States culture.

A major problem for such migrants is their willingness or resistance to fitting into a prescribed social role (i.e., "seeker of freedom from Communism") which may or may not fit into their identity needs. Therefore, if Americans with whom the migrant comes into contact treat him and expect him to play the "freedom seeker game," then the migrant must either fit into the pattern and adapt, which may be uncomfortable, or reject the pattern and receive behaviors which may enforce social distance and alienation.

Too frequently, as social scientists and laymen, we pay little attention to the reciprocal relationship between actors. The importance of examining and understanding this reciprocity is essential when examining minority-majority relations. What is perceived by the stationary population as acceptable role behaviors for migrants and what migrants perceive as acceptable behaviors for self in relation to migrants and others is frequently divergent. When one moves from a culture in which he was born and reared, where he has maintained a comfortable, adaptive set of behavior patterns at cognitive, affective, and cultural levels, old ways of feeling, thinking, and acting remain the only criteria upon which the migrant can validate new experiences. When trying to validate new experiences with self as a social barometer, comfortable, old ways of behaving do not provide the same fulfillment as they did in the old world, culture, or social group, because feeling states, interaction patterns, and cultural mechanisms are less successful at tension reduction in this new environment.

By virtue of moving and seeking to establish territorial rights, one sets up within the organism's psychology, a need to feel and display, among other things, dominance and security over some aspects of one's life. Overdetermined fears of rejection, alienation, and just plain discomfort increase the possibilities for carrying out a self-fulfilling prophecy (Thomas and Znaniecke, 1927). Migrant-nonmigrant interaction, by virtue of manifest differences as well as the more subtle behavioral and feeling nuances, often decreases communication validation. Therefore, with social interaction there is a potential for the migrant to receive and communicate a fear of role nonacceptance which, in itself, when it is recognized at a conscious or subliminal level in the other participant creates a distrust which forces the maintenance of social distance to provide safeguards against uncomfortable feelings of strangeness between the two participants.

These psychodynamics become intensified and of greater importance when one examines the social dynamics for migration. Some expressed reasons for migration are: (a) social factors over which the individual feels he has little or no control (i.e., revolution, economic depression, acts of God, e.g., floods, hurricanes, fires and other disasters); (b) external social forces over which he feels he may have some control yet he is unable to cope with these in his present environment (i.e., loss of job, extrusion of close family member from society to prison, hospital, etc., death of a loved one, unexpectedly poor return on self-employed endeavors, e.g., farming, light manufacturing, etc.); (c) social forces which perpetuate the appearance that the new land of the immigrant has greater opportunities than where he presently lives. These and other migratory forces are not mutually exclusive and become weighted determinants in family and individual decision-making processes. Therefore, it seems reasonable to conclude that any understanding of differences between migrants and nonmigrants must be examined in terms of both psychodynamic and social reasons for the migratory patterns. For example, one can hypothesize that the length of time a migrant family maintains the old cultural identity will be inversely related to social-psychological traumas (i.e., unanticipated culturally shocking experiences created by moving and the quality and quantity of subsequent experiences) which force sanctuary in one's cultural identity. Also it can be hypothesized that those persons who perceive their migration as forced or maladaptive, yet have a desire to return to the donor culture, will participate in less effective social-psychological interplay between migrant and nonmigrant (Scotch, 1963).

The following interview brought to my attention the many feeling problems engendered during migration. A seventeen-year-old Mexican-

[278] BEHAVIOR IN NEW ENVIRONMENTS

American youth had arrived in East Los Angeles at the age of thirteen with his father, mother, three brothers and two sisters. He stated:

> The ride from Mexico to Los Angeles was long, hot and troublesome. With each stop for gas and truck repairs people acted funny. When they smiled I did not know why. When they frowned I was puzzled and hurt. Even though I understood English, the tongue was strange and not easy to follow. People's faces made me feel strange and unwanted. We ate very little and I was hungry all the way. I did not know what people wanted or how to act. I was scared. I did not trust anyone. It was this way everywhere. People, land and smells, all were strange. I remember feeling lost, helpless and unhappy yet my father said he was happy to be going to a new job in Los Angeles.

This paper compares the attitudes and behaviors of lower class Mexican American migrants and nonmigrants. It is assumed that recent migrants will have attitudes affected by Mexico and their more recent nomadic type experiences while second- and third-generation migrants will have adapted to a ghettoized more or less supported life of the host culture. Although the data presented here do not adequately validate the aforementioned hypotheses and assumptions, some insights are provided.

Mexican American

Terminology difficulties are inherent within minority populations who are citizens of the United States of America (Heller, 1966: Chap. 1). All members of minorities, either born or naturalized in the United States, are legally considered, tend to be, but frequently are not responded to as American citizens first. Aside from American citizenship, however, there are certain ethnic characteristics which promote or inhibit human interaction and relationships within the United States culture (Marden, 1952). These differences are based upon physiognomy, i.e., hair structure and color, nasal index, height and weight, eye color and eye fold and skin color. There is no conclusive evidence that physiognomic characteristics of ethnic categories determines their behavior. However, determinants of behavior are related to gross cultural differences, linguistic nuances of persons from varying backgrounds and societal reactions to physiognomic differences. When these differences interact with the dominant white, Protestant, "capitalistic," Anglo-Saxon "democratic" culture of the United States, prejudicial and discriminatory behavior often occur. In other words, differences in behavior among ethnic groups is the result of at least two major phenomena: (a) the "right" and "wrong" of behavior due to socialization supporting the dominant values within one's culture (Williams, 1960), and (b) these behaviors in interaction with the United States culture which views minority persons and their behavior as "alien."

Mexican American, without the hyphen, is used to identify the minority population with whom this project is related. Although we recognize numerous identity conflicts exist within the community of East Los Angeles, it is not the intention of this research to alienate, determine concensus, or ameliorate feelings of group identification. Although other Spanish-speaking populations, i.e., Cubans, Puerto Ricans, and other South and Central American nationalities, reside in East Los Angeles, statistical analysis of the community indicates that the bulk of the population originally migrated from Mexico. Overtones of group identity problems are not dismissed. Self-reference labels of Latin Americans, Americans of Spanish descent, Americans of Mexican descent, Mexican Americans, or Spanish-speaking Americans and others are frequently heard in East Los Angeles (Weaver, 1968). Although the lack of consensus over semantic identification problems is recognized, I use the abbreviated form without the hyphen for identifying Mexican Americans. The persons described in this research are Americans who by virtue of their Mexican ethnicity have received differential and inferior treatment. However, one teenager, searching for an equal or superior identity, stated, "Mexican American means bicultural and bilingual a little more than other Americans."

The East Los Angeles Mexican American

A brief subjective view of the area where data were gathered may help to interpret the findings (Derbyshire, 1968: 75-80).

East Los Angeles, the unincorporated area, in which this study was carried out, lies east of Boyle Heights, south of City Terrace and Alhambra, west of Monterey Park and north of Commerce and Montebello. Those census tracts in East Los Angeles, which supply the greatest amounts (in Los Angeles County) of family disorganization, juvenile delinquency, crime, drug addiction, dilapidated housing, poverty, and other indicators of community pathology, are surrounded by four huge concrete and steel freeways. Apparently these freeways tend to limit ecologic mobility. The Long Beach Freeway on the east, Santa Ana Freeway on the south, San Bernardino Freeway on the north, and Golden State Freeway on the west are the "Anglo" curtains segregating the most economically deprived Mexican Americans.

On a clear day, when smog neither settles in this basin nor nestles against the San Bernardino Mountains, snow-capped Mount Baldy is seen as it overlooks Los Angeles County. On humid days yellowish, eye-burning, mucous-generating smog devastatingly increases the health hazards of this community.

Although a California state college and junior college are located in or near this area, few residents join its student body. Fewer East Los Angeles residents are members of the faculties. Statistics on junior high and high school dropouts are the highest in the County.

The "Serape Belt" as East Los Angeles has been designated, was at one time a predominately Jewish neighborhood. Today, however, on main thoroughfares, Brooklyn Avenue, Third Street, and Whittier Boulevard one notes advertisements in Spanish, cantinas, corner fruit stands, motion picture theaters with Spanish-speaking movies and Anglo merchants who display signs reading "hablo Espanol." Many merchants and some physicians' offices have televisions tuned to Spanish language programs during the day. Tortillas and tacos are sold at small sidewalk carry-out shops. Signs designating the office of a "curendero" (a folk curer or medicine man) and "abogado" (a false lawyer) are visible.

From second-hand furniture and clothing stores filter the sounds of Spanish language radio programs. On the streets Spanish can be heard more frequently than English. Dark complexioned faces of older males reveal the rugged outdoor manual labors of a lifetime. While women are frequenting the shops and stores, groups of young men are visible on several corners. Clothing is worn, tattered, yet neat; hands, face and eyes appear fatigued; a massive pride is characterized by the erectness of body posture and lightness of gait.

Houses in East Los Angeles are small, colorfully painted, with dilapidated fences and small flower gardens. Much of the housing is rented, not owner occupied. In many neighborhoods the interdependent atmosphere of the barrio (small community) exists. Females, particularly mothers and wives, are often busy with local church and neighborhood affairs. Children frequent the streets for companionship and peer activities. Local settlement houses as well as more recent poverty programs provide some recreational and social facilities.

The sheriff and his deputies, and the state police are avoided when possible. Church and its responsibilities are taken most seriously by females, while males pay lip service. Immigration officers are well known and avoided. Unknown Anglos asking questions in the neighborhood are viewed suspiciously. Middle- and upper-class "chicanos" (slang for Mexican Americans) are also viewed with trepidation for fear they are investigators from Anglo institutions.

Although families in this ghetto appeared to be patriarchal and matricentric, there was excellent evidence that male strength and dominance exists only because females feel it is best to play the subordinate role. Females frequently verbalized dissatisfaction with their husbands' and fathers' positions in the family but give and promote family respect out of deference to his loss of status in Anglo culture. Among families interviewed, there was an apparent role deception by females to provide a foundation for the emasculated if not lost "machismo" of their Mexican heritage. This was most frequently revealed through the wives' lack of condemnation for their husbands' extramarital sexual exploits. Praise was often given for the male's lack of fear and his physical combativeness outside of the house. Although sexual and aggressive behavior was seldom condemned, most mothers and wives were able to verbalize to the interviewer that these actions were dysfunctional for getting ahead in Anglo land. "The poor man fights so many problems every day that the least I can do is support him in the ways of his father."

Child-rearing practices are punitive and severe. Girls are protected while boys are encouraged to be aggressive. Gang life is most usual for boys. "Pachuco" gangs have their distinct dress and speech patterns (Meeker, 1964). Gangs frequently attack isolates, dyads or triads of their rivals. Large gang warfare is seldom encountered although it does exist. These gang disagreements most frequently involve territory or females. Most youngsters in these gangs are not delinquent.

Although East Los Angeles, no doubt, produces a variety of family belief systems, most older persons give lip service to the Mexican family prototypes. This has been described by Oscar Lewis (1949).

> According to the ideal pattern for husband-wife relations in Tepoztlan, the husband is viewed as an authoritarian patriarchal figure who is head and master of the household and who enjoys the highest status in the family. His prerogatives are to receive the obedience and respect of his wife and children as well as their services. It is the husband who is expected to make all important decisions and plans for the entire family. He is responsible for the support of the family and for the behavior of each member. The wife is expected to be submissive, faithful, devoted, and respectful toward her husband. She should seek his advice and obtain his permission before undertaking any but the most minor activities. A wife should be industrious, frugal and should manage to save money no matter how low her husband's income. A good wife should not be critical, curious, or jealous of her husband's activities outside the home.

Adolescents, in this study, of both a recent and not so recent migratory background have interacted within this milieu and each has been uniquely affected by these transactions.

Adolescent Migration Research

The problem of growing up as a first-generation Mexican migrant in the United States was discussed as early as 1931 (see Gamio, 1931; Humphrey, 1946; Leonard and Loomis, 1941; McWilliams, 1949; Sanches, 1943; Taylor, 1932). Children born to the migrant Mexican and reared in East Los Angeles, present unique United States adaptation patterns. Conflicts due to diversity of cultures create adaptive problems within American institutional life in areas of values, education, religion and occupation.

During the initial data gathering phase of this experiment I was fortunate to receive aid and support from an East Los Angeles teen center and its leader. One Friday evening while approximately thirty teenagers were completing the questionnaire, a rival gang harassed members of the teen center. Not only was the experiment interrupted during this encounter but many of the forms were destroyed. This incident extended the data gathering period several months because these adolescents were brought in two at a time to complete the 34 page questionnaire.

Purpose

The purpose of this research is to compare the attitudes of Mexican American adolescents who were born and reared or whose parents were born and reared in the United States with those adolescents who migrated or whose parents migrated from Mexico to the United States.

In an earlier work Seward (1958) suggested "the paradoxically sounding assumption that the more firmly an individual is embedded in his primary ingroup the better integration he may be expected to make with the dominant culture." Peak (1958) writes, "The great differences in philosophy of acceptance and resignation, passivity, dependency, etc. between Mexicans and Anglos have been the cause of much misunderstanding of Mexicans by the Anglo population who misinterpret Mexican philosophy and see these people as indifferent, lazy or unambitious." These differences appear to be exaggerated early in the migratory process. Since Mexican values are the antithesis of values cherished by American adults, then unless Mexican American adolescents can find congruence in, and lack ambivalence toward, their subcultural values, transition into adulthood may be filled with conflict over the necessity of giving up one's sacred beliefs in order to be accepted by a hostile environment, which results in role strain and social deviance.

The dominant pattern in Mexican familes is to develop in the growing child, obedience, humility and respect for elders (Guerrero, 1955). Adolescent crisis is turbulent with problems of authority and self-determination. An intensification of ambivalence toward obedience, humility and respect is encountered at this time for most Mexican American adolescents. It is hypothesized that role strain is particularly evident for youngsters of Mexican parentage who, having been born and reared in the United States, have not adequately internalized the dominant Mexican values.

Associated with these difficulties is the humility of the older generation. A lower-class Mexican lady indicated, "when my teenage children do wrong I blame nobody but myself for not being more strict." In response to being asked how she felt when the children lived up to her expectations she stated. "I thank God for making them that way." The Mexican American adult, with his extreme humility associated with success and self-punishment associated with failure, establishes a psychosocial situation within which the developing child, when in contact with the Anglo world, finds difficulty integrating the concepts of responsibility, aggressiveness, authority and independence. To manifest a world view that "man can do only bad" and "God can do only good" is functional only with strong (Mexican) cultural support and a lack of interference from outside (United States) cultures. Culture shock for Mexicans is apparent when the Anglo world suggests that man is responsible for his "good" and his "bad" deeds. This, no doubt, establishes for the growing child a conflicting situation, creating uncertainty toward the Mexican value system, while concomitantly supporting acceptance of the Anglo culture which is more functional for upward mobility, yet it has only limited usefulness for self-preservation (Derbyshire, 1966, 1964; Brody, 1966).

Since adolescence is a time for breaking away from authority, both parental and religious, the Mexican American youngster who has uncertain feelings toward his culture may strike out at his parents through deviant acts, without offending God. Mexican American informants in East Los Angeles indicate that youthful gangs are most frequently those boys with extreme feelings of uncertainty toward Mexican culture.

According to Erikson's assumption (1950), the mastering of future life experiences (for adolescent Mexican Americans) depends upon the success with which their own culture is mastered (and later the culture of the Anglo world). It may be hypothesized that if one successfully gains "wholehearted" cultural support for his Mexican identity, then later success at coping with incongruent Anglo modes of behavior may be functional for political and socioeconomic success, yet not necessarily dysfunctional for ego identity. If, as it has been suggested by other investigators (Spiegel and Kluckhohn, 1954), the Mexican American family has been, during the last generation, moving away from overt father dependence to the American pattern of individualism, then the strong dominant father and passive, submissive mother of the Mexican cultural pattern is no longer functional as a "successful varient of group identity" (Erikson, 1950). As stated by Seward (1964), "This change has had a disorganizing influence on family structure and adversely affected the personalities of family members, often resulting in antagonism between eldest and younger sons, and becoming manifest in delinquent behavior."

Information gathered in this study describes differences in attitudes between two groups of Mexican American adolescents. It is the hypothesis of this research that the values, attitudes, and behavioral patterns of teenagers who were born in Mexico, or who are children of parents who were born in Mexico, are significantly different from those teenagers in the same socioeconomic situations who were born to parents who have resided in the United States for two or more generations.

Methodology and Social Characteristics of Sample Population

The sample consists of eighty-nine adolescents of Mexican American background who lived in a low-income area of East Los Angeles. These young persons, forty-two males and forty-seven females, were between the ages of thirteen and nineteen. Each youngster anonymously completed a thirty-four-page questionnaire. This included ten pages of information covering a personal and family history as well as subjective feelings and attitudes toward persons and values significant in the life of adolescents.

Together with these data was a series of twenty-four concepts followed by nineteen Osgood semantic differential scales. These scales were selected to reveal concept differences for male-female roles and American and Mexican value orientations, (e.g., proud-humble, dark-light, etc.). Each concept concerns itself with attitudes related to persons who may, for these adolescents, present problems of ego integration during adolescence (e.g., Mexican, Father, Mother, bullfighter, etc.).

The eighty-nine adolescents who contributed data to this study were divided into two categories: (a) Mexican American migrants who had moved to East Los Angeles within their lifetime or whose mother or father moved from Mexico to East Los Angeles; (b) Mexican Americans whose father and mother as well as themselves were born and reared in the United States. Forty-one adolescents were defined in the migrant category, twenty-six male and fifteen female. The nonmigrant or what is also labelled as the "established" category consisted of forty-eight adolescents, twenty-one male and twenty-seven female. When comparing the median age (fifteen years) of the migrants with the established category (sixteen years) there is only a one year difference. However, the mean age of the established category is two years older than the migrants. Only fifteen migrants were born in Mexico while twenty-six of these adolescents had one or both parents who were born in Mexico. An overwhelming majority of both categories professed Catholicism as their religious preference. However, church attendance was significantly more frequent among migrants.

When comparing migrants with nonmigrants, the fathers of nonmigrants were significantly more highly educated (formally by three years) while there is no difference between mother's education for the two groupings. The majority of mothers in both categories were unemployed and remained at home, (according to these youths) because their mothers "desire to be with family."

Twelve adolescent nonmigrants had forty-six arrests while seven migrants were involved in twenty arrests. There was no difference in school failures between these two categories. Nonmigrants were further ahead in their formal education, but this was, no doubt, due to the fact that they were somewhat older. School dropouts, however, were most frequent for the established grouping.

Only one nonmigrant is an only child. Established families average two children more than the migrant families. Although migrants speak Spanish significantly more than nonmigrants, there is no difference in the reading of Spanish newspapers. Reading appears to be more influenced by social class than ethnicity since there is also very little reading of English newspapers by both categories of adolescents.

Findings and Discussion

Although significant differences between categories do not appear on attitudes concerning family, father, mother and Mexican cultural patterns (e.g., "control over environment," "sickness as punishment," and "present-time orientation") large differences eixst in attitudes toward premarital sex. While migrant adolescents, somewhat more strongly, feel premarital sexual relations are appropriate for boys, nonmigrants are significantly more willing to suggest premarital sexual behavior for girls. The lower class established adolescents do not have a dual standard for premarital sexual relations while the immigrant category maintains a dual standard.

Nonmigrants view formal education as a means for upward mobility while migrant adolescents less frequently see education in this manner. "Getting even" for perceived injustices and "turning to authority" when difficulties arise are significantly more important to migrants than established adolescents. Established adolescents also see employment as a means to success. Significantly fewer migrants see work in terms of future orientation but work is viewed by these adolescents as a present-oriented phenomena.

[286] BEHAVIOR IN NEW ENVIRONMENTS

There is also a surprising sophistication within both categories concerning mental illness. Most of these youngsters view mental illness as a "misfortune," an illness which "can be cured," and something for which "one should not be punished."

Significantly more migrants than nonmigrants desire to live in Mexico. Somewhat fewer migrants than nonmigrants want to continue to live in Los Angeles. The gang as a mechanism of identity is significantly more important to established adolescents than to migrants.

When there is a "problem" both migrant and nonmigrant adolescents see their parents seeking assistance in the following order: (a) family; (b) God; (c) priest, policeman and doctor; (d) friends and social worker; (e) curandera. However, when seeking help with mental illness migrants feel their parents would turn to: (a) God; (b) physician; (c) priest; (d) family and psychiatrist; (e) friend; (f) social worker; (g) policeman. Nonmigrants, on the other hand, see their parents using: (a) physician; (b) God; (c) family; (d) psychiatrist; (e) priest; (f) social worker; (g) friend; (h) policeman. These data indicate that migrants are less likely to use traditional American mental health resources than nonmigrants. Apparently Mexican cultural heritage plays a more important role than social class when dealing with disordered behaviors designated as mental illness.

Table 1 compares migrants and nonmigrants as each category evaluates self with reference to other meaningful Mexican and American roles. Each plus and minus indicates the direction of the significant difference between self and other. Squares which are blank indicate no statistically significant difference. Migrants more frequently (twenty-four) than nonmigrants (seventeen) differentiate themselves from other role categories on potency scales. These data indicate the self-determination concerns of migrants. Evidently nonmigrants are less concerned, deny or have dealt with their feelings of control over their lives.

Nonmigrants are more concerned than migrants with evaluative responses. Viewing the world in terms of good and bad is important to both categories; however, nonmigrants are more likely to differentiate between self and others in evaluative terms.

Activity scales do not differentiate migrant adolescents from the established category.

While migrants see little difference between their fathers as they are and as they wish them to be, they view their mother as extremely different from what they wish her to be. The established category, on the other hand, reverses this perception by indicating high congruence between what mother is and how they wish her to be while fathers lack this congruence to a high degree.

These data raise an interesting question as to sex role differentiation and confusion for migrants and nonmigrants. Migrant adolescents who

TABLE 1

SIGNIFICANT DIFFERENCES (T > 2.0) ON NINETEEN ADJECTIVE SCALES WHICH COMPARE THE
MANNER IN WHICH MIGRANTS AND NONMIGRANTS "SEE MYSELF" AND THE MANNER IN WHICH THEY
VISUALIZE A SERIES OF MEXICAN AND AMERICAN ROLES

			Migrants (N = 41) Females (N = 15) Males (N = 26)													Nonmigrants (N = 48) Females (N = 27) Males (N = 21)												
P = Potency E = Evaluation A = Activity	(+) or (−)		Mexican	Anglo	Negro	Bullfighter	Police	Priest	Crazy Person	God	Babies	Father	Father as I like him to be	Mother	Mother as I like her to be	Mexican	Anglo	Negro	Bullfighter	Police	Priest	Crazy Person	God	Babies	Father	Father as I like him to be	Mother	Mother as I like her to be
P	Brave or Cowardly		+			+	+	+	−	+		+	+	+	+			−		−	−	−		−		+		
E	Unfriendly or Friendly							−	+	−	+					+	+	+		+	−	+		+			+	+
E	Sober or Drunk							+	−	+	+				+	−		−			+	+			+		+	+	+
E	Sinful or Virtuous								−	−	−		−		+	+				−	+	−		+	+			+	+
E	Dark or Light		+	+	+											+	+	+									−	
A	Rash or Cautious						+		+	−						+	+	+		+		−		+				
E	Proud or Humble									−							−	+	−		−		−		+			−	
A	Excitable or Calm		+			+		+	+										−				+		+				
P	Weak or Strong					−			+	−	+	+	+	+	+	+					−		+	+	+				−
A	Following or Leading		−	−				−			+		−	+	−			−			−			+	+			−	
A	Competitive or Cooperative		+	+			+	−	+	−							+	+		+		+	−	+					
E	Inferior or Superior						−	−	+						−										+				
E	Beautiful or Ugly		+					+		+	+	+	+	+	+	−	−	−					+	+	+		+	+	+
P	Unintelligent or Intelligent		−					−	+	−		−	+	+	−								+				+		−
E	Sensitive or Insensitive					+			−	+		+	+											−					
A	Active or Passive		−	+					−				+	+	+		−	+			−		−	+	−	−	+		
E	Wise or Foolish						+	+	−	+	+		+		+		+	+		−	−	+	−	+		+	+		
P	Hard or Soft		−			+		−	+		−														−	+			
E	Kind or Cruel		+	−					−	+					+						−		−	+	−		+		+
Responses			9	7	1	5	4	10	16	16	6	6	7	3	11	4	6	9	1	9	6	12	6	15	3	10	6	7

see their fathers as they desire them to be but do not view their mothers in a similar fashion may be responding to the Mexican mother's "lack of fit" in terms of behavioral prototypes in American culture. Fathers meet the required behavior of these adolescents. Apparently, fathers of migrants are able to carry out role expectations of their children while mothers are not.

Migrant adolescents view themselves as most highly different from "crazy person," "God," "priests," and "Mexican." Adolescents of the established category see themselves as highly different from "babies," "crazy person," "father as I would like him to be," "police," and "Negro."

Differentiation of self from others for nonmigrants appears to be based on an age-masculinity complex, strangeness, father's desired behaviors and lower-class American culture stereotypes. However, migrants differentiate themselves on the basis of strangeness, religion, Mexican culture.

While nonmigrant adolescents highly identify themselves as being least different from bullfighter, father and Mexican, the migrant category see themselves as most similar to Negro, mother, and police. Nonmigrants see little difference between self and those whose role includes highly aggressive, masculine and proud behaviors while migrants identify more closely with less potent, supportive yet aggressive categories.

During this process of acculturation, migrant Mexican American adolescents in search of self and adaptively meaningful behavior, utilize father, religion and Mexican culture as positive value orientations. Migrants not only place high value on these items but they also view representatives of these values as being different from self in a culturally positive direction. As the acculturation process extends through two or more generations, a major shift in value orientation takes place. Nonmigrants in their search for a positive identity utilize inadequacy and strangeness as self-differentiating factors. In other words, babies, crazy persons, father as I would like him to be, police, and Negro are all viewed by the established adolescent as people not like himself. Adolescent migrants utilize culturally positive value orientation representatives in differentiation of self, while nonmigrant adolescents adhere to culturally negative value orientation representatives as a comparative reference group. This major switch from a positive value orientation reference group to a negative value orientation reference group probably has significant behavioral adaptive functions. Seeing oneself as different from, yet working toward, behavior which has positive meaning in the culture increases interest in behavior and behavioral alternatives which support sociocultural expectations. On the other hand, viewing oneself as different from, yet focusing upon, representatives of negative value orientation, does not provide role models which support acculturative behavior with sociocultural expectations.

Although intrapsychic mechanisms assisting successful adaptation are not delineated, this research indicates that most recent migrant Mexican American adolescents tenaciously hold Mexican ideals and values by viewing them as desirable goals. However, the established category see themselves as extremely similar to the role categories displaying Mexican American values. The impact of United States society either reduces the importance of Mexican values as a personal goal or more established adolescents view themselves as a part of these goals for adaptive reasons. In other words, after one or two generations in the United States it may be adaptively necessary for Mexican American adolescents to overly view themselves as highly Mexican in order to defend against the "cultural stripping" process of American society.

To maintain the values and seek the goals of Mexican culture while learning the culture of the United States may be the most adequate mechanism for adapting. However, as American culture consistently denies the importance of Mexican values, established adolescents are forced to overidentify and to some degree identify with the most visible or masculine aspects of the culture (bullfighter, father and Mexican). This overidentification has in the past been a deterrent to upward mobility, acculturation and assimilation. According to others (Meeker, 1964) moving up for Mexican Americans has meant leaving the ethnic community, changing names and assimilating into the Anglo-white community. Apparently, the overidentification with Mexican values is functional to identity maintenance for established Mexican American adolescents, but not necessary for new immigrants. However, it appears as though this overidentification has been dysfunctional to upward mobility and acculturation.

For these Mexican American adolescents, adaptive behavior learned at the onset of migration becomes maladaptive during succeeding steps in the migratory process. If adolescents, their families, or society's institutions cannot provide adaptive techniques for rapid change during the several generations of the migratory process, then the ghettoes of East Los Angeles with their social disorganization will be perpetuated.

REFERENCES

BRODY, E. B. (1966) "Cultural exclusion, character and illness." American J. of Psychiatry 122: 852-858.

DERBYSHIRE, R. L. (1968) "Adolescent identity crises in urban Mexican Americans in East Los Angeles." Pp. 73-110 in E. B. Brody (ed.) Minority Group Adolescents in the United States. Baltimore: Williams & Wilkins.

――― (1966) "Cultural exclusion: implications for training adult illiterates." Adult Education (October): 3-11.

――― and E. B. BRODY (1964) "Marginality, identity and behavior in the American Negro: a functional analysis." Sociology and Social Research 48 (April): 301-314.

ERIKSON, E. (1950) Childhood and Society. New York: W. W. Norton.

GAMIO, M. (1931) The Mexican Immigrant: His Life Story. Chicago: Univ. of Chicago Press.

GUERRERO, R. D. (1955) "Neurosis and the Mexican family structure." American J. of Psychiatry 112: 411-417.

HELLER, C. S. (1966) Mexican American Youth: Forgotten Youth at the Crossroads. New York: Random House.

HUMPHREY, N. D. (1946) "The housing and household practices of Detroit Mexicans." Social Forces 21 (4): 433-437.

LEONARD, O. and C. P. LOOMIS (1941) "Culture of a contemporary rural community, El Cerrito, Mexico." Rural Life Series, No. 1. Washington, D.C.: U. S. Department of Agriculture, Bureau of Agricultural Economics.

LEWIS, O. (1949) "Marriage and the family: husbands and wives in a Mexican village: a study of role conflict." American Anthropology 51: 602-610.

MARDEN, C. F. (1952) Minorities in American Society. New York: American Book.

McWILLIAMS, C. (1949) North From Mexico. (People of America Series) Philadelphia: J. P. Lippincott.

MEEKER, M. (1964) Background for Planning. Los Angeles: Welfare Planning Council.

PEAK, H. M. (1958) "Search for identity by a young Mexican American." In G. Seward (ed.) Clinical Studies in Culture Conflict. New York: Ronald Press.

SANCHES, G. I. (1943) "Pachucos in the making." Common Ground 4 (Autumn): 13-20.

SCOTCH, N. A. (1963) "Social change and personality: the effects of migration and social mobility on personality." Pp. 323-330 in N. J. Smelser and W. T. Smelser (eds.) Personality and Social Systems. New York: John Wiley.

SEWARD, G. (1964) "Sex identity and the social order." J. of Nervous and Mental Disease 139: 126-136.

――― [ed.] (1958) Clinical Studies in Culture Conflict. New York: Ronald Press.

SPIEGEL, I. P. and F. R. KLUCKHOHN (1954) "Integration and conflict in family behavior." Group for Advancement in Psychiatry, monograph 27.

TAYLOR, P. S. (1932) An American-Mexican Frontier. Chapel Hill: Univ. of North Carolina Press.

THOMAS, W. I. and F. ZNANIECKE (1927) The Polish Peasant in Europe and America. (2nd. ed.) New York: Alfred A. Knopf.

WEAVER, T. (forthcoming) "Sampling and generalization in anthropological research on Spanish speaking groups." Paper presented at the American Ethnological Society, Detroit, May 3, 1968.

WILLIAMS, R. (1960) American Society. New York: Alfred A. Knopf.

Chapter **13**

Immigration, Migration and Mental Illness: *A Review of the Literature with Special Emphasis on Schizophrenia*

VICTOR D. SANUA

Since World War II, large population movements have occurred throughout the world. This has revived an interest in the effects of immigration on the psychological adjustment of immigrants, migrants, and refugees.

Two major theories have been proffered to explain the apparently high admission rates of immigrants to mental hospitals. One theory assumes that change of environment, with the ensuing problem of social and cultural adaptation, may cause psychological stresses which are reflected in hospital statistics. This is part of what Murphy (1961) has called the "general hazard theory." The other major hypothesis pertains to what is commonly referred to as the "self-selection theory," which presumes that those who are predisposed to mental illness are prone to immigrate.

During the early part of this century, most hospital statistics in the United States have shown that the rate of admission of the foreign-born was two to three times higher than the admission rate for the native-born. This led to the prevailing belief that mental illness is "racially determined." However, improved statistical procedures, which consider the age and sex distribution of the population, and later, social class and occupational structure, have narrowed the differences between the foreign-born and native-born.

To illustrate the prevalent "scientific" convictions of the time, we shall quote Salmon (1913) on the subject of the relationship between mental illness and race:

This is particularly true of mental diseases, for if racial characteristics profoundly affect political, social and religious ideals, we must look for a similar influence upon the individual makeup which so largely determines trends in mental disease. All those who are familiar with mental diseases among the Japanese in California testify to the remarkable tendency to suicide in that race, not only in depressed conditions but in conditions in which suicidal tendencies, in other races, are not frequent. This is in accordance with the general attitude of the Japanese toward self-destruction. The strong tendency to delusional trends of a persecutory nature in West Indian Negroes, the frequency with which we find hidden sexual complexes among the Hebrews and the remarkable prevalence of mutism among Poles, even in psychoses in which mutism is not a common symptom, are familiar examples of the influence of racial traits upon mental diseases.

J. V. May, the Superintendent of the Boston State Hospital, and Chairman of the Committee on Statistics of the American Psychiatric Association, had the following to say about immigration to the United States, in a book published in 1922:

While it must be conceded that we are indebted to European countries for much that has been contributory to the welfare and success of American institutions, it is equally true that the tremendous increase in mental diseases and defects here is to be attributed to no small degree to immigration.... Section 19 (Act of Sixty-Fourth Congress, 1917) provided that any alien "who within five years after entry becomes a public charge for causes not affirmatively shown to have arisen subsequent to landing" shall, upon warrant of the Secretary of Labor, be taken into custody and deported. The act also made provision for the first time for a literacy test which has been a subject of discussion for years. These amendments are of far-reaching importance and will eventually undoubtedly afford the hospitals considerable relief. The fact still remains, however, that the individual states are expending millions of dollars annually for the care and maintenance of an alien population which should have been excluded by the federal government. Under these circumstances it would seem nothing more than fair that the states should be reimbursed for the cost of carrying a burden for which they are in no way responsible.

All population movements can be subsumed under the general concept of geographical mobility. The following is an effort to classify the different types of geographical mobility:

(1) *Residential mobility* – This represents a change of home within a circumscribed geographical location, such as a county, city, town, village, etc. (2) *Internal migration* – Some arbitrary criteria are used to indicate internal migration, usually movement from one state or province, to another. Under this rubric urban-rural and rural-urban movement could also be included. (3) *External migration* – This could be movement from one county to another, usually in the same continent. Or, it could be international migration from one continent to another, usually involving the crossings of oceans, such as Europeans settling in Australia, or coming to the United States.

A very rigid adherence to these definitions may lead to methodological problems in research. For example, as pointed out by Malzberg (1967) there may be some question about calling an individual from New Jersey who moves to New York an in-migrant, while such a label could not be given to an individual coming from rural up-state New York to New York City. The problems of adaptation in the move of the latter individual involve a much greater degree of social change than for an individual who simply crosses the Hudson River. Another methodological problem in trying to define a migrant or immigrant is raised by the fact that some may be moving voluntarily, others involuntarily. Nevertheless, they are often included in the same category.

Most of the studies in this review use admission rates to mental hospitals as the criteria to measure degree of adjustment. Since there are so many diagnostic categories, we have decided to consider the overall total hospital admission rates and one major psychosis, schizophrenia. It was felt that the use of this diagnostic category would be more revealing, in view of the usual tendency to connect immigration with schizophrenia. We expect also to include a number of clinical studies in this review.

RESIDENTIAL MOBILITY

It has been hypothesized that prepsychotic individuals tend to leave their own community in order to settle in certain areas of the city. Such geographical mobility would naturally inflate the figures on hospital admissions coming from these areas. This was one of the major issues raised in Faris and Dunham's study (1939) which was based on incidence rates in Chicago. They discovered that the downtown rooming-house districts provided a greater share of schizophrenia. They also found that schizophrenics tended to reside in central areas while manic depressives were more widely distributed. Both geographical and social mobility may be involved in what has been called the "drift hypothesis."

However, at that time, Faris and Dunham felt that the nature of living conditions in the neighborhood, particularly extended isolation, produces the abnormal traits of behavior and mentality and thus, that the urban conditions may have been the cause of a higher rate of schizophrenia, while manic depressive psychosis was caused by psychological factors of family life.

One of the early studies in the United States, which used the survey approach, and included a recognition of the relationship between mental illness and mobility, was conducted by Tietze, Lemkau, and Cooper (1942). While the Faris and Dunham studies relied exclusively on hospital admissions, and used the "area" as an ecological variable, Tietze et al. used prevalence rates based on records of public and private hospitals, clinics, social agencies, with the household used as a unit of analysis. They observed that intracity migrants rather than in-migrants from other communities had an excess in the prevalence of personality disorders. They suggested two interpretations for such a relationship: either families on the move have adjustment difficulties, or families with a tendency toward mental deviation change residence oftener than more stable families. They concluded that the truth is probably somewhere between these extremes.

In the following studies, which are more clinical in their approach, an effort was made to examine the mobility, prior to their committment, of schizophrenic patients in particular hospitals.

Gerard and Houston (1953) conducted a study at the Worcester State Hospital which revealed that certain central urban areas "attract" schizophrenics, rather than "breed" them. The investigators divided their cases into those who were "in a family setting" and those who were "out of family setting." While the first group was distributed in the city at random, the latter came largely from the central zones. They suggested that schizophrenics tend to drift into the deteriorated zones where they lodge in cheap rooming houses, and thus seem to segregate themselves.

Hare (1956a, 1956b) in Bristol, England, also found evidence in favor of the "drift hypothesis." He analyzed the life histories of 64 schizophrenics in "out of family settings in Bristol. In almost fifty percent of the cases he found difficulties in interpersonal relations had caused the *separation*, while *isolation* was considered an important causal factor in one-fourth of the cases. Thus Hare believes that both prepsychotic drift and social isolation play a part in the increase of schizophrenia.

Lapouse et al. (1956) studied the residental mobility of first admission schizophrenics from the city of Buffalo to see whether they had the characteristic of either having drifted from a higher status or having recently moved to their present neighborhood prior to hospitalization. Comparing these schizophrenics to a control group, they found no evidence that "drift" could account for the concentration of schizophrenics in a low-income neighborhood.

Contrary evidence regarding higher residential mobility of schizophrenics was obtained by Lystad (1957) in a study of the hospital population in New Orleans. In comparing a group of schizophrenics and nonschizophrenics, she found that the schizophrenics, besides having a low upward mobility, *show less geographic mobility*. The conclusion was that the nonmentally ill, who are in the middle class, are white, younger, and have more than a grammar school education, show more mobility than do schizophrenic patients having these same characteristics.

Plank (1959) also studied geographic mobility of schizophrenics among a group of U.S. veterans. He used a control group that was slightly maladjusted, with a ten percent or less service-connected disability for psychoneurosis. He included a consideration of residential and social mobility in his study. Plank clearly demonstrated that schizophrenics do not tend to move into lower class neighborhoods; they (seventy-five percent) tend to remain in the same residential level. On the other hand, in the control group, approximately forty-five percent had changed in residential status, mostly upwards. In this study, residential status was considered at two points in time; residence at the time of induction and at the time when study was undertaken. Thus, schizophrenics tend to remain on the residential level on which they are, while other people move upwards.

Freeman et al. (1960) challenged the interpretation that mobility of the families of the mentally ill is "pathological." In the families of a group of psychiatric patients, they found that forty percent had lived in their present residence for at least ten years; twenty percent had moved once during ten years. Clausen and Kohn (1960) in Hagerstown, found that schizophrenics changed their residence and their jobs more frequently than did normal people. However, they considered this to be a symptom of the illness rather than an antecedent.

It would seem that while Tietze et al., Gerard and Houston, and Hare's data strengthened the hypothesis of residential movement of mental patients within the city, Lapouse, Lystad, Plank, Freeman, and Clausen and Kohn did not find the schizophrenic more geographically mobile; if anything, he was found to be less mobile.

Two recent large-scale epidemiological studies tend to weaken the hypothesis relating to the positive relationship between residential movement and mental illness.

Hollingshead and Redlich (1958) obtained residential histories of 428 patients. Their finding clearly shows that schizophrenia patients committed to mental hospitals from slum areas do not "drift" there as a result of their illness.

The other study by Leighton et al. (1963) controlled for the age at which residential movement occurred, arbitrarily using residential movement prior to the age of twenty, and after the age of twenty. They found that there was a strong association between the prevalence of psychiatric disorders in the normal sample of the population and the number of moves prior to the age of twenty (at least three moves). No relationship was found when the moves took place after the age of twenty. Thus, the data would seem to indicate that residential movement during childhood and adolescence may have some deleterious effect, but that "self-initiated" moves after the age of twenty are not related to mental disorders. Assuming that the moves in the majority of the cases less than age twenty were not voluntary, the conclusion may be drawn that during one's formative years several changes of environment represents a mental health risk.

Thus on the basis of present evidence, no conclusive generalization has emerged on the relationship between residential mobility and mental illness. Probably some of the inconsistencies in these findings are attributable to the lack of uniformity in the methodologies used in these studies.

MIGRATION, IMMIGRATION, AND MENTAL ILLNESS IN THE UNITED STATES

This section will deal with moves that go beyond a specific, narrow locale; that is, migration from one state to another, and migration between countries. New York State represents an ideal place to study morbidity differentials, in view of the large number of in-migrants and foreign-born. For more than forty years, Malzberg has been providing us with statistics on admission rates in the State of New York. His papers have become classics in the field of epidemiology.

Studies on the White Populations

Comparing the morbidity rates of native-born migrants in the State of New York with native-born, nonmigrants, and also foreign-born, Malzberg and Lee (1956) found that the migrant, in general, had higher rates of admission. Among the nonwhite, the differences were even higher. The following table gives the annual rate of admissions for the three white groups.

TABLE 1

RATES, PER 100,000, OF FIRST ADMISSIONS TO ALL MENTAL HOSPITALS IN NEW YORK STATE, 1939-1941, FOR THE NATIVE-WHITE POPULATION

	Born in N.Y. State		Born Outside N.Y. State		Foreign-Born	
	Male	Female	Male	Female	Male	Female
Schizophrenia	33[a]	30	52	57	51	48
Total Admissions	137[b]	115	216	199	162	154

[a]The standardized rates for schizophrenia are based on N.Y. population aged 20-59.

[b]Standardized on basis of N.Y. population age 20 and over. Both apply to morbidity rates given by Malzberg in the rest of the review.

Controlling for years of residence, they found that the migrant's rate of psychoses was twice as high as the nonmigrant within five years of his arrival in New York State. For those who had moved earlier than the five year period, the rate was one-fourth higher. It would seem that the first years after migration are the most difficult, since the data reflects the stress of recent change. Correction for age and sex reduced the differences, but still revealed that the migrant suffered from higher rates of morbidity. It should be noted that in these studies the total rates for the foreign-born are well below those for native-born in-migrants to New York State.

Malzberg (1962c) repeated the study of migrants and nonmigrants in comparing admission rates during the 1949-1951 period. Internal migrants had a total admission rate of 171.9 per 100,000, nonmigrants a rate of 135.4. The rates for schizophrenia were 53.9 and 35.4 per 100,000 respectively. For the admission rates in 1960-1961, Malzberg (1967b) found the figures to be 242.9 for internal migrants, and 158.0 for nonmigrants, and 64.8 and 40.1 for schizophrenia. Malzberg showed that the over-all rate has increased, with schizophrenia rates showing a marked upward trend.

In the 1939-1941 studies, Malzberg corrected for sex, age, urban-rural residence. He found that differences in rates between native born, nonmigrant or in-migrant, and foreign-born were quite reduced, except for schizophrenia where there was a substantial excess among the foreign-born males. No differences were found between foreign-born and native-born in the other psychoses. However, foreign-born females continued to show higher rates than native-born females in all psychoses. Other studies, Lemert (1948) in Michigan and Clark (1948), found that the foreign-born in general were overrepresented in the morbidity rates. Malzberg points out that prior to the forties it was believed that immigrants, because of poor biological stock, were more prone to mental illness. In other words, it was believed that Europe was dumping its defectives through immigration to the United States. However, when it was later found that native-born Americans of foreign-born parents had a morbidity rate closer to the total population, it was assumed that because immigrants had been subjected to more social stress in their adjustment to the new world, they were more likely to suffer breakdowns.

Since New York State is atypical, with the largest city in the United States, it is conceivable that special conditions in the state make for a wide difference between migrants and nonmigrants and a narrower one for foreign-born and native-born.

We shall now examine extensive studies that were conducted in Ohio and California. Locke et al. (1960), like Malzberg and Lee, found in Ohio that migrants into the state had higher rates than nonmigrants, and foreign-born had higher rates than native-born. The following table summarizes their findings for total admission rates.

TABLE 2

FIRST ADMISSION RATES PER 100,000, TO OHIO PUBLIC MENTAL HOSPITALS, FOR THE WHITE POPULATION, 1948-1952

	Ohio-Born	Other States	Total Native-Born	Foreign-Born
Male	93.3	103.7	98.2[a]	119.4[a]
Female	70.5	78.2	77.6[a]	111.3[a]

[a]For urban areas.

Comparing the admission figures between New York and Ohio, several differences become evident. First, rates in New York seem to be higher than in Ohio. While the largest discrepancy in New York is between nonmigrant and migrant, we find that in Ohio, the largest discrepancy is between foreign-born and native-born. However, Locke et al. indicate that even if the rates for the foreign-born were similar to the native-born, this would not necessarily be evidence that there are no differences. According to Locke et al., there are a number of forces that would tend to reduce the rates of the foreign-born, and other forces that would tend to increase them. Lack of family, limited language facility, economic problems, etc. would tend to increase the rates. On the other hand, it might be possible that foreign-born attitudes toward mental hospitals would tend to keep some away, unless they were extremely ill. However, such variables cannot enter into the discussion when rates of morbidity for native-born migrants are considered.

Malzberg (1964c) conducted a similar analysis with the admission rates in Canada for the years 1950-1952. He likewise found a significant difference between migrants and nonmigrants. The rates were 167.9 and 109.3 (total admissions) per 100,000, and 40.4 and 23.4 (schizophrenia) per 100,000, respectively. These discrepancies remained even when the native population of a state was compared with former residents who had moved elsewhere. With regard to the foreign-born in Ontario, Canada, Malzberg found that the over-all rate for the foreign-born is four percent higher than the native-born rate. More recent immigrants seem to have fewer problems than the old immigrants. Immigrants from the British Isles have rates which are less than the average native-born Canadians. The rate was high amoung those who had emigrated during 1931-1940. Malzberg assumes this is due to severe stresses incurred by refugees prior to their emigration. In contrast to the total rates, the rates for schizophrenia went up with the more recent arrivals. Malzberg attributes this increase to the higher percentage of non-English-speaking elements in recent immigration, compared to past immigrants who came primarily from the British Isles, and whose adjustment to Canada presented fewer problems.

In a somewhat different research approach, Wilson, Saver and Lachenbruch (1965) found that there was two and one-half times more mental illness among 322 migrants who had established new residence in Los Angeles County. The rate was particularly high for those migrants who had changed residence within a year. This study, although conducted on a small scale, confirmed Malzberg's findings in New York. However, Wilson et al. defined a migrant as anyone who had moved into Los Angeles County, while in the New York study a migrant was anyone coming from another state. In the Los Angeles study the criterion used for morbidity was seeking psychiatric help, while in New York the criterion was hospital admission. These differences may explain the different ratios found between the migrant and the nonmigrant in the New York and Los Angeles studies.

In his original data, Malzberg did not control for education and socioeconomic status; it was believed that a control of these variables would greatly influence the rates. The following represent efforts to conduct comparative studies with a reanalysis of the data.

In 1963, under the sponsorship of Milbank Memorial Fund, three studies on migration differential in New York, Ohio and California, by Lazarus et al. (1963), Thomas and Locke (1963) and Lee (1963) were published. Lazarus et al. analyzed Locke's data for Ohio on migration status by age, sex, and color or race, in relation to similar data for New York and California. The Thomas and Locke study controlled for socioeconomic differentials in mental disease, marital status, education by age and sex, for Ohio and New York. The Lee study cross-classified the New York admissions—population data by marital status, and education (by sex) and occupation of employed males—each by age, color or race, and migration status, and thus tested the importance of intervening variables.

Lazarus et al. pointed out that the ethnic, cultural and experiential differences of the three states are offset by certain basic structural similarities, such as large populations, urbanness, high degree of industrialization, and relatively high average income level. They found that for all "disorders" the relative rates differ only slightly, state by state, for foreign-born compared with native whites, and that the difference is not always toward the same direction. In Ohio and California the male foreign-born have a slightly favorable differential, their rates being ten to thirteen percent below those of native whites, whereas among females in New York and Ohio, a differential of the same order favors the native-born. However, if schizophrenia is considered, the foreign-born white, in five out of six possible comparisons, shows a marked unfavorable differential. The data on migration and nonmigration was more consistent, but color differentials showed the most consistent and striking findings.

Thomas and Locke in Ohio and New York, found that in spite of the controls, there was a clear inverse relationship between the rates of first admission for mental disease and occupation, in-migration, and status as measured by education.

Lee reached the conclusion that a control of the socioeconomic composition of the migrant and nonmigrant in New York State did not eliminate the differences between them. Here again, the foreign-born had intermediate rates compared to the nonmigrant and the migrant, the latter having the highest rates. However, there was a marked difference between the sexes. Foreign-born white females were closer to the native migrant in rates of morbidity.

While there seems to be general consistency in the foregoing studies, a study by Jaco (1960) in Texas, does not seem to agree. In general, he found that the migrant had approximately the same incidence of schizophrenia as the nonmigrant (26 per 100,000). In general, female migrants had higher rates than nonmigrant females (34 and 29 per 100,000, respectively) while the opposite held for the males (20 and 23 per 100,000, respectively). The rate of schizophrenia for the migrant Spanish American was found to be lower than that of the native-born Anglo American. Thus, it would seem that migration, particularly in the case of the Spanish American male, does not lead to greater morbidity rates. Jaco proposes a number of hypotheses to explain the fact that his study did not support Malzberg's conclusions, or those of subsequent studies:

(1) Malzberg included "transients" with the migrants. Rates might be lower if only migrants were included in the statistics.

(2) The 1939-1941 period studied by Malzberg included migrants who were the product of the depression. Jaco's own data was collected in 1950, when circumstances were more secure.

(3) Malzberg's rates were based upon indirect estimates of the age and sex distribution of the migratory population of New York, whereas Jaco conducted a direct adjustment of his rates.

(4) Conditions of migration in the two states are quite different. Migration is much higher, and more upwardly mobile members in-migrate to New York. Jaco feels that internal migration within the State of Texas might be possibly more an important factor than in-migration from another State.

Malzberg (1967b: 184-191) also tried to explain the discrepancy between his findings and those of Jaco. Jaco was able to identify the origin of only sixty-five percent of his patients, and eliminated the others. It is possible that a relatively larger percentage of cases whose origin could not be identified belonged to the migratory population. Furthermore, Jaco included foreign-born among the migrants, and this may have confounded the data.

The higher rates for migrants may be attributed to two sets of hypotheses, according to Malzberg: (1) that a higher proportion of schizoid types tend to move, and ultimately end up in mental hospitals, and (2) that the stresses incurred in migration may precipitate psychological disturbance. Certainly both of these factors may play a role. However, in view of the contrary evidence found in the Scandinavian studies (reviewed below), these hypotheses will be elaborated upon further.

As for differences between native-born and foreign-born, the trend shows higher rates for the foreign-born. Insofar as total rates are concerned, it would seem that control for socioeconomic variables tends to narrow the discrepancies, which sometimes disappear. In most instances, the foreign-born rates for schizophrenia exceed the rates for the native-born. In a later section of this paper we shall see that an analysis of the foreign-born according to national origin shows an inconsistent pattern of first admission rates, depending on the origin of the subgroups.

Studies of the Negro Population

The following table shows the incidence rates found by Malzberg and Lee (1956) in the Negro population of New York State.

TABLE 3

AVERAGE ANNUAL ADMISSION RATES TO ALL MENTAL HOSPITALS, PER 100,000 FOR THE NEGRO POPULATION (1939-1941)

	Born in N.Y. State Male	Female	Born Out-of-State Male	Female	Foreign-Born Male	Female
Schizophrenia	77	65	90	78	102	113
Total Rate	304	207	455	316	342	304

The out-of-state Negro has the highest rate of hospitalization. Compared with admission rates for the white population Negro rates were approximately twice as high. Malzberg (1963a) provided us with further admission rates for native, migrant and foreign-born Negroes for the years 1949-1951. The total standardized admission rates are respectively 256.0, 327.4, and 252 per 100,000. For schizophrenia, the figures are 77.1, 108.5, and 88.3 respectively. Malzberg (1965a) confirmed these findings in the most recent analysis of admission rates for 1960-1961. In general, there seems to be an over-all decrease of morbidity rates among Negroes, possibly because of an improvement of their economic status. Malzberg attributed the higher rates among the Negro migrants to the probability that they had a higher proportion of schizoid personalities and to an association of such characteristics with a drive towards geographic mobility. However, the foreign-born had moved from their place of birth, and still had a lower rate than the native-born Negro, although their rate for schizophrenia was higher.

Malzberg believes that the foreign-born Negroes, mostly from the British West Indies, have a background environment superior to the migrant Negroes, are better educated and experience less racial prejudice while growing up in their native land. This may explain their lower total overall incidence rates. The fact still is that their incidence rate of schizophrenia is very high compared to the other groups and therefore there is some justification to relate immigration to a higher rate of schizophrenia but not necessarily to a higher overall rate of morbidity. It is also to be noted that females in practically every instance have lower incidence rates than males. In two instances, however, the rates for females are higher than males in schizophrenia. This is found among white females in-migrants and Negro female immigrants. Thus it would seem that migration and immigration represent a more difficult process of adjustment for certain types of females.

The explanation given by Malzberg regarding the lower rates for foreign-born Negro males may be questioned. If the Negro from the West Indies was brought up and raised with little discrimination, his experience in the United States should be traumatizing; coming from a culture which accepts him, he nonetheless finds himself sharing the onus of the prejudice which prevails in the United States. The relatively high rates for Negroes immigrating to England as compared to the native white English population is attributed to the fact that they suffer from discrimination in England.

Having in mind the differences in ethnic composition of the population in England and the United States, a comparison of the adaptation of the West Indian Negro living there would be of great interest. Different selective factors, and further differences in identification would tend to operate. The Negro from the West Indies may not identify himself with the native-born American Negro, but may feel a strong sense of identity with his own group. Another factor not to be neglected is the attitude of the dominant white group, which may be less prejudiced towards a Negro coming from the Islands than toward a native-born Negro.

In Ohio, Locke et al. (1960), like Malzberg, found that the Negro born out of the state had higher admission rates. The following table gives us incidence rates.

TABLE 4

TOTAL ADMISSION RATES PER 100,000, OF NEGROES,
MIGRANTS AND NONMIGRANTS, TO MENTAL HOSPITALS
IN OHIO, 1948-1952

	Born in Ohio State	Born Outside Ohio
Male	178.0	218.3
Female	124.0	155.8

We have already mentioned Jaco's study (1960) which found contrary evidence about the higher admission rates for the migrant. Jaco also found that nonwhite migrants had a lower rate of psychoses than nonwhite, native-born in Texas (39 and 44 per 100,000 respectively). Another striking difference between the rates in Texas and in New York is the higher rates found in the latter state, despite the fact that Jaco included patients seen by private psychiatrists and in private hospitals. For example, Jaco's incidence rates of schizophrenia among native-born nonwhite males is 25 per 100,000. Among the nonwhite migrants it is 21 per 100,000. These figures contrast sharply with the New York incidence rates which are three to four times higher.

Another study which does not corroborate Malzberg's general findings is one conducted by Kleiner and Parker (1959), Parker, Kleiner and Taylor (1960), and Parker and Kleiner (1966) among Negroes in Philadelphia. They found that the Southern-born Negro migrant was underrepresented in hospital statistics, while migrant Negroes from the Northern parts of the United States and Negroes Philadelphia-born were overrepresented. Furthermore, Negroes coming from other States in the North had more mental illness than Negroes born in Pennsylvania. To account for these differences, they hypothesized that they might be attributed to the discrepancy between the level of aspiration and goal achievement. In other words, the Southern-born Negro may be satisfied with less desirable jobs, whereas among native-born Negroes, the level of aspiration was found to be higher than their level of accomplishment. Kleiner and Parker also found that the mentally ill had a greater number of intracity residential moves.

Kleiner and Parker (1968) conducted further analysis of their data, and found that the Northern Negroes had relatively higher levels of upward and downward social mobility, besides having an intensive goal-striving behavior and weak and ambivalent attitudes towards their Negro group membership. They felt, in a sense, that the Northern Negro was more "marginal" than the Southern Negro, according to the Lewinian hypothesis.

Thus, while Malzberg found that out-of-state migrants have higher rates of admissions for mental illness, Jaco found the opposite. In Pennsylvania, when the early environment is controlled (the first seventeen years of socialization), the results show that the point of origin of the migrant is important. The findings with the migrant Northern-born Negro coming to Philadelphia agree with Malzberg's findings in New York State, while the findings with the Southern-born Negro migrant moving to Philadelphia agree with Jaco's findings in Texas. We expect to discuss these discrepancies further. Malzberg is inclined to explain his findings on the basis of selective migration and "cultural shock" while Kleiner and Parker espouse social-psychological variables, such as the discrepancy between the level of aspiration and the level of accomplishment, to explain the discrepancies in morbidity rates.

Studies on the Puerto Rican Population

Puerto Ricans who come to the continental United States are technically in-migrants, but in view of the disparity between the cultural environments, their move to the mainland could be considered as international migration. Malzberg (1956b) found that fifty-eight percent of Puerto Rican patients admitted to New York City mental hospitals (1949-1951) were diagnosed as schizophrenics, as contrasted to twenty-nine percent for non-Puerto Ricans. This is the highest percentage of schizophrenia of all groups in the United States. One of the explanations for this high percentage of schizophrenia is that the Puerto Rican population is relatively young. The corrected incidence rate was found to be 99.4 per 100,000 (males, 114.3 and females, 85.6) as compared to 55.6 per 100,000 (males, 56.5 and females, 54.5) for the rest of the population. The total admission rates are respectively, 239.3 (males, 262.7 and females, 214.5) and 185.5 (males, 183.4 and females, 180.7) per 100,000.

In 1960-1961, Malzberg (1965a) found approximately the same rates—230.2 and 188.4 respectively. It should be noted that among Puerto Ricans, females had relatively a much lower rate than males, compared to non-Puerto Rican male and female patients. Standardized rates for Puerto

Rican males in New York City was 284.0 and for females it was 188.1 per 100,000, the excess being fifty percent as compared to only seven percent excess for the non-Puerto Rican male and female.

When the Puerto Rican rates were broken down according to nativity, the second generation, born in the United States, had a higher total rate than their parents (429.7 and 266.8 respectively, for males, and 220.3 and 179.3 for females). This is in contrast to other large ethnic and nationality groups in New York State, where most of the native-born had less mental illness than their immediate forebears. The Puerto Rican rates are also similar to the native-born and foreign-born Negro. However, in the latter case, they represent groups of different origin. While education and better economic background may tend to lessen mental illness among the foreign-born Negro (mainly from the West Indies, it would seem that better education and better socioeconomic status of the native-born Puerto Rican does not prevent a high rate of hospital admissions. These advantages, however, may be offset by problems of identification of the native mainland Puerto Rican and racial prejudice which may be less acceptable than for a Puerto Rican born on the island. Jaco (1960) in Texas, likewise found that the native-born Spanish American had a higher rate of schizophrenia than the Spanish American migrant, and that mental illness was positively correlated with education in this particular group. Furthermore, in Philadelphia Parker and Kleiner (1966) found that the Negro native-born had more mental illness than the Southern-born Negro migrant, despite the former's higher education. Thus, it would seem that a higher education for certain minority groups may be conducive to more mental illness, as in the case of Puerto Ricans, Spanish Americans in Texas and the more educated Negro in Philadelphia. On the other hand, the Negro coming to New York from the West Indies who has more education has less mental illness than the Negro in-migrant.

Comparing the rates of white and nonwhite Puerto Ricans, Malzberg (1965b) found, as expected, that the admission rates for the nonwhite Puerto Rican were higher than for white Puerto Ricans, 611.4 and 225.2 per 100,000 respectively. (Nonwhite Puerto Rican male, 730.3, female, 479.6 and white Puerto-Rican male, 269.4 and female, 180.9) It seems that the Negro Puerto Rican is worse off than any other group since he has three problems to cope: his color, the fact that he is Puerto Rican, and the stresses of in-migration or 'foreign birth.' The language handicap probably compounds the problems of adjustment.

The foregoing rates pertain to total admission rates for Puerto Ricans. However, when we compare admission rates for schizophrenia, we see some inconsistencies compared to the total rates of admission. The ratio of schizophrenia in nonwhite, male Puerto Ricans is 5.15, as compared to 1 for white males. There does not seem to be a difference between the combined rates of male and female Puerto Ricans born on the Island and those born on the mainland. A breakdown of the figures in New York State, based on sex, reveals that the male born in Puerto Rico has a higher rate of schizophrenia than the Puerto Rican born on the mainland (160.9 and 126.7 per 100,000, repectively), while the incidence rates for the females go in the reverse direction (125.0 and 151.4 respectively).

How could these discrepancies be explained? Earlier studies, particularly by Hollingshead and Redlich (1958) showed that schizophrenia is more prevalent among lower classes. This may explain the higher rates for the migrating Puerto Rican who belongs primarily to the lower socioeconomic group, compared to the mainland Puerto Rican. On the other hand, the better educated mainland Puerto Rican male has a higher total rate of admission, than the Puerto Rican Island-born male, but he has a lower rate of schizophrenia. One other possible explanation is that difficulties in communicating with only Spanish-speaking Puerto Ricans from the Island may make the psychiatrist more inclined to label his patients as schizophrenic. However, this does not apply to females. Mehlman (1961) has discussed the "Puerto Rican syndrome," a collection of various disease processes that tend to be superficially and deceivingly similar for different illnesses in the Puerto Rican culture.

With the female migrant from Puerto Rico, her initial adjustment may be facilitated by her job security, particularly in the needle trades, where she usually finds quick employment. Regarding the native-born Puerto Rican woman, a combination of high aspiration in a free society, and the old Spanish tradition of female subservience to the male, may present serious problems leading to schizophrenia.

Another variable which should enter into the interpretation of the data is the factor of "selectivity." Males and females may leave the Island for different reasons, and possibly different reasons may also prompt their return to the Island. It may be possible that females who are not capable of adjusting to mainland life may return more frequently to Puerto Rico than males. A clarification of the role that this selectivity factor plays on admission rates would require intensive longitudinal and follow-up studies. Rogler and Hollingshead (1965) in a study of Puerto Rican schizophrenics on the Island, reported that schizophrenia may bring about a reversal of roles. Males abandon the "machismo" pattern and become mild, meek, and defenseless, whereas the female becomes hostile, intolerant, and assaultive, which is the opposite of the expectation of the traditional culture. A similar phenomenon was reported by Schooler and Caudill (1964) when they compared the symptomatology of the Japanese schizophrenic with his American counterpart. The Japanese is more likely to be physically assaultive, while the American suffers a disruption of reality. They hypothesized that in Japanese society, which insists on decorum, a breach of the norm leads to hospitalization, while in United States' society, a lack of reality testing, which seems to be important, leads to hospitalization.

Regarding the relationship of geographic mobility within the Island and schizophrenia, Rogler and Hollingshead found no connection whatsoever between any kind of movement, either from one area to another, or residential movement and schizophrenia. We do not know of any studies which tried to find a relationship between mental illness and residential mobility of the Puerto Rican on the mainland.

Studies on the Jewish Population

Malzberg (1960) published a book on mental illness among Jews, based on the 1939-1941 admissions. As expected, he found that foreign-born Jews accounted for a disproportionate percentage of admissions. However, since he could not correct for age and education, this difference cannot be considered to be conclusive. Of the mentally ill, more than half (55.2 percent) were foreign-born, although it is estimated that foreign-born Jews represent only thirty-five percent of the total Jewish population in the United States. In general, it was found that Jews in New York State had a lower rate of first admissions for mental disease than non-Jews to all hospitals.

An analysis of admissions rates between 1949-1951 (Malzberg, 1963e) confirmed the 1939-1941 results. With respect to nativity among the Jewish patients, foreign-born Jews again appeared in more than their expected percentage of total first admission rates. However, in the case of schizophrenia, the native-born Jew, compared to the foreign-born, was overrepresented (seventy-six percent and twenty-four percent). The difference, according to Malzberg, is due to the fact that schizophrenia is more prevalent when the population is younger. However, no corrected rates are possible, since they cannot be computed in the absence of appropriate age statistics of Jews.

Malzberg (1963g) analyzed the admission rates (1959-1962) of Jews in Canada. The total admission rates for Jews and non-Jews were found to be 109.6 and 130.3, respectively. Malzberg believes that these figures are authentic. It is noted, however, that for Canada as a whole Jews have more admissions for schizophrenia than non-Jews. This is explained by the fact that schizophrenia is more prevalent among urbanized groups, and Jews are highly urbanized. When limited to Ontario, where the degree of urbanization is approximately the same for all groups, no difference in schizophrenia was found between Jews and non-Jews. Higher rates for Jews occur principally with repsect to manic depressive psychosis and psychoneurosis.

Studies on Other Nationality Groups
In the United States

The following tables represent a compilation of admission rates for a number of major nationality groups in the State of New York, as studied by Malzberg.

TABLE 5

AVERAGE TOTAL ANNUAL STANDARDIZED RATES OF FIRST ADMISSIONS TO ALL HOSPITALS FOR MENTAL DISEASES IN NEW YORK STATE, PER 100,000 OF DIFFERENT NATIONALITY GROUPS, NATIVE AND FOREIGN-BORN, 1949-1951

	Native-Born			Foreign-Born			
	Male	Female	Total	Male	Female	Total	
English	158.9	142.3	157.3	138.8	137.2	138.3	(1964e)
Irish	228.4	194.3	220.2	240.7	216.9	231.7	(1963d)
Norwegian	183.2	119.8	155.0	147.7	135.0	143.2	(1962a)
Swedish	131.2	152.0	140.4	229.0	144.6	188.2	(1962b)
Italian	119.2	102.8	114.6	146.2	130.4	141.3	(1963b)
French Canadian	175.7	160.2	175.4	141.4	91.2	117.1	(1966a)
Polish	151.9	146.1	155.5	167.3	207.6	191.3	(1963c)
Russian	160.2	146.9	157.1	169.8	153.0	164.1	(1963f)
German	144.2	147.2	152.4	157.3	175.5	169.4	(1964g)
All Foreign				168.2	180.5	178.7	
Native White	157.1	141.8	152.0				

In general, the total rates for the foreign-born were found to be higher than for the native-born. However, only in the case of the English, Norwegian, and French Canadians were the foreign-born rates lower than for the native-born. A similar pattern is found with rates in schizophrenia. Comparing the different nationality groups, the rates for the Italians and

TABLE 6

AVERAGE ANNUAL STANDARDIZED RATES OF FIRST ADMISSIONS FOR SCHIZOPHRENIA TO ALL HOSPITALS FOR MENTAL DISEASE IN NEW YORK STATE, PER 100,000 OF DIFFERENT NATIONALITY GROUPS, NATIVE AND FOREIGN-BORN, 1949-1951

	Native-Born			Foreign-Born		
	Male	Female	Total	Male	Female	Total
English	45.5	41.4	44.6	27.3	26.3	26.6
Irish	63.	52.1	59.0	71.2	50.1	60.2
Norwegian	53.8	35.1	45.4	24.4	29.3	26.7
Swedish	40.8	37.4	40.3	75.8	51.6	63.4
Italian	35.0	35.6	36.3	56.9	40.2	48.4
French Canadian	19.9	32.3	27.1	29.4	33.3	31.6
Polish	56.4	52.0	55.7	66.7	78.3	73.1
Russian	41.8	47.4	46.	40.9	41.1	65.4
German	40.1	41.8	42.0	58.0	58.3	58.2
All Foreign				57.2	50.3	52.7
Native White	41.8	40.8	41.3			

French Canadians were the lowest, with the Irish having the highest rates among both the native-born and foreign-born. The fact that English is the native language of the Irish does not seem to help them in their adjustment, while the Italians, despite their lack of familiarity with the English language, and their low socioeconomic status, seem to be better off than most other nationality groups, including the English and United States native-born.

Comparing the rates between males and females, in the majority of cases, females seem to have less mental illness. This seems to be in contradiction to earlier studies by Odegaard and Malzberg, who found that females usually seem to have a more difficult time when they immigrate. In the foregoing table, this generalization applies only to Polish and German-born females for total rates, and to Norwegian, French Canadian and Polish-born females for schizophrenia rates. Since Malzberg did not break down the foreign-born according to nationality in his early studies, it is possible that unequal numbers of females in New York State confounded the results.

The disparities between different nationality groups, native-born and foreign-born, indicate that combined rates for the foreign-born or even the second generation may be confounding, since it would appear that some foreign groups have less mental illness than United States native-born, and sometimes the rates are higher. Malzberg (1964a) calculated the rates of admission of New York hospital population according to nativity. The group with the lowest rate of mental illness was one of native parentage and the group with the highest rate was one of mixed parentage—that is, one parent foreign-born, and the other native-born. The rates when both parents were foreign-born fell between these two groups.

Other Major Studies in the United States

Roberts and Myers (1954) analyzed the distribution of mental illness among religious and nationality groups of the New Haven Study. Since the foreign-born tended to be older than the native-born, their analysis of morbidity rates was limited to those over 21 years of age. They found a significant difference in the distribution of native and foreign-born: there was a higher rate of affective disorders, illnesses of senescence, and organic ailments in the latter group. Psychoneurosis was found to be more prevalent among the native-born. They believe that acceptance of psychiatry by the native-born accounts for their high rate of psychoneurotics. It should be noted that morbidity rates in New Haven were based on treated prevalence and included patients seen by psychiatrists in private practice. The foreign-born population was broken down according to nationality. Italians were high on affective psychoses and illnesses of senescence; Irish were high on illnesses of senescence and addiction, but devoid of any psychoneurotic disorders; Poles and Roumanians were high on affective disorders and schizophrenia.

Gibbs (1962) tried to relate what he called noninstitutionalized indicators of psychopathology (suicide, death from alcoholism, narcotics violation, etc.) with rates of hospitalization in forty-eight states. In his article, he provided correlations for three states—California, Texas, and Mississippi. He found that the rates of hospitalization are not closely related to the actual amount of mental illness in the population as measured by noninstitutionalized indicators of psychopathology. However, using another index which he called Index of Socio-Cultural Alienation, he found a positive relationship between the latter, and rates of hospitalization. The Index of Socio-Cultural Alienation was arrived at by the addition of three criteria—percentage of nonwhite population, percentage of foreign-born, and percentage in lower-class occupations. It would be of interest for the purpose of this paper to partial out how the percentage of foreign-born had contributed to the total relationship.

Most of the data reviewed here shows a trend toward a higher admission rate for foreign-born. However, with control of demographic variables, such as socioeconomic status, education, urban-rural status, the differences are reduced, and sometimes, even disappear. On the other hand, it seems apparent that in-migration in the United States, with some exceptions, is related to higher morbidity rates, particularly in schizophrenia.

At this point, it would be of interest to review a discussion by Mintz and Schwartz (1964) on the differences between the morbidity of native-born and foreign-born. While many studies have shown that there is a positive relationship between social class and mental illness, studies on the relationship between ethnic density and psychosis have received scant attention. Furthermore, they feel that lumping all foreign-born into one group is inappropriate, that separate analysis for each ethnic or nationality group, as conducted by Malzberg in the last ten years, is more appropriate. Faris and Dunham (1939) had found that Negroes and whites not living in their own racial neighborhoods had a higher rate of mental illness than Negroes and whites living in their own neighborhoods. However, when Mintz and Schwartz reanalyzed Faris and Dunham's data, this reciprocal relationship did not hold for the native-born and the foreign-born. They found that the foreign-born, while having a higher rate of schizophrenia in a native-born area, had less mental illness than the foreign-born in all the other areas in the city of Chicago. To explain this inconsistency, Mintz and Schwartz indicated that all foreign-born cannot be considered as a homogeneous group. They do not see why the foreign-born Irish would benefit psychologically by living in an area predominantly inhabited by Puerto Ricans. What is to be considered is that each foreign-born population experiences its ethnicity differently, depending on clannishness, retention of native mores, integration with the dominant culture, etc. Mintz and Schwartz, in studying the incidence of schizophrenia of Italians in Boston, found that communities of higher "Italian" density tended to have fewer Italians admitted for schizophrenia. Communities which have a large number of foreign-born are usually low on a socioeconomic index. While there may be high rates for the combined ethnic population, some ethnic groups or nationality groups may have more favorable rates than others because of density, or other factors of integration which may help in maintaining the psychological health of the group. However, Mintz and Schwartz believe that in a homogeneous city with a stratified society socioeconomic factors may play a more important role, but that lower class community integration achieved by ethnic groups can relegate socioeconomic status to secondary importance.

AREAS OF RECENT HIGH IMMIGRATION

Following World War II, there were large scale movements of populations in different parts of the world, with persons either fleeing from politically undesirable conditions in the country of origin, or fulfilling the need to settle in a new country, to start a new life. Countries which accepted new immigrants were England, Australia, Canada, the Scandinavian countries, France, and Israel. We shall now review some of the research conducted in these countries.

England

West Indians

In recent years, England has been faced with such a serious influx of immigrants from the West Indies (former British colonies) that immigration limitations were recently introduced by the British government. A number of studies dealing with psychiatric illness among West Indians in London have been published.

Pinsent (1963), examining service carried out in general practice in Birmingham, found a higher rate of incidence for mental disorders among the English as compared with West Indian men. On the other hand, the rate of mental illness for Jamaican women was twice as high as the rate for English women.

Kiev (1964b, 1965) reported that, while the rates for severe psychiatric illness were not high among West Indians, he found, in a six-month survey in a group general practice, a high rate of psychiatric morbidity in both sexes of the lesser disorders. In general, schizophrenics among them displayed religious delusions accompanied by hallucinatory commands to preach and heal. Their beliefs were strongly related to the beliefs of normal West Indian migrants (Kiev, 1963, 1964a). Kiev believes that because of his social and cultural anomie, the West Indian turns to the Pentecostal sects which provide social acceptance for everyone and a socially acceptable way of releasing suppressed emotions and frustrations, which is very therapeutic for this particular group living under stressful conditions.

Gordon (1965) found a low incidence of psychiatric disorder prior to migration in a group of West Indian patients—only seven percent had been hospitalized prior to migration. He concluded that environmental factors were contributory to the illness. He found a low incidence of neurosis and personality disorder (six percent) and a prominence of schizo-affective illness (28.8 percent) usually of the reactive type, and schizophrenia (38.4 percent) with an excess of paranoid tendencies. This finding was in marked contrast to the patterns of admission of English patients in mental hospitals. Tewfik and Okasha (1965) likewise found a relative excess of paranoid illnesses among West Indians compared with English patients.

The differences between the findings of Pinsent, Kiev, Gordon, and Tewfik and Okasha may be due to the fact that the first two authors relied on psychiatric data in general practice, while the latter examined all mental patients admitted in mental hospitals. Hemsi (1967) using the 1961 census, found that in two boroughs in London the rates for first admissions for all psychiatric illnesses was considerably higher among West Indians than among the native population (9.5 males, 12.2 females per 10,000, and 31.8 males and 30.4 females, respectively). Regarding the diagnostic distribution, schizophrenia among the West Indian males was 13.1 per 10,000, and 5.1 for females. On the other hand, females suffered more from affective disorders, 16.8 per 10,000 (males 11.5). Contrary to Gordon's findings, West Indians had a higher rate of character disorders but fewer organic syndromes, drug or alcohol dependence than the native population. It is to be noted that twenty-five percent of the West Indian patients were ill on arrival or within three months of arrival. While Hemsi has some reservations about comparing morbidity rates in the West Indies and England, he found that the rates in England were much higher. According to Hemsi, lack of consistency in the findings of these studies may be due to differing criteria in the selection of patients and to differing diagnostic criteria. For example, Hemsi included first admissions only, whereas Gordon included admissions and readmissions. This is probably why Gordon found less personality disorder and more schizophrenia, as compared to Hemsi's data.

A most comprehensive study on mental illness among immigrant groups is presently being conducted by Bagley (1968a, 1968b) at the Institute of Psychiatry, in London. The following represent preliminary results, the most striking of which are the low rates of prevalence among immigrants from Cyprus and Malta, (9.9 per 10,000) and the high rates among immigrants from Africa, (71.01 per 10,000). In contrast to previous studies, the rate for Caribbean immigrants (16.21 per 10,000) was found to be similar to that for the British population. Higher rates were found among immigrants coming from the "old" Commonwealth (69.23 per 10,000). The rates for Indians and Pakistanis (33.85) and Irish (24.5) were relatively low.

Controlling for age, social class, and sex, it was found that the Irish display a marked dearth of schizophrenia, but were high on alcoholism and drug dependency. The author suggested that alcoholism may mask schizophrenia among the Irish. Both the Caribbean and African patients showed an excess of schizophrenia and a dearth of depression. It was suggested by Copeland (1968), who studied mental illness among African students in England, that paranoid beliefs may be fostered by objectively experienced racial discrimination. Copeland found that a large number of African students seen at Maudsley and Bethlehem Hospitals suffered from schizophrenia or "paranoid state." Four variables were considered by Bagley in relation to mental illness: (1) community integration, (2) status isolation, (3) color discrimination, and (4) selective factors. The hypothesis offered to explain low prevalence among the Cypriots and Maltese is that they have a well-developed community, do not suffer from racial discrimination, and have few selective factors in immigration. The Africans may be the victims of status isolation, color discrimination, lack of community integration, and the West Indians, while suffering from color discrimination have better community integration. The Irish seem to have been ill prior to migration and possess little community organization.

Some comparisons can be made between the morbidity rates of the West Indian immigrating to England and the foreign-born Negro, mainly from the West Indies, immigrating to New York. While most of the English investigators emphasize the difficulties of adjustment of the West Indian in England and his high morbidity rate as compared with the native white population, the rates of the black immigrant in the United States, when compared with the native-born black and the migrant black, are found to be lower except for schizophrenia. The assumption is made that these immigrants have fewer difficulties in adjustment to life in New York than the native-born and migrant Negro. However, the rate of morbidity of the black immigrant in the United States is still much higher than the morbidity rate of the white. This discrepancy found in the United States is much wider than the discrepancy in England between the West Indians and the English-born. Either self-selective factors in immigration or different environmental stresses in the two countries makes the West Indian who settles in New York less well adjusted than the West Indian who settles in London.

Hungarians

In Great Britain, Mezey (1960a, 1960b, 1960c) conducted a most intensive study of Hungarian refugees from the 1956 uprising. He found that the schizophrenics in the group he examined were different from other diagnostic groups among Hungarians. They had a higher record of migration from rural/provincial areas to urban Budapest, and a higher frequency of nonpolitical causes for leaving Hungary. Thus, the stresses of immigration did not play a causal role in schizophrenia. Mezey's data seems to support the "selection hypothesis" in the interpretation of differential incidence of schizophrenia among immigrants. As in the case of Hungarians who immigrated to Canada, Mezey found that somatic manifestations and aggressive outbursts were dominantly represented in the symptomatology. Again, the most frequent form of schizophrenia was the paranoid type. However, Mezey feels that adaptive difficulties are etiologically important in affective disorders.

Other Groups

Regarding the length of hospitalization, Brown (1960) reports that only twenty-six percent of Polish soldiers who had been committed had been discharged after ten years. The reason given is that those soldiers had no relatives. Murphy (1955) who was the first to study admission rates of refugees in England, found that their rates were much higher at each age period than for the British population, the increase being due primarily to schizophrenia. Female refugees suffered from depression and a high rate of of suicide.

In a study in West Sussex, Lucas, Sainsbury and Collins (1962) studied the types of delusions present in a group of schizophrenics. They found that paranoid delusions were more frequent among an "immigrant" group than among "natives." This consistency of paranoid content in the immigrant was also reported by Kino (1951) and Prange (1959). The hypothesis is that cultural isolation often leads the foreigner to what has been called "alien's paranoid reaction." It is interesting to note that Singer and McCraven (1962) in a study of daydreaming among various subcultural groups, found that the more insecure or the recently emigrated groups (Jews, Italians and Negroes) showed considerably more daydreaming than the more assimilated groups, Irish, Anglo-Saxon, and German.

Studies in Canada

Another country within the British Commonwealth which experienced a large influx of immigrants is Canada. Studies conducted in Canada, however, were done primarily by psychiatrists who dealt with mental patients from foreign countries. Early in 1951, Tyhurst discussed forty-eight displaced persons under psychiatric care. They were mostly of Baltic, Polish and Ukrainian background. He noted common features in the symptomatology—suspiciousness and paranoid trends, coexistence of anxiety and depression with somatic reactions. Tyhurst points out that the displaced person who left his country of origin involuntarily may express a more negative attitude towards his emigration than the person who left voluntarily.

Koranyi et al. (1958) reported on the difficulties encountered by fifty-three Hungarian immigrants in Canada. With the exception of three cases, all showed marked neurotic conflicts in their earlier lives. Conscious and unconscious motivations for leaving Hungary were described, and the investigator noted that the more the unconscious motives outweighed the realistic ones, the more likely was the appearance of adaptive difficulties.

In a later study, Koranyi et al. (1963) pointed out that the hard core of Hungarians who could not adapt to their new land had been morbidly ill, psychotic, or functioning on the borderline of psychosis in their native Hungary. They carried out verbal thrusts, fought, or indulged in heavy drinking. One special feature of this hostility is a culturally determined anti-Semitism where they equated foreigners with Jews.

Terashima studied second generation Japanese in Canada (1958) and presents evidence which supports the etiological relevance of disturbed relations between parents and children to the etiology of schizophrenia.

However, in the province of Saskatchewan, where 561 cases were studied in one hospital, Ward (1961) found that differences in rates could only be attributed to age. Migration and ethnicity did not seem to be important. Ward believes that the frontier-type society of Saskatchewan allowed for expression of individual belief and aspiration, as opposed to the highly structured societies of provinces with large cities such as Ontario and Quebec.

Malzberg (1968a) analyzed admission rates in Canada on the basis of country of birth of the foreign-born. Rates for the whole of Canada, the Province of Ontario, and the City of Toronto are given in his numerous papers. To control for urban-rural differences, Malzberg calculates the rates for Toronto. In the following tables we have compiled the admission rates for the City of Toronto. This limitation will give us a more accurate picture of the differences between native-born and foreign-born, since these admission rates pertain only to urban patients.

TABLE 7

AVERAGE ANNUAL STANDARDIZED RATES OF FIRST ADMISSIONS FROM TORONTO TO ALL HOSPITALS FOR MENTAL DISEASE IN CANADA 1950-1952, PER 100,000 POPULATION, CLASSIFIED ACCORDING TO SEX AND NATIVITY GROUPS

Nativity	Male	Female	Total	
Native-born	158.3	141.1	149.2	(1968b)
Foreign-born	158.4	152.1	155.4	(1968b)
United Kingdom	109.1	111.1	109.7	(1968c)
United States	216.3	194.2	206.8	(1968d)
Italy	137.7	87.6	111.3	(1968a)
Scandinavia	218.0	140.3	175.1	(1962b)

TABLE 8

AVERAGE ANNUAL STANDARDIZED RATES OF FIRST ADMISSIONS FOR SCHIZOPHRENIA FROM TORONTO TO ALL HOSPITALS FOR MENTAL DISEASE IN CANADA, 1950-1952, PER 100,000 POPULATION, CLASSIFIED ACCORDING TO SEX AND NATIVITY GROUPS

Nativity	Male	Female	Total
Native-born	41.2	38.3	40.1
Foreign-born	52.4	46.2	49.8
United Kingdom	23.3	25.7	24.7
United States	63.4	28.7	46.6
Italy	57.1	46.2	49.8
Scandinavia	103.8	46.1	75.4

On the basis of his analysis, Malzberg reaches two major conclusions. First, when the foreign-born and native-born population are compared on most demographic variables, the similarity in first admission rates is greater. Second, the continued significant differences in schizophrenia, in spite of the statistical adjustment, implies some fundamental difference. Foreign-born tend to suffer more from schizophrenia. However, these generalizations do not hold for every group. The English-born and the Italian-born have lower total rates, and the English-born have a lower rate in schizophrenia than native-born Canadians. The most striking difference between males and females are for the Italian-born, and Scandinavian-born on total admission rates, and for the United States-born and the Scandinavian-born on schizophrenia.

It is to be noted that for the total rates, the United States-born have almost twice the admission rates of any foreign-group in Canada. Likewise, in New York State the in-migrant has a much higher admission rate than the native-born New Yorker. We could assume that there is a similarity between the United States-born who goes to Canada, and the in-migrants who come to New York.

Another major finding by Malzberg (1964d) in Canada, is the differential rate of mental disease among those of English and French origin. While the two groups are not immigrants, they have in general maintained their ethnic identity. It was found that the total admission rate for the Canadian of French origin is lower than the rate of the Canadian of English origin. However, when French Canadians live in areas where they are a minority, they have a higher rate than the English Canadian, and likewise, when the English Canadians live in areas where they form a minority, their hospital admission rates are higher. Also, French Canadians living in Ontario had higher rates than the French Canadians living in Quebec, and the British living in Quebec had higher rates than those living in Ontario. It would seem that minority or majority status has some influence on hospital admission rates.

Murphy (1965), reporting on Chinese hospitalization rate in Canada, indicates that the lowest hospitalization rates are to be found in areas which have a real Chinatown.

It should also be noted that in New York State (Malzberg 1966a) the native-born Canadian of French origin has lower rates than other native-born Canadians of different parentage. The same applies to the foreign-born French Canadian when compared to the foreign-born Canadian of other than French origin.

Murphy (1965) reporting on unpublished data for the Canadian Department of Citizenship and Immigration, indicates that an analysis of the 1958 admission rates show the immigrant population of Canada producing proportionally less mental hospital admissions than native-born. We shall later discuss Murphy's hypothesis regarding the differences in hospitalization rates of migrants and nonmigrants around the world.

Recently Murphy (1968) analyzed hospital statistics in Canada for various "origin" groups, immigrants and native-born. He found that the French, Dutch, Polish and Asiatic male and female immigrants had higher total admission rates than the native-born of French, Dutch, Polish and Asiatic origin. The German male and female immigrant had less mental illness than the native-born of German origin. British and Italian male immigrants had less mental illness than the native-born of British and Italian descent, while Ukrainian, Russian and Scandinavian male immigrants had more mental illness than their native-born counterpart. British and Italian female immigrants had more mental illness, Ukrainian, Russian and Scandinavian female immigrants had less mental illness than their Canadian native-born counterparts. Murphy's single main conclusion is that 'origin' is associated with hospitalization. However, he is very cautious about drawing etiological implications. He is uncertain whether these differences are due to genetic or cultural factors among the various "origin" groups.

Studies in Australia

Following the unsuccessful threat of Japanese invasion, the Australian government after World War II opened its doors to immigrants in order to Italian descent, while Ukrainian, Russian and Scandinavian male immigrants had more mental illness than their native-born counterpart. British and Italian female immigrants had less mental illness, Ukrainian, increase its scant population and thus meet any future threat of invasion. Since 1947, total immigration has been two million. A study conducted with manual workers (Richardson, 1957) revealed that those with the greatest difficulties of adjustment tended to feel strongly attached, or identified with the social order in Britain. Australia thus became a scapegoat for all difficulties, real or imagined. The satisfied immigrants came with a greater readiness to change their behavior to fit Australian conditions. However, a study by the Australian Commonwealth Immigration Advisory Council (1961), using hospital admissions, revealed that the average first admissions rate of immigrants (1948-1952) was lower than among Australian-born. If displaced persons are eliminated from the statistics, the rate for immigrants is even lower, 0.38 as opposed to 0.70 per 1000 for native-born Australians. This low rate is attributed to the fact that Australian laws governing the admission of immigrants are very stringent; screening procedures generally succeed in eliminating the potentially ill applicants.

Another possible reason for a lower admission rate is the criterion for an immigrant, which is five years' residence in Australia. This means that immigrants who have resided in Australia more than five years were included with the majority group. A similar explanation is given by Srole et al. (1962) regarding the favorable mental health rating of immigrants to the United States following the introduction of the general quota system in 1922, when applicants received intensive screening.

However, a number of studies were conducted by Australian psychiatrists, in which admission figures were broken down on the basis of the immigrant's country of origin. These statistics were collected at a *later* period, after Australia permitted large-scale immigration from countries outside the British Commonwealth. Cade (1956) found that there was a higher incidence of schizophrenia among non-British European males and females than among Australian and British-born. Listwan (1956), studying 244 new patients of the psychiatric outpatient department of Sydney Hospital, found that twice as many migrants as native-born were paranoid. He feels that migrants in Australia develop paranoid states and paranoia-like reactions, either because of the social or cultural make-up developed in the country of origin, or because of what he calls "migration stresses." Since most of the paranoidal patients were from Eastern Europe, he believes that these mental patients tend to develop schizophrenia.

Cade and Krupinski (1962) conducted a study in one psychiatric hospital in Victoria; they found that the rate of alcoholism was highest among males born in Australia, Britain, and Eastern Europe, and lowest in males originating from Southern Europe. Schizophrenia occurs much more frequently among migrants from countries other than Britain, particularly those of Eastern and Southern Europe. Other affective and schizo-affective diseases show a much higher overall incidence among males from Eastern Europe, and females from Eastern and Southern Europe. Females from Germany and Holland also had a high rate of schizo-affective reactions.

Cade and Krupinski believe that migrants face manifold stresses of war, migration, and assimilation, and even British-born, despite a common language with Australians, suffered from depression. The most notable finding of their study was that schizophrenia is seven times more frequent among Eastern and Southern Europeans than among native-born Australians. Cade and Krupinski did not correct these statistics for socio-economic status. It should be recalled from other studies that such a correction decreases the discrepancies found.

In a later publication, Krupinski and Stoller (1965) reported the following admission rates, both in-patient and out-patient, by major groups for the whole state of Victoria:

TABLE 9

INCIDENCE OF MENTAL ILLNESS, CORRECTED BY AGE, IN-PATIENT AND OUT-PATIENT, PER 100,000, ACCORDING TO ORIGIN OF GROUP IN VICTORIA, 1962

	Australia M	Australia F	Britain M	Britain F	Western Europe M	Western Europe F	Eastern Europe M	Eastern Europe F	Southern Europe M	Southern Europe F
Schizophrenia	21.2	28.6	24.9	21.7	56.3	51.4	121.9	159.9	52.5	66.7
Total	259.2	255.6	330.9	271.8	232.8	145.1	424.8	375.2	149.3	190.0

As can be seen from the above figures, migrants from Eastern European countries had the highest rates in total admissions, and also for schizophrenia. The total rate for Southern Europeans was relatively low, compared to British and Australian-born, but the rate of schizophrenia remained higher. Females from Eastern Europe and Southern Europe had higher rates of schizophrenia than the males. It should be noted that migrants from Western Europe and from England can return to their homelands if they wish, and are therefore relatively better off than migrants from Eastern Europe, who are least likely to return to their homelands, or migrants from Southern Europe, who are mostly poor. It is interesting to note that Southern European migrants, despite their lower class status, and the financial difficulty of returning to their homeland, suffered less overall mental disorders than any other group, including the native-born Australian.

Krupinski, Schaechter and Cade (1965) in a clinical study of a number of mental patients, attribute the high incidence of schizophrenia in Australia to the migration of single, unstable men who break down in the first year of their economic and financial struggle. High incidence of schizophrenia among professionals can be related to their inability to get recognition of their qualifications. Breakdown in female migrants, usually after seven to fifteen years residence, is assumed to result from lack of assimilation, leading to isolation within their own families. Wartime experience played an important role in the causation of mental illness, particularly with the Eastern Europeans.

Schaechter (1962) found a large percentage of non-British female migrants suffered from paranoid schizophrenia and indicated that this is related to the difficulties of assimilation. In a later study (1965) she conducted intensive interviews with one hundred non-British female migrants. The existence of mental illness was established in twenty-seven cases, and suspected in ten cases. Early breakdown occurs mostly among persons who have already been ill in their own country, while the later breakdowns appear to be more closely related to effects of migration.

It is to be noted that there is a difference in the findings between the Commonwealth Immigration Advisory Council's report and the work conducted by Krupinski et al. in Victoria. The first study included the whole of Australia. However, Krupinski and Stoller (1965) criticized the study because of its arbitrary definition of a migrant as a person who had spent less than five years in Australia. This elimination of migrants beyond the five year period, besides the fact that displaced persons were eliminated, and considering that the earlier wave of immigrants were British, may explain the discrepancies in the results of the two studies.

Studies in the Scandinavian Countries

Because of the stable population and a central registry of patients, statistics on hospitalization and follow-up studies in the Scandinavian countries are quite developed. Ødegaard (1932) in his classic and often-quoted study, found that admission rates among the Norwegian-born in Minnesota were about thirty to fifty percent higher than those of American-born, and higher than among the Norwegians living in Norway proper.

Schizophrenia was found to be more than twice as high among Norwegians in Minnesota than in Norway. The preponderance of schizophrenia in the immigrant may be caused either by life conditions that favor the development of schizophrenic symptomatology when one lives in an "alien" milieu, or by the "schizothymic" element in the population which immigrates. Ødegaard believes that the schizoid character of an individual predisposes him to migration. He found even more insanity among those Norwegians who returned to Norway after staying awhile in the United States. He felt that this latter finding strengthens the hypothesis that there was more disturbance prior to immigration.

In a later paper (1945), Ødegaard found that in Norway the migrant has lower admission rates than the nonmigrant. However, the admission rates are high when the migrant settles in Oslo, the capital. In a replication of the study, Astrup and Ødegaard (1960) provide us with more detailed categories of internal migration. Again, they found that all migrants, except those that go to Oslo, have a lower rate of mental hospitalization than nonmigrants. When migrants to Oslo were divided into urban/rural origin, the rural migrants to the city, contrary to expectation, had less mental illness than migrants coming from other urban areas. Ødegaard believes that there are different selective factors between inland and overseas migration. Inland migration represents a continuous progress in life and work; the international migration represents a complete break with the past. He even found that immigrants from Sweden and Denmark to Norway have a significantly higher morbidity rate than the native Norwegian population, but not as high as the Norwegian immigrants to the United States.

On the other hand, in a study of hospital admissions in Helsinki, Sternback and Achte (1964) found that schizophrenia was statistically more frequent in males born in Helsinki, than in migrants. Female patients with presenile and senile psychoses were found to be born outside Helsinki more than twice as frequently as those born in Helsinki. It is possible that ecological conditions, both outside and inside the capitals of Norway and Finland, make for these differences. In an earlier study by Kaila (1942) in Finland, it was found that the part of the country with the greatest inward migration had the lowest hospitalization rates, thus corroborating Ødegaard's finding in Norway, except for Oslo.

In a recent study in Iceland, Helgason (1964) found no difference in the incidence of mental illness among nonmigrant males and those migrating to Reykjavik, while for the men migrating to places other than Reykjavik, it was found to be lower than for nonmigrating men by 7 percent. These are somewhat similar findings to those of Ødegaard in Norway, except that individuals migrating to Oslo have higher morbidity rates than the local population. Psychosis, in general, was found to be higher in the following groups: nonmigrants; inhabitants of rural communities; those of low socioeconomic level; and unmarried persons.

From the data presented, it would seem that studies in the United States and the Scandinavian countries are contradictory. Migration in the United States is related to greater hospitalization, while in Norway it is related to less hospitalization, with the exception of migrants to Oslo. However, more recent data, Malzberg, 1962a) shows that Norwegians born in New York State had lower rates of psychoses (1949-1951) than native-born in the United States, except for paresis and alcoholic psychosis. United States-born population of Norwegian parentage shows a higher rate of schizophrenia than their parents (20.7 per 100,000, and 45.4, respectively).

How can these differences be explained? We have already indicated that Norway has a stable population; therefore, migration would represent a greater sense of initiative, possibly reflected in the individual's greater feeling of adequacy and resourcefulness. Thus, in Norway, a need to move and a desire to change is probably an indication of mental health. In the United States, on the other hand, the freedom of geographical mobility represents an escape for any individual facing personal difficulties. Thus, people on the move in the United States may be more predisposed to mental illness. However, how can we explain the greater hospitalization rate of Norwegians going to Oslo? Our assumption is that possibly the Norwegian going to Oslo may seek isolation in a big city, which he cannot find in other areas of the country, and thus may be a different type of migrant than the Norwegian moving other parts of the country. Sundby and Nyhus (1963) have found that half of the schizophrenics in the downtown and central western sections of Oslo (areas of one-room apartments and boarding houses) were migrants to the town, indicating that those without established social and family relations live there.

On the basis of the above findings, with some exceptions, we could hypothesize that countries where opportunity for geographic mobility is limited, and where the population tends to be stable, as in the Scandinavian countries, it is the more resourceful and capable who migrate; those with much better mental health stamina, who not only are capable of overcoming the stresses of adjustment in a new environment, but even show less tendency toward mental illness than the native-born around them. However, they do not seem to migrate to the capital cities. There are no statistics in these studies regarding the social class of those individuals who migrate. Therefore, it is not the variable of migration alone which should be related to mental health, but the fact that possibly people from the upper classes are more prone to change residence within the country in the Scandinavian countries.

Eitinger (1958; 1959) studied one group of migrants to Norway who had sustained serious stress: refugees, primarily manual laborers of Polish and Czech origin, who had settled in Norway during the postwar period. Psychosis was about ten times higher among these refugees than in the long-settled population. There are three reasons for this according to Eitinger: (1) a higher rate of psychosis in the native land of the refugee; (2) premorbid personality; (3) mental and physical stress incurred by the refugee. In his study of these refugees, Eitinger was convinced of the importance of the second and third factor, with the first being least important. Refugees who broke down seemed to have had limited psychological resources in their premorbid condition. Both isolation and the feeling of insecurity were found to be the psychodynamic elements determining the illness. A combination of these elements resulted in persecutory paranoid delusions.

In a later paper, Eitinger (1963) pointed out that, on the basis of Jasper's theory that psychogenic disturbance is bound to disappear after the provoking cause has vanished, he sees no "scientific" reason for former concentration camp inmates to claim their present mental disturbance was caused by suffering during war time. It would seem that Eitinger belongs to the school which believes in the "hereditary taint."

Studies in Israel

As far as we can see in reviewing the literature on the relationship between emigration and mental illness, Israel has provided us with most interesting data. The fact that a large number of psychiatrists were among these emigrants has undoubtedly enriched the literature. We shall consider first admission rates in Israel, and later discuss local clinical and epidemiological studies.

Halevi (1963) reviewed major studies on the frequency of mental illness among Jews in Israel. Israel, within a few years, had accepted more than a million refugees. Most of them represented the remnants of European Jewry who had survived the concentration camps, or Oriental Jews who were transferred from a conservative, somewhat primitive culture into an industrialized society.

In a census by Herman (1931 reported in Halevi, 1963) he found that Jews had more mental illness than Christians and Muslims, and that while females among Jews had higher rates than males, the trend is reversed with

Christians and Muslims. The differences could be attributed, possibly, to a fuller coverage of Jewish patients.

Another survey by Halpern (1938) showed that while seventy-nine percent of the Jewish cases suffered from endogenous psychoses, only two-thirds of the Christians and Muslims suffered from the same type of illness. The third survey carried on by Halevi (1963) represents a study of admissions, while the two previous studies were studies of prevalence. Halevi points out that in Israel the "selective factor" does not play an important role. Prior to 1958, whole communities from Yemen, Iraq, and Libya were transplanted to Israel, thus leaving little room for emigration to be influenced by schizoid personalities.

In considering total admission, Halevi found that Jews coming from Oriental countries had higher overall rates than Jews coming from European countries; in considering manic depressives, he found European Jews had higher rates. In unspecified psychoses, Jews from Turkey and Iran showed a high incidence.

Native-born Israelis are closer to Oriental Jews in overall incidence than European Jews. However, the most interesting finding is that the incidence of schizophrenia among native-born Israelis is the highest of any group in Israel (Israel-born, 80.8; Jews from Eastern Europe, 34.2; Yemenite Jews, 56.8). Rates among male Israelis and Oriental Jews are higher than among females, while the trend is reversed for European Jews. The lowest rates for combined psychoses were found among Jews from Yemen (123.3) and immigrants from Eastern Europe (124.3).

It is to be noted that the percentage of schizophrenia increases among immigrants who arrived recently, and reaches maximum among native-born (58.2 percent). Halevi points out that emigration to Israel has a different meaning to Jewish refugees. Even in cases of economic difficulties, there is a strong feeling of coming "home" which provides a heightened feeling of security.

Miller (1966) tries to explain differences in the incidence rates of schizophrenia. He believes that the rates for Oriental Jews (46 to 65) are somewhat overestimated, since diagnosing patients from cultures they are unfamiliar with presents difficulties for the Western-trained psychiatrists, particularly when it comes to acute reactions among the new immigrants. Further explanation is necessary to account for the high rate of schizophrenia among native-born Israelis. This, according to Miller, could be attributed to rapid social changes, and possibly to the fact that forty percent of the Israel-born are of Oriental stock, which further complicates the picture. However, he tends to believe, since a number of studies found that schizophrenic symptomatology is more frequent among technologically-advanced people, that the native-born Israeli would therefore be most primarily affected by this condition.

While schizophrenia increases with the more recent emigrants, manic depressive disorders decrease with more recent arrivals. The ratio of the rates of schizophrenia to manic depression is 3:1 among Oriental Jews and native-born Israelis, but among European Jews there is no preponderance of schizophrenia.

Hes points out (1960) that, contrary to expectations, Jews suffer less from manic depressive disorders. He found a low morbidity rate of .4 per 1000, as compared to the findings of Mayer Gross in England (see Hes, 1960), where the rate is 3.4 per 1000. Jews emigrating from Africa and Asia were found to have a lower incidence of the disease than those coming from Europe and the Americas. Hes suggests it would be interesting to study the rates after a few decades, and see how they are affected by acculturation.

Hes (1958) conducted a comparative study of Oriental Jews and European Jews with regard to hypochondriacal symptoms. He found that a large percentage of Oriental Jews expressed hypochondriacal complaints. He believes that schizophrenia manifests itself in the form of hypochondriasis more often in Oriental Jews than in occidental Jews.

Palgi (1963), discussing the psychiatric problems of so many different nationality groups, believes that Israeli Jews will in time lose the minority feelings acquired in the Diaspora, and through a sense of community feeling of being Israeli, all cultural differences will tend to disappear. An extensive program of education in the Army for all ablebodied men and women will certainly contribute to this coalescence. Nevertheless, Weinberg (1955), having studied the adaption of voluntary immigrants, still feels that they ran some dangers of stress. Therefore, he espouses preventive mental health measures for everyone. In an extensive research, Weinberg (1961) makes a series of practical suggestions regarding immigrants, such as developing a sense of belonging in an effort to integrate the immigrant as soon as possible, and providing social services prior to and after arrival in Israel.

We shall now review studies, clinical, rather than epidemiological in approach, conducted by psychiatrists in three Israeli communities. Hoek et al. (1965) compared the prevalence among various ethnic groups in a lower class suburban community. They found more mental illness among older females than among younger ones. The trend was reversed with Jewish females from Asia and Africa. The authors believe this is due to the fact that older Oriental Jewish females have a clear function within the extended family which the European Jewish female does not have. There were no differences in psychosis or psychoneurosis, and little difference among males of different origin.

The rates were higher for European women, with respect to psychophysiological conditions, but were lower on the sociopathic side. This did not hold with the males.

Surprisingly, and contrary to expectation, the Afro-Asian patients were much readier to talk about themselves than the European women, but were less satisfied with the results. Sanua (1961, 1966) has reviewed the literature concerning the problems of acceptance of psychotherapy, and reaction to illness among different cultural groups. Although previous studies indicated less willingness on the part of lower-class patients to talk about their problems, it would appear that this does not apply to lower-class Afro-Asian Jewish patients in Israel, It should be noted that in the clinical study by Hoek et al. findings clearly indicated that adjustment to the new country contributes to the frequency of psychiatric disorders, particularly with the middle-aged Afro-Asian immigrant male who shows especial stress.

This is somewhat different from the general finding that there is more schizophrenia among the native-born Israelis. It may be possible that, while Oriental Jews complain more frequently of physical and mental problems, which are not very serious, the Israeli-born appears more frequently in the hospital statistics in regard to severe breakdown.

For the first time, Maoz et al. (1966) conducted a complete epidemiological study of a whole community covering all facilities. Like Hoek et al., Maoz found that the rate of personality disorders was higher among the Afro-Asian groups. Females from Poland contributed a larger percentage of patients, while those from Rumania contributed least. According to the authors, this is probably related to the fact that Poles suffered more severely under the reign of Nazism. However this discrepancy in rates does not apply to males. European patients exhibited the more classic forms of psychoneurosis, the Afro-Asian the more psychophysiological complaints. Females in the latter group described their complaints "all over the body," while males tended to complain about abdominal pains.

Psychoses not falling into any standard classification occurred especially among recently arrived Oriental Jews. Nonpsychotic disturbances usually labelled "transitional stress reaction" occurred, especially in those who came from Africa.

In the overall admission pattern, immigrants who had reached Israel between the years 1948-1951 had a higher morbidity rate than those who had immigrated in the years 1957-1961. Immigration in the earlier period, 1948-1951, was notable for the masses who came from Iraq, Yemen, and Iran, which may explain the high morbidity rate. Females from Iran and Iraq had the highest rates. Antisocial violence was more common among the males from Africa. Total prevalence of psychoses was 12.1 per 1,000, which is considered to be low in comparison to other surveys, and this in spite of constant stress of adaptation and threats from the Arab world.

The ratio of 3:1 between schizophrenia and manic depressive reactions, as indicated earlier by Halevi, and which is characteristic of lower-class strata, was confirmed for the first time by a field investigation. Moses and Shanan (1961) in a study of out-patients of the Hadassah Hospital, found no relationship between socioeconomic level and psychosis, contrary to expectation. There were also more psychoses among the sabras (native-born Israelis).

The general picture of Jewish immigrants to Israel appears somewhat favorable in relation to surveys of Jewish immigrants to other countries. A study by Sunier and Mairlot (1962) revealed that the most frequent illness found among Jewish immigrants to Belgium and Holland was schizophrenia. They had a relatively higher rate of prevalence than the rest of the population. According to Sunier and Mairlot, the traumatic experience they had undergone had probably caused permanent psychological damage.

Kabaker and Azoulay (1962) in a study in Paris, reached the same conclusion as Sunier and Mairlot. Almost sixty percent of the Jewish patients had lived in Nazi-occupied European countries, and thus had been exposed to various degrees of Nazi persecution. Since this was not evident in Israel, we could assume that whatever trauma was incurred by Jewish refugees was tempered by the feelings of returning to Zion.

STUDIES IN OTHER COUNTRIES

Studies in Asia

We now turn our attention to the Asian countries. Psychiatry has not yet become well developed in countries which are engaged in a basic struggle for survival. Hospitals are few in number; therefore, incidence and prevalence studies are rather sparse. There are, however, a number of studies which have been conducted by a group of Chinese psychiatrists on the Island of Formosa, and by other psychiatrists in the South Pacific Islands and Hawaii.

In Formosa, Rin, Wu, and Lin (1962) compared the context of the hallucinations and delusions of native-born Formosans and mainland Chinese who settled on the Island. Despite the fact that both are of the same stock, there are great differences between the two groups: (a) Mainlanders who had experienced much social mobility migrated to the island under chaotic conditions and suffered severe stresses; (b) Formosans had been largely Japanese educated; (c) There was economic decline for the mainlanders; (d) Belief in magic and superstition is more prevalent among the Formosans.

Grandiose ideas were more frequent among the Formosan males, possibly accounted for by their lower social and cultural prestige. Formosan males had less sexual content in the hallucinations and delusions than mainlanders. The latter's delusions were filled with political affairs. Mainland females seemed to be more concerned with neighborhood affairs.

In another study, Rin and Lin (1962) compared the prevalence rates of four subgroups. They found that the catatonic type was more prevalent among the Formosan aborigines than among the Chinese. Schizophrenia differed in other aspects. The aborigines had more acute psychomotor excitement, visual hallucinations, less elaborate delusions. It was mostly of a benign nature. The authors attribute this favorable prognosis to a less complicated village life. The community itself, which affords an abundance of opportunities for group and community participation in respect to daily life, may act as a therapeutic agent.

Tzuang and Rin (1961) noted that paranoid psychosis was observed most frequently among female migrants in an out-patient clinic in Formosa. Manic depressive psychosis was found to be decreasing for migrants and nonmigrants in Formosa.

In an unpublished study, Murphy (1961) found that rates among the native-born in Singapore exceeded the rates among immigrants at most ages, contrary to the general findings that immigrants suffer from greater stress.

Schmitt (1959) found that most of the hospital admissions in Sarawak, Borneo, were Chinese schizophrenics of the hebephrenic or simple type.

Medlicott (1961) indicated that Chinese in Indonesia, because of their ambitious nature and higher economic level, have more mental illness than Malays, who are presumably protected by fatalism, a philosophic aspect of their Islamic faith.

Studies in Hawaii

Regarding the adjustment of various ethnic groups in Hawaii, Finney (1963) found that the Portuguese had the highest admission rates—more than four times their number in the total population. Using the MMPI with out-patients, he found the anxiety was highest with the Portuguese, and lowest with the Chinese, who had the lowest admission rates. A study by Schmitt (1956, 1957a) found that schizophrenia was lowest with the Chinese and Caucasians, and highest with the Hawaiian and Japanese. It is to be noted that Finney found total admission rates highest for the Portuguese; we could therefore assume that the Portuguese are underrepresented as far as schizophrenia is concerned.

With the Japanese group, Okinawans had the higher rate of admission than mainland Japanese, despite similar ethnic background (Ikeda et al. 1962). The assumption is that Okinawans have not been able to maintain all of their traditional institutions. Okinawan women have a low rate of schizophrenia, which the author assumes is due to the fact that they marry outside the group.

In an earlier study by Wedge (1952), he formulated the hypothesis that the intensive mothering of the Okinawan mother would afford protection against psychosis in later life. While the mothering was continued by the immigrant Okinawan mother in Hawaii, this particular group showed a high rate of psychosis. It was suggested that the influence of cultural practices on personality development should be considered in the context of the total social setting to which psychological adjustment must be made.

Schmitt (1957b) tried to test the hypothesis that high rates in low economic level sections are the result of recent migration into the areas of men who live alone. His data suggest that the relationship between mobility and mental health may be less important in some communities since he found no such correlation in Hawaii.

Zunin and Rubin (1965) found that foreign students from non-Western countries, who became psychotic, had a strong paranoid element. The uniqueness of the stress and the similarity of the acute illness which developed in the students suggested a causal relationship. They found high parental expectations of success, and authoritarian, controlling fathers to be characteristic of these students. These are some of the usual premorbid stresses of the paranoid in our own culture. As a result of these parental pressures, individuals tend to become shy, experience difficulty in close personal relationships, and are hypersensitive to criticism. With what Zunin and Rubin have called a cultural shift, there is an explosive breakdown.

Studies in Europe and Latin American Countries

In Europe, social psychiatry or research conducted to determine the correlation between social variables and mental illness is somewhat rare outside of the Scandinavian countries. It is possible that this dearth of research is due primarily to the strong tradition in Europe that all mental illnesses can be accounted for in organic terms. In view of the weakness of the teaching of the social sciences in European universities, social psychiatry cannot have a fertile soil in European countries. However, we shall review a few of the sparse studies which could be subsumed under the name of social psychiatry.

After World War II, there was an enormous increase in the number of Italian laborers in Switzerland. In 1964 (Risso & Boker, 1964) they numbered approximately 500,000 in an indigenous population of five million. Many of these migrants had to be hospitalized in mental hospitals; it became evident that they developed a symptomatology very different from that of Swiss patients. Among the peculiar characteristics of these patients was that their delusions bore the stamp of the archaic magical vision of the world in which they grew up in Southern Italy. The main features were depression with anxiety, an inhibition of the will, vivid feelings of bodily transformation, and the overpowering delusion that they had been bewitched.

Violent emotional outbursts accompanied by delusions of being poisoned and feelings of bodily transformation would be diagnosed by a Swiss psychiatrist as a severe, acute schizophrenia. However, to the families of the patients these were natural reactions to being bewitched. Inquiries showed that these patients, who had never left their villages prior to their migration to Switzerland, suffered from the most extraordinary misunderstandings in interpersonal relations in their new environment, particularly in their relationships with women. Accustomed to the rigid segregation of the sexes which existed in their native villages, they did not understand the ease which characterized the manners of women in Switzerland. However, Laffranchini (1965) noted that in most instances, Italian workers seemed to have been predisposed to mental illness prior to immigration, and that problems of assimilation may have precipitated the illness. Even in Italy proper, it was found (Rose, 1964) that mobility is probably linked to mental disorders. Of the mental patients in Rome, half originated from the other areas of the country, particularly from Southern Italy, while the figure for the general population is twenty-nine percent. Likewise, in the province of Lazio, migrants accounted for a high percentage of mental patients, a large proportion of whom suffered from paranoid schizophrenia.

Pflanz et al. (1967), in Germany, studied Greek workers. They found that their functional complaints, such as feelings of fatigue, unrest and sleep disturbances, supported the hypothesis that the difficulties of adaption were the cause. Poeck (1964) reported that Italian workers in Germany tended to have neurotic depressions with hypochondriacal features.

A study in Poland by Szwedkowicz et al. (1965) found that hospitalization was higher in the migrant population, with schizophrenia and senile psychoses particularly high. They believe that psychic disturbances are produced by difficulties of adaptation and of assimilation, not by constitutional personality features, and that women, in particular, are more affected by these difficulties. Crocetti et al. (1964a, 1964b) in Yugoslavia found that some of the communities with the highest rates of hospitalized psychotics and schizophrenics are communities which have declined in population. Thus those who do not move have greater risks of mental illness than those who move, a finding similar to Odegaard's in Norway.

A number of short articles on North African immigrants who had settled in the Paris region of France have been published. Daumezon et al. (1958) found in 1945 that the percentage of Africans in mental hospitals was 2.18 percent; in 1957, the percentage was 14.9. Alliez and Mayaud (1960) studied a group of repatriates of European stock, originally from Tunisia. They found a high degree of difficulties of adjustment among them, and a good deal of psychosomatic illness. Frey (1961) found that Muslim workers returning to Algeria after spending some time in France suffered from all kinds of physical complaints.

It is interesting to note the differences in diagnosis of the Muslim patients. Assicot et al. (1961) pointed out the relative rarity of schizophrenia among them, and the unusual ratio of catatonic to paranoid forms. If there are any paranoid delusions in Muslims, they tend to be more associated with short-lasting episodes in females, the context relating especially to witchcraft, sorcery, or poisoning. This becomes all the more surprising in view of the paranoid tendencies of the "normal Arab." This is somewhat similar to the finding among Mexicans (Meadow and Stoker, 1965) who, despite their reputation for jealousy, do not develop the paranoid types of schizophrenia. Reading newspapers in the Middle East, particularly in the case of their dealings with Israelis and the Western world, the content of Muslin thinking could only be labelled as being of the paranoid type. Sanua (forthcoming) has reviewed the literature on the psychology of the Arab, which shows this great tendency.

A number of studies have been conducted with migratory Indians in Latin America. Seguin (1956) found that Indians from the Andes migrating to the coast of Peru usually develop psychosomatic ailments when, in addition to the normal social stresses, are added those such as being unable to find jobs, or receiving bad news from home.

Brody (1966) showed that the dark-skinned, less educated, less skilled recent migrants were more anxious, showed more distrust, and had more ideas of reference than their more advantaged counterparts.

Likewise, Fried (1959) discussed the problems of adjustment of Peruvian Indian migrants to the coast. To all the stresses, he adds the physiological factor which produces somatic disturbances. The migrant, high altitude man working in a low coastal region, is subject to another pathogenic set of forces. Thus, the author concludes that the psychodynamic and sociocultural framework used with immigrants in Europe and America could be applied to Peru. The difference between the culture of the Indians and the coastal population is extreme enough to magnify the dimensions of change required by the Indian. It is to be noted that neither Seguin nor Fried mention schizophrenia as resulting from the migration.

SUMMARY AND CONCLUSIONS

In this paper, we have brought together the findings of a number of studies, spanning many countries, which have tried to relate migration with mental illness. Most of the research is based on hospital admissions, but a few limited studies based on clinically selected populations have been included.

Over the course of the years, there has been a refinement in statistical procedures. Originally, crude admission rates for native and foreign-born populations were compared, and the latter group was found overrepresented in United States statistics. In addition to controlling for age, sex, and race, some recent studies have controlled other demographic variables, such as occupation, social class, urban-rural residence, ethnicity, cultural background, and even concentration of the migrant population. In addition to using hospital admission rates, statistics have included patients seen in out-patient clinics and by private practitioners.

A general conclusion which we can draw from this review is that some migrations are related to greater risks in mental health and some migrations are related to favorable mental health. However, in most of the research, findings are drawn from a comparison of the migrant population to the nonmigrant population. Except for Malzberg and Odegaard, no study has compared the migrant population with the nonmigrant population that remained at the original point of departure.

While the "general hazard theory" has been weakened by the data, there is no adequate study which has tried to test the "self-selection theory." The only effort to date is the study by Keeler and Vitols (1963) in which it was found that forty percent of a Negro hospital population had histories of migration, as compared to four percent in the general population. Sixty-five percent of them were deemed to have been free of schizophrenia prior to migration on the basis of interviews conducted with relatives of patients who still resided in the areas from which the patients had migrated. While this study seems to indicate some relationship between schizophrenia and migration, the sample is too small to make any generalization regarding the cause and effect relationship.

The complexity of the problems in conducting such research and the lack of comparability of the research available makes the formulation of any kind of generalization highly tentative. In the introduction to Malzberg and Lee (1956), Thomas states:

> "Closer examination of both generalizations and exceptions show so many inconsistencies in definition, so few adequate bases of controls, so many intervening variables, so little comparability as to time and place, that the fundamental cause of the discrepancies may well be merely the non-additive nature of the findings of the different studies."

Nevertheless, we shall make an effort to explain the differences found among the various studies and suggest that some of the discrepancies are due to intervening variables. We shall discuss (a) the characteristics of the migrant or immigrant, (b) the motivation behind his migration, and (c) the type of environment in which the migrant resided. Finally, we shall try to see how these may be related to his adaptation in his new environment.

Malzberg found that while differences between foreign-born and native-born are disappearing, differences between migrants and non-migrants in the State of New York still remain high. Such findings are also supported by his Canadian data. Consistently, however, schizophrenia seems to be characteristic of both the foreign-born and the migrant. This lends support to his general hypothesis that "schizoid" individuals are more likely to change residence. A number of recent findings do not seem to support this relationship between geographic mobility and schizophrenia. For example, intensive clinical studies of schizophrenics did not disclose that they tended to move prior to their illness; if anything, they might be less mobile.

Srole and Langner (1962) seem to have weakened the hypothesis of "self-selection" in their study of a representative sample of a normal population living in one section of New York. However, their criterion for morbidity was a mental health rating arrived at by psychiatrists looking at interview records. They found no differences between individuals who came to the United States as children and those who came as unaccompanied adults. They found that the sick-well ratio was higher for the foreign-born than for the native-born, but when they corrected for socioeconomic status and age, differences were eliminated. A further reanalysis of the data which compared the old immigrant with the new immigrant who came after 1921 revealed a higher percentage of impaired among old immigrants. According to Srole and Langner (1962) this reflects the "different magnitude of the role discontinuity bridged in the transition from their respective environment." The older generation of immigrants came mainly from villages or towns, while those arriving after the twenties, when immigration screening was introduced, came from big or medium-sized cities. Thus, the latter were better prepared to live in a highly technological society. In the post-war period, environmental changes and social changes also occurred in the United States, which facilitated adaptation of the immigrant in the American metropolis.

Thus, both variables, different types of immigrants, and an environment more conducive to the adaptation of the immigrant, may explain the consistent trend found by Malzberg, namely, the decrease in differences between the foreign-born and native-born in the United States and Canada. This may also explain why Norwegians who immigrated to Minnesota approximately seventy years ago had high rates of mental illness as compared to Norwegians who came to the State of New York during the last few decades. The rates are reversed. The Norwegian-born has less mental illness than the native-born, and even has less than second generation Norwegian Americans.

We have indicated that the indiscriminate grouping together of all "foreign-born" tends to confound the results. It was shown that Italians, in the United States, despite language and cultural differences, have less mental illness than the Irish or native-born Americans. The same applies to Italians in Canada and Australia. The explanation for the relatively better adaptation of the Italian immigrant must probably be sought in the cohesiveness of their family units.

Malzberg also indicated that the foreign-born female seems to have greater difficulties with immigration, since higher over-all rates of admission support this hypothesis. However, when separating the morbidity rates according to nationality, females among the Poles and Germans had a much higher rate of hospital admission than males. Since the Poles and Germans are among the most numerous immigrants to the State of New York, the overall female rate was affected. Rather than saying that immigration is more difficult for women in general, it would be more precise to say that immigration may cause more stress in females from one nationality group than from another group.

Studies by Ødegaard in Norway showed that the migrant, except for those who migrate to Oslo, have a lower mental health risk. It would seem, therefore, that the individuals who in-migrate to New York are of a different type from those who migrate within Norway. In the United States, perhaps a move represents an escape, or possibly that less resourceful people are more inclined to move, while in Norway, migration may represent a well-thought-out move and improved perspectives. Furthermore, the homogeneity of the Norwegian population as compared to the heterogeneity of the New York population may affect the adaptation of the migrant. In Norway, however, the migrant who goes to Oslo may seek anonymity. However, the two studies in the United States and Norway are not exactly comparable. In the United States, the migrant comes from another state, but we have no indication whether he moved to an urban or rural community in New York State, while in Norway, movement is within one relatively small country having one major city, Oslo.

A recent study by Locke and Duval (1964) in Ohio may throw further light on these contradictory findings. They reported in 1958-1961 that, contrary to their previous findings, the foreign-born had lower total admission rates than the native-born, but schizophrenia rates still remained higher among the foreign-born. Again, it is plausible that the foreign-born may make less use of mental hospitals, either because the family prefers to care for its disturbed members, or simply because they may not be therapy-oriented, except in the case of a very serious mental disturbance, such as schizophrenia.

In the previous census of Locke et al. (1964), dealing with the period 1949-1952, the migrant had higher rates than the native-born Ohioan. In their new census, Locke and Duval (1964) divided the nonmetropolitan counties into three groups, based on the population mobility measured by the percentage of individuals who had been living in the county all their lives; they found that the counties with the least mobility had the highest admission rate to mental hospitals. In the metropolitan areas of Ohio, the rates were reversed. From this data, it would appear that the individual faces some mental health risks when moving to a metropolitan area from a nonmetropolitan area, or even a change from one metropolitan area to another. While neither Locke and Duval, nor Ødegaard, have given us exact details of the moves, the above study by Locke seems to support the Norwegian findings. Thus, more fruitful results are obtained if the migrant group is broken down according to the point of origin and the point of destination.

We mentioned that motivation on the part of the immigrant and his attitude towards the change of home may affect his mental health status. Eitinger in Norway found that the rate of psychoses among a group of Polish refugees was ten times higher than the rate among the native-born population. Studies in England, Australia, France, Switzerland, Formosa, showed that foreign-born have problems of adjustment in their new environment. Among a number of exceptions to the above generalization were studies conducted in Israel, Singapore, and Canada.

Murphy (1965) tries to explain these differences by introducing the influence of two variables, although he is aware that they do not explain all of the discrepancies. He feels that if the immigrant population, as in the case of Israel and Singapore, constitutes a large proportion of the total population, their hospitalization rate is relatively lower. In the case of Israel there is the added factor of total community acceptance of the immigrant. He suggests another variable to explain the low rates for the foreign-born in Canada, compared to the rates for immigrants in Australia and the United States. In both Australia and the United States, there is strong, official support for the "melting pot" policy; immigrants are urged to adopt the language and customs of the dominant group, which include an emphasis on individuality and independence. Canada has a different attitude towards its ethnic minorities; because of historical forces, the French in Canada have been able to maintain their culture, and dependency has never been considered an indication of immaturity. Possibly the rest of Canada may have been influenced by such an attitude. Therefore, there is less pressure to conform on the part of the foreign born. This permits the immigrant a more gradual and less conflicting process of adaptation into the new society.

Regarding mental illness of Negroes and Puerto Ricans in New York State, the rates were found to be the highest for this group. Contrary to expectation, the foreign-born Negro had lower rates than the native-born

Negro, while the mainland Puerto Rican had higher rates than the Island-born Puerto Rican. In Pennsylvania, native-born Negroes had higher rates than Negro migrants from the South, but lower rates than Negroes coming from elsewhere in the North. Hollingshead and Redlich (1958) found that lower class groups have approximately eight times more psychoses than upper class groups. This hypothesis, however, cannot reconcile the discrepancies between foreign-born and native-born. While social class may be a factor, it is too narrow a determinant to explain the vagaries of the research. It would seem that social class, plus other sociopsychological variables, such as sense of identity, family life and composition, attitude of the dominant group, aspiration level, etc. may throw further light on this elusive problem of the relationship between mental illness and social class.

While morbidity rates can be corrected for age, sex, education or social class, there are other variables which are too complex to control, particularly in studies which use admission rates as a basis for differential morbidity. A discussion of methodological problems in epidemiological research was provided by Sanua (1963b). A number of studies have shown that availability and proximity of mental hospitals may affect the morbidity rates. There is further the problem of diagnostic standards. Differences in theoretical orientations in psychiatric diagnosis prevent a uniformity in assessing the need for hospitalization. This could be reflected both in the total morbidity rates as well as the distribution of the percentages of the various diagnostic categories. Special biases of psychiatrists may be reflected in their reluctance to label upper-class patients as psychotics. Other sources of variations are the degree of tolerance towards mental illness shown by various ethnic, religious, racial and social class groups. Hospital admission rates include only "treated mental illness," while the "untreated mental illness" is not included in hospital admissions. Variations in statistical procedures or lack of uniformity in reporting morbidity rates, errors in estimating the population at risk, all these factors should necessarily make a comparison of these rates highly tentative. Thus migration or immigration and their relation to mental illness as measured by hospital statistics are too gross variables. Research would be more fruitful in elucidating this relationship if both the characteristics of the migrant and immigrant and the conditions under which they move are fully considered. Scientific answers can be obtained through longitudinal studies of subjects who, over extended periods of time and space, are facing different types of stresses and cultural demands. I wish to suggest that a concerted effort to coordinate research in mental illness in various states or countries could provide invaluable information to those who are concerned with prevention and treatment of mental illness. It would be possible to establish research stations and programs in various areas under the auspices of an international agency directed by multilingual scientists. Replication of studies in several countries would make for more valid generalizations.

REFERENCES

ALLIEZ, J. and MAYAUD, R. (1960)"Role of displacement and of habitat in the psychopathological reactions of a group of repatriates of European stock originally of Tunisia." Annales Medico-Psychologiques 118: 918-923.
ASSICOT, J. et al. (1961) "Main causes of psychiatric morbidity among Algerian Muslims." L'Hygiene Mentale 50: 261-286.
ASTRUP, C. and ØDEGAARD, O. (1960)"Internal migration and mental disease in Norway." Psychiatric Quarterly, Supplement, 34: 116-130.
Australian Commonwealth Immigration Advisory Council (1961) Report on the Incidence of Mental Illness Among Migrants. Canberra.
BAGLEY, C. (1968a) "Migration, race and mental health: a review of some recent research." Race 9: 343-356.
———(1968b) "A comparative study of mental illness among immigrant groups in Britain." Paper presented at the Center for Studies of Multi-Racial Groups, University of Brighton, September.
BASTIDE, R. (1965) La sociologie des maladies mentales. Paris: Flammarion.
———(1964) Table ronde sur l'adaptation des Africains en France. Paris: Ecole Practiques des Hautes Etudes.
BLUMENTHAL, K. (1948) "Problems of social psychiatry in Palestine." American Journal of Psychiatry 104: 563-568.
BRODY, E. B. (1966) "Recording cross-culturally useful psychiatric interview data: experience from Brazil." American Journal of Psychiatry 123: 446-456.
BROWN, G. W. (1960) "Length of hospital stay and schizophrenia." Acta Psychiatrica Scandinavica et Neurologica 35: 414-430.
BRY, I. [ed.] Mental Health Book Review Index. New York: Council on Research in Bibliography.
CADE, J. F. (1956) "The etiology of schizophrenia." Medical Journal of Australia 43: 2.
CADE, J. F. and J. KRUPINSKI (1962) "Incidence of psychiatric disorders in Victoria in relation to country of birth." Medical Journal of Australia 49: 400-404.
CLARK, R. E. (1948) "The relationship of schizophrenia to occupational income and occupational prestige." American Sociological Review 13: 325-330.
CLAUSEN, J. A. and M. L. KOHN (1960) "Social relations and schizophrenia." In D. D. Jackson (ed.) Etiology of Schizophrenia. New York: Basic Books.
———(1959) "The relation of schizophrenia to the social structure of a small city." Pp. 69-94 in B. Pasamanick (ed.) Epidemiology of Mental Disorders. Washington, D. C.: American Association for the Advancement of Science.
COPELAND, J. R. M. (1968) "Aspects of mental illness in West African students." International Journal of Social Psychiatry 3: 7-13.
CROCETTI, G. M., Z. KULCAR, B. KESIC, and P. V. LEMKAU (1964) "Differential rates of schizophrenia in Croatia, Yugoslavia." American Journal of Public Health 54: 196-206.
———(1964b) "Selected aspects of the epidemiology of schizophrenia in Croatia, Yugoslavia." Milbank Memorial Fund Quarterly 42: 9-37.

DAUMEZON, G. et al. (1958) "Increases in the psychiatric hospitalization in metropolitan France of immigrant native males from North Africa for the year 1957." Annales Medico-Psychologiques 116: 1031-1033.
EITINGER, L. (1963) "Preliminary notes on a study of concentration camp survivors in Norway." Israeli Annals of Psychiatry and Related Disciplines 1: 59-67.
———(1960) "A clinical and social psychiatric investigation of a 'hard core' refugee transport in Norway." International Journal of Social Psychiatry 5: 261-275.
———(1959) "The incidence of mental disease among refugees in Norway." Journal of Mental Science 105: 326-338.
———(1958) Psykiatriske Undersokelser Blant Flyktninger i Norge (with an English summary). Oslo: Universitetsforlaget.
FARIS, R. L. and H. W. DUNHAM (1939) Mental Disorders in Urban Areas. Chicago: University of Chicago Press.
FINNEY, J. C. (1963) "Psychiatry and multiculturality in Hawaii." International Journal of Social Psyciatry 9: 5-11.
FREEMAN, H. E. et al. (1960) "Residential mobility inclinations among families of mental patients." Social Forces 38: 320-324.
FREY, F. (1961) "Development after returning to Algeria of mental disorders contracted in the metropolis by Moslem workers." L'Hygiène Mentale 50: 244-249.
FRIED, J. (1959) "Acculturation and mental health among Indian migrants in Peru." Pp. 119-140 in M. K. Opler (ed.) Culture and Mental Health. New York: Macmillan.
GERARD, D. L. and L. G. HOUSTON (1953) "Family setting and the social ecology of schizophrenia." Psychiatric Quarterly 27: 90-101.
GIBBS, J. P. (1962) "Rates of mental hospitalization: a study of societal reaction to deviant behavior." American Sociological Review 27: 782-798.
GORDON, E. B. (1965) "Mentally ill West Indian immigrants." British Journal of Psychiatry 111: 877-887.
HALEVI, H. S. (1963) "Frequency of mental illness among Jews in Israel." International Journal of Social Psychiatry 9: 268-282.
———(1960) Mental Illness in Israel. Jerusalem: Ministry of Health.
HALPERN, L. (1938) "Some data of the psychic morbidity of Jews and Arabs in Palestine." American Journal of Psychiatry 94: 1215-1222.
HARE, E. H. (1956a) "Mental illness and social conditions in Bristol." Journal of Mental Science 102: 349-357.
———(1956b) "Family setting and the urban distribution of schizophrenia." Journal of Mental Science 102: 753-760.
HELGASON, T. (1964) "Epidemiology of mental disorders in Iceland." Acta Psychiatrica Scandinavica Supplementum 173: 1-258.
HEMSI, L. (1967) "Psychiatric morbidity of West Indian immigrants." Social Psychiatry 2: 95-100.
———(1966) "Aspects of psychiatric morbidity among West Indian immigrants." Proceedings No. 4 of the World Congress of Psychiatry, Madrid.
HES, J. P. (1960) "Depressive illness in Israel." American Journal of Psychiatry 116: 1082-1086.
———(1958) "Hypochondriasis in Oriental Jewish immigrants." International Journal of Social Psychiatry 4: 18-23.

HOEK, A., R. MOSES and L. TERRESPOLSKY (1965) "Emotional disorders in an Israeli immigrant community: a prevalence study." Israeli Annals of Psychiatry and Related Disciplines 3: 213-228.

HOLLINGSHEAD, A. B. and F. C. REDLICH (1958) Social class and Mental Illness. New York: John Wiley.

IDEDA, K., H. V. BALL and D. S. YAMAMURA (1962) "Ethnocultural factors in schizophrenia: the Japanese in Hawaii." American Journal of Sociology 68: 242-248.

JACO, E. G. (1960) The Social Epidemiology of Mental Disorders. New York: Russell Sage Foundation.

KABAKER, J. and J. AZOULAY (1962) "Psychiatric care problems within the Jewish population in the Paris region." Study conducted by the Centre Medico-social, O.S.E., Paris, with the cooperation of the Jewish Distribution Committee Medical Department, Geneva.

KAILA, M. (1942) "Uber die Durchschnittshaüfigkeit der Geisteskrankheiten und des Schwachsinns in Finnland." Acta Psychiatrica et Neurologica 17: 47-67.

KEELER, M. H. and M. M. VITOLS (1963) "Migration and schizophrenia in North Carolina Negroes." American Journal of Orthopsychiatry 33: 554-557.

KIEV, A. (1965) "Psychiatric morbidity of West Indian immigrants in an urban group practice." British Journal of Psychiatry 111: 51-56.

——— (1964a) "Psychotherapeutic aspects of Pentacostal sects among West Indian immigrants to England." British Journal of Sociology 15: 129-138.

——— (1964b) "Psychiatric illness among West Indians in London." Race 5: 48-53.

——— (1963) "Beliefs and delusions of West Indian immigrants to London." British Journal of Psychiatry 109: 356-363.

KINO, F. F. (1951) "Alien's paranoid reaction." Journal of Mental Science 97: 589-594.

KLEINER, R. J. and S. PARKER (1968) "Social-psychological aspects of migration and mental disorder in a Negro population." Paper presented at the Conference on Migration and Behavioral Deviance, San Juan, Puerto Rico, November 4-8.

——— (1959) "Migration and mental illness: a new look." American Sociological Review 24: 678-690.

KORANYI, F. K., A. B. KERENYI, and G. S. SARWER-FONER (1963) "On adaptive difficulties of some Hungarian immigrants. part II: clinical considerations and the process of acculturation." American Journal of Orthopsychiatry 33: 760-763.

——— (1958) "On adaptive difficulties of some Hungarian immigrants: a sociopsychiatric study." Medical Services Journal, Canada 14: 383-405.

KRUPINSKI, J. (1967) "Sociological aspects of mental ill-health in migrants." Social Science and Medicine 1: 267-281.

KRUPINSKI, J., FRIEDA SCHAECHTER and J. F. CADE (1965) "Factors influencing the incidence of mental disorders among migrants." Medical Journal of Australia 52: 269-277.

KRUPINSKI, J. and A. STOLLER (1965) "Incidence of mental disorders in Victoria, Australia, according to country of birth." Medical Journal of Australia 52: 265-269.

LAFFRANCHINI, S. (1965) "Psychiatric and psychotherapeutic problems of Italian workers in Switzerland." Praxis 54: 786-795.
LANGNER, T. S. and S. T. MICHAEL (1963) Life Stress and Mental Health: The Midtown Manhattan Study. New York: Free Press.
LAPOUSE, R., M. MONK and M. TERRIS (1956) "The drift hypothesis and socioeconomic differentials in schizophrenia." American Journal of Public Health 46: 978-986.
LAZARUS, S., B. Z. LOCKE, and D. S. THOMAS (1963) "Migration differentials in mental disease: state patterns in first admission to mental hospitals for all disorders and for schizophrenia: New York, Ohio and California as of 1950." Milbank Memorial Fund Quarterly 41: 25-42.
LEE, E. S. (1963) "Socioeconomic and migration differentials in mental disease, New York State, 1949-1951." Milbank Memorial Fund Quarterly 41: 244-268.
LEIGHTON, D. C., J. S. HARDING, D. B. MACKLIN, A. M. McMILLAN, and A. L. LEIGHTON (1963) The Stirling County Study of Psychiatric Disorder and Sociocultural Environment. New York: Basic Books.
LEMERT, E. M. (1948) "An exploratory study of mental disorders in a rural area." Rural Sociology 13: 48-64.
LISTWAN, I. A. (1959) "Mental disorders in migrants: further study. Medical Journal of Australia 43: 776-777.
LOCKE, B. Z. and H. DUVAL (1964) "Migration and mental disease." Eugenics Quarterly 11: 216-221.
LOCKE, B. Z., M. KRAMER and B. PASAMANIK (1960) "Immigration and insanity." Public Health Reports 75: 301-306.
LUCAS, C. J., P. SAINSBURY, and J. G. COLLINS (1962) "A social and clinical study of delusions in schizophrenia." Journal of Mental Science 108: 747-758.
LYSTAD, M. H. (1957) "Social mobility among selected groups of schizophrenic patients." American Sociological Review 22: 282-292.
MALZBERG, B. (1968a) Migration in Relation to Mental Disease. Albany: Research Foundation for Mental Hygiene.
---(1968b) "Mental disease among native and foreign-born in Canada, 1950-1952." Pp. 1-37 in B. Malzberg (ed.) Migration in Relation to Mental Disease. Albany: Research Foundation for Mental Hygiene.
---(1968c) "Mental Disease in Canada, 1950-1952, among those born in the United Kingdom." Pp. 38-77 in Malzberg, 1968a.
---(1968d) "Mental Disease in Canada, 1950-1952, among those born in the United States." Pp. 78-123 in Malzberg, 1968a.
---(1968e) "Mental disease in Canada, 1950-1952, among those born in Italy." Pp. 124-161 in Malzberg, 1968a.
---(1967a) "Mental Disease in New York State, 1910-1960. Albany: Research Foundation for Mental Hygiene.
---(1967b) "Internal migration and mental disease among the white population of New York State, 1960-1961." International Journal of Social Psychiatry 13: 184-191.
--- (1966a) "Mental disease among French Canadians and French Canadian native-whites of French Canadian parentage in New York State, 1949-1951." Pp. 267-281 in Malzberg, 1966b.

——— (1966b) Ethnic Variations in Mental Disease in New York State, 1949-1951. Albany: Research Foundation for Mental Hygiene.
——— (1965a) New Data on Mental Disease among Negroes in New York State, 1960-1961. Albany: Research Foundation for Mental Hygiene.
——— (1965b) Mental Disease among the Puerto Rican Population of New York State, 1960-1961. Albany: Research Foundation for Mental Hygiene.
——— (1964a) "Mental disease among native-White and foreign-born White in New York State, 1949-1951." Mental Hygiene 48: 478-499.
——— (1964b) Internal Migration and Mental Disease in Canada, 1950-1952. Albany: Research Foundation for Mental Hygiene.
——— (1964c) "Mental disease among foreign-born in Canada, 1950-1952 in relation to period of immigration." American Journal of Psychiatry 120: 528-532.
——— (1964d) Mental Disease in Canada, 1950-1952: A Study of Comparative Incidence of Mental Disease among those of British and French Origin. Albany: Research Foundation for Mental Hygiene.
——— (1964e) "Mental disease among English-born and native-White of English parentage in New York State, 1949-1951." Mental Hygiene 48: 32-54.
——— (1964f) "Mental disease among native-White in New York State, 1949-1951, classified according to parentage." Mental Hygiene 48: 517-536.
——— (1964g) "Mental disease among German-born and native-White of German parentage in New York State, 1949-1951." Mental Hygiene 48: 295-317.
——— (1964h) "Mental disease among native and foreign-born White in New York State, 1949-1951." Mental Hygiene 48: 478-499.
——— (1963a) The Mental Health of the Negro: A Study of First Admissions to Hospitals for Mental Disease in New York State, 1949-1951. Albany: Research Foundation for Mental Hygiene.
——— (1963b) "Mental disease among the Italian-born and native-White of Italian parentage in New York State, 1949-1951." Mental Hygiene 47: 421-451.
——— (1963c) "Mental disease among Polish-born and native-Whites of Polish parentage in New York State, 1949-1951." Mental Hygiene 47: 300-337.
——— (1963d) "Mental disease among Irish-born and native-White of Irish parentage in New York State, 1949-1951." Mental Hygiene 47: 12-42.
——— (1963e) The Mental Health of the Jews in New York State: A Study of First Admissions to Hospitals for Mental Disease, 1949-1951. Albany: Research Foundation for Mental Hygiene.
——— (1963f) "Mental disease among Russian-born and native-born of Russian parentage of New York State, 1949-1951." Mental Hygiene 47: 649-678.
——— (1963g) The Mental Disease among Jews in Canada: A Study of First Admissions to Mental Hospitals, 1950-1952. Albany: Research Foundation for Mental Hygiene.
——— (1962a) "Mental disease among Norwegian-born and native-born of Norwegian parentage in New York State, 1949-1951." Acta Psychiatrica Scandinavica 38: 48-75.
——— (1962b) "Mental disease among Swedish-born and native-born of Swedish parentage in New York, 1949-1951." Acta Psychiatrica Scandinavica 38: 79-107.
——— (1962c) "Migration and mental disease among the white population of New York State, 1949-1951." Human Biology 34: 89-98.

——— (1960) Mental Disease among Jews in New York. New York: International Medical Book Corporation.
——— (1959) "Mental disease among Negroes: analysis of first admissions in New York State, 1949-1951." Mental Hygiene 43: 422-459.
——— (1956a) "Mental disease among native and foreign-born Negroes in New York State." Journal of Negro Education: 175-181.
——— (1956b) "Mental disease among Puerto Ricans in New York City, 1949-1951." Journal of Nervous and Mental Disease 123: 262-269.
——— (1955) "Mental disease among native and foreign-born white populations of New York, 1939-1941." Mental Hygiene 39: 545-563.
——— (1936) "Migration and mental disease among Negroes in New York State." American Journal of Physical Anthropology 21: 107-113.
——— and E. S. LEE (1956) Migration and Mental Disease, 1939-41. New York: Social Science Research Council.
MAOZ, B., S. LEVY, N. BRAND, and H. S. HALEVI (1966) "An epidemiological survey of mental disorders in a community of newcomers to Israel." Journal of the College of General Practitioners 11: 267-284.
MAY, J. V. (1922) Mental Disease: A Public Health Problem. Boston: Richard G. Badger.
MEADOW, A. and D. STOKER (1965) "Symptomatic behavior of hospitalized patients." Archives of General Psychiatry 12: 267-277.
MEDLICOTT, R. (1961) "Psychiatric impressions of South East Asia." Transcultural Research on Mental Health Problems 10: 33-35.
MEHLMAN, R. D. (1961) "The Puerto Rican Syndrome." American Journal of Psychiatry 118: 328-332.
MEZEY, A. G. (1960a) "Psychiatric illness in Hungarian refugees." Journal of Mental Science 106: 628-637.
——— (1960b) "Psychiatric aspects of human migrations." International Journal of Social Psychiatry 5: 245-260.
———(1960c) "Personal background, emigration and mental disorder in Hungarian refugees." Journal of Mental Science 106: 618-627.
MILLER, L. (1966) The Social Psychiatry and Epidemiology of Mental Ill Health of Israel. Jerusalem: Ministry of Health.
MINTZ, N. L. and D. T. SCHWARTZ (1964) "Urban ecology and psychosis: community factors in the incidence of schizophrenia and manic depression among Italians in Greater Boston." International Journal of Social Psychiatry 10: 101-118.
MISHLER, E. G. and N. A. SCOTCH (1963) "Sociocultural factors in the epidemiology of schizophrenia." Psychiatry 26: 315-351.
MOSES, R. and J. SHANAN (1961) "Psychiatric outpatient clinic: an analysis of a population sample." Archives of General Psychiatry 4: 60-73.
MURPHY, H.B.M. (1968) Mental Hospitalization Patterns in Twelve Canadian Subcultures. Montreal: McGill University Department of Psychiatry.
——— (1965) "Migration and the major mental diseases." Pp. 5-29 in M.B. Kantor (ed.) Mobility and Mental Health. Springfield, Ill.: Charles C. Thomas.
——— (1961a) Culture, Society and Mental Disorder in South East Asia. (unpublished).

——— (1961b) "Social change and mental health." Pp. 280-340 in Causes of Mental Disorders: A Review of Epidemiological Knowledge, 1959. New York: Milbank Memorial Fund.
——— (1955) Flight and Resettlement. Paris: Unesco.
MYERS, J. D. and B. H. ROBERTS (1959) Family and Class Dynamics in Mental Illness. New York: John Wiley.
NATHAN, T. S., L. EITINGER, and H. Z. WINNIK (1964) "A psychiatric study of survivors of the Nazi holocaust: a study in hospitalized patients." Israeli Annals of Psychiatry and Related Disciplines 2: 47-76.
ØDEGAARD, O. (1961) "L'épidémiologie des troubles mentaux en Norvège." Evolution Psychiatrique 26: 193-253.
——— (1957) "The epidemiology of schizophrenia in Norway." Pp. 49-51 in Vol. 3 of the Report of the Second International Congress in Psychiatry, Zurich.
——— (1945) "Distribution of mental diseases in Norway: contribution to ecology of mental disorder." Acta Psychiatrica et Neurologica Scandinavica 20: 247-284.
——— (1932) "Immigration and insanity: a study of mental disease among the Norwegian-born population in Minnesota." Acta Psychiatrica et Neurologica Scandinavica, Supplementum 4: 1-206.
OPLER, M. K. (1967) Culture and Social Psychiatry. New York: Atherton Press.
——— (1959) Culture and Mental Health: Cross-Cultural Studies. New York: Macmillan.
PALGI, P. (1963) "Immigrants, psychiatrists and culture." Israeli Annals of Psychiatry and Related Disciplines 1: 43-58.
PARKER, S. and R. J. KLEINER (1966) Mental Illness in the Urban Community. New York: Free Press.
PARKER, S., R. J. KLEINER and H. G. TAYLOR (1960) "Level of aspiration and mental disorder: a research proposition." Annals of the New York Academy of Science 84: 878-886.
PFLANZ, M., O. HESENKNOPF, and P. COSTAS (1967) "Blood pressure and functional complaints in guest workers: a transcultural comparison." Arbeitsmedizin, Socialmedizin, Arbeitshygiene, Heft 5: 1881-1885.
PINSENT, J. (1963) "Morbidity in immigrant population." Lancet 1: 437-438.
PLANK, R. (1959) "Ecology of schizophrenia: newer research on the drift hypothesis." American Journal of Orthopsychiatry 29: 819-826.
POECK, K. (1964) "Social and cultural factors in mental illness of immigrant workers." World Medical Journal 11: 75-76.
PRANGE, A. J. (1959) "An interpretation of cultural isolation and alien's paranoid reaction." International Journal of Social Psychiatry 4: 254-263.
RICHARDSON, A. (1957) "Some psycho-social characteristics of satisfied and dissatisfied British immigrant skilled manual workers in Western Australia." Human Relations 10: 235-248.
RIN, H. and T. Y. LIN (1962) "Mental illness among Formosan aborigines as compared with the Chinese in Taiwan." Journal of Mental Science 108: 134-146.
RIN, HSIEN, KWANG-CHUNG WU, and CHING-LING LIN (1962) "A Study of the content of delusions and hallucinations manifested by the Chinese paranoid psychotics." Journal of the Formosan Medical Association 61: 47-57.

RISSO, M. and W. BOKER (1964) "Delusions of being bewitched: a contribution to the understanding of delusional disorders in Southern Italian labourers in Switzerland." Bibliotheca Psychiatrica et Neurologica (Basel) Fasc. 124: 1-79.

ROBERTS, B. H. and J. K. MYERS (1954) "Religion, native origin, immigration." American Journal of Psychiatry 110: 759-764.

ROGLER. L. A. and A. B. HOLLINGSHEAD (1965) Trapped: Families and Schizophrenia. New York: John Wiley.

ROSE, A. M. (1964) "The prevalence of mental disorders in Italy."International Journal of Social Psychiatry 10: 87-100.

SALMON, T. W. (1913) "Immigration and the mixture of races in relation to the mental health of the nation." In Vol. I of A. White and S. E. Jeliffe (eds.) The Modern Treatment of Nervous and Mental Diseases. Philadelphia: Lea & Febiger.

SANUA, V. (forthcoming) "The psychology of the Egyptian fellahin." In E. L. Margetts (ed.) The Mind of Man in Africa: An Integrated Anthology Relating to Psychiatry and Mental Health. London: Pergamon Press.

--- (1969) "Sociocultural aspects of schizophrenia." Pp. 256-310 in L. Bellak and L. Loeb (eds.) The Schizophrenic Syndrome. New York: Grune & Stratton.

--- (1967) "The sociocultural aspects of childhood schizophrenia: a discussion with special emphasis on methodology." Pp. 159-176 in G. H. Zuk and I. Boszormenyi-Hagy (eds.) Family Therapy and Disturbed Families. Palo Alto, Calif.: Science and Behavior Books.

--- (1966) "Sociocultural aspects of psychotherapy and treatment: a review of the literature." Pp. 151-190 in L. Abt and B. Riess (eds.) Progress in Clinical Psychology. New York: Grune & Stratton.

--- (1963a) "The sociocultural aspects of schizophrenia: a comparison of Protestant and Jewish schizophrenics." International Journal of Social Psychiatry 9: 27-36.

--- (1963b) "The etiology and epidemiology of mental illness and problems of methodology." Mental Hygiene 47: 607-621.

--- (1962) "Comparison of Jewish and Protestant paranoid and catatonic patients." Diseases of the Nervous System 29: 1-7.

--- (1961) "The sociocultural factors of families of schizophrenics: a review of the literature." Psychiatry 24: 246-265.

SCHAECHTER, F. A. (1965) "Previous history of mental illness in female migrant patients admitted to the psychiatric hospital, Royal Park." Medical Journal of Australia 52: 277-279.

--- (1962) "A study of psychoses in female migrants." Medical Journal of Australia 49: 458-461.

SCHMIDT, K. E. (1961) "Management of schizophrenia in a Sarawak mental hospital." Journal of Mental Science 107: 157-160.

--- (1959) "The racial distribution of mental hospital admissions in Sarawak." Annual Report of Sarawak Mental Hospital.

SCHMITT, R. C. (1957a) "Population densities and mental disorders in Honolulu." Hawaii Medical Journal 16: 396-397.

--- (1957b) "Areal mobility and mental health on Oahu." Sociology and Social Research 42: 115-118.

--- (1956) "Psychosis and race in Hawaii." Hawaii Medical Journal 16: 144-146.
SCHOOLER, C. and W. CAUDILL (1964) "Symptomatology in Japanese and American Schizophrenics." Ethnology 3: 172-178.
SEGUIN, C. A. (1956) "Migration and Psychosomatic disadaptation." Psychosomatic Medicine 18: 404-409.
SINGER, J. L. and V. G. McCRAVEN (1962) "Daydreaming patterns of American subcultural groups." International Journal of Social Psychiatry 7: 272-282.
SROLE, L., T. S. LANGNER, S. T. MICHAEL, M. K. OPLER, and T.A.C. RENNIE (1962) Mental Health in the Metropolis: The Midtown Manhattan Study. New York: McGraw-Hill.
STERNBACK, A. and K. ACHTE (1964) "An Epidemiological study of psychiatric morbidity in Helsinki." Acta Psychiatrica Scandinavia 40, Supplementum 180: 287-307.
SUNDBY, P. and P. NYHUS (1963) "Major and minor psychiatric disorders in males in Oslo." Acta Psychiatrica Scandinavica 39: 519-547.
SUNIER, A. and F. MAIRLOT (1962) A Survey of Psychiatric Patients in the Jewish Communities of Belgium and Holland. Paris: Medical Department of the Jewish Distribution Committee.
SZWEDKOWICZ. H. et al. (1965) "The effect of migratory movements on the hospitalization indices of some psychic diseases." Neurologia et Neurochirurgia Psychiatrica Polska 15: 255-262.
TERASHIMA, S. (1958) "Schizophrenic Japanese-Canadians and their sociocultural backgrounds." Canadian Psychiatric Association Journal 3: 53-62.
TEWFIK, G. I. and A. OKASHA (1965) "Psychosis and immigration." Postgraduate Medical Journal 41: 603-612.
THOMAS, D. S. and B. Z. LOCKE (1963) "Marital status, education, and occupation differentials in mental disease: state patterns in first admission to mental hospitals for all disorders and for schizophrenia in New York and Ohio as of 1950." Milbank Memorial Fund Quarterly 41: 145-160.
TIETZE, C., P. LEMKAU, and M. COOPER (1942) "Personality disorder and spatial mobility." American Journal of Sociology 48: 29-39.
TSUANG, M. T. and H. RIN (1961) "An evaluation of social and migration factors among psychiatric out-patients in the University Hospital." Journal of the Formosan Medical Association 60: 30-36.
TUCKMAN, J. and R. KLEINER (1962) "Discrepancy between aspiration and achievement as a predictor of schizophrenia." Behavioral Science 7: 443-447.
TYHURST, L. (1951) "Displacement and migration: a study of social psychiatry." American Journal of Psychiatry 107: 561-568.
WARD, T. F. (1961) "Immigration and ethnic origin in mental illness." Canadian Psychiatric Association Journal 5 and 6: 323-332.
WEDGE, B. M. (1952) "Occurrence of psychosis among Okinawans in Hawaii." American Journal of Psychiatry 109: 255-258.
WEINBERG, A. A. (1964) "On comparative mental health research of the Jewish people." Israeli Annals of Psychiatry and Related Disciplines 2: 27-39.
--- (1961) Migration and Belonging: A Study of Mental Health and Personal Adjustment in Israel. The Hague: Martinus Nijhoff.
--- (1955) "Mental health aspects of voluntary migration." Mental Hygiene 39: 450-454.

WILSON, A. W., G. SAVER, and P. A. LACHENBRUCH (1965) "Residential mobility and psychiatric help-seeking." American Journal of Psychiatry 121: 1108-1109.

ZUNIN, L. M. and R. T. RUBIN (1965) "Paranoid psychotic reactions in foreign students from non-Western countries." Paper presented at the American Psychiatric Association Meeting, New York.

Chapter **14**

Social-Psychological Aspects of Migration and Mental Disorder in a Negro Population

**ROBERT J. KLEINER
and SEYMOUR PARKER**

The present paper will examine the relationship between migration and mental disorder. It will attempt to explain differences in rates of mental disorder between a population of Negroes who migrated from the South to the North and Negroes native to the North, in terms of such social and psychological factors as social mobility, goal-striving behavior, and ethnic or racial identity. Before dealing directly with these issues, however, we shall outline the research problems and findings that led to the present endeavors.

In previous research we attempted to replicate the findings of Malzberg and Lee (1956), who reported that Negroes moving to New York State from the South had higher rates of mental disorder than those born in that state. This attempted replication (which was carried out in Pennsylvania) yielded surprising results; rates of first admissions (1951-1956) of Negroes to Pennsylvania public mental hospitals were significantly higher for *natives* or that state than for southern migrants.[1] Rates for migrants from other northern communities, constituting only seven percent of our total sample, fell between those of the southern migrants and natives (Parker and Kleiner, 1966: 210).

Our interest in pursuing this matter was stimulated by studies indicating that juvenile delinquency in the Philadelphia Negro community was more characteristic of adolescent children of natives than for children of migrants (Savitz, 1960), and that "broken homes," as measured by divorces, separations, and desertions were also more frequent among natives than migrants in the same city (Kephart and Monahan, 1952). A possible explanation for these findings emerged from work done by Tuckman and Kleiner (1962). When they controlled for occupational level, they found that Negroes native to Philadelphia had significantly higher educational levels than migrants. Assuming that educational achievement was positively correlated with level of aspiration, this finding indicated the possibility that, compared to migrants from the South, the native Negro population was characterized by a higher discrepancy between level of aspiration and achievement.

The notion of the aspiration-achievement discrepancy suggested the relevance of Merton's means-ends paradigm (Merton, 1957: 131-194) and Lewin's level of aspiration theory (Lewin et al., 1944) for understanding mental disorder. We predicted that in an anomic situation (i.e., one in which there was a disjunction between means and ends) a mentally ill population, in contrast to a community population, would

(1) perceive a more open opportunity structure;
(2) experience higher levels of goal-striving stress;
(3) evaluate their own achievements as lower than those of their informal reference groups; and
(4) have lower self-esteem.

Within this frame of reference, we conducted a large sample study in the Philadelphia Negro community to see how mental disorder related to these social-psychological factors and to other selected social-structural variables. This investigation allowed us to derive rates of mental disorder for various population groups, and also to determine the degree of association between these groups and the social-psychological factors included in our theoretical approach. Differences obtained between the mentally ill and community populations supported the four predictions.

The social-psychological variables noted above were then utilized in an attempt to explain the differences in rates of mental disorder found for native Negroes and southern migrants. We reasoned that a Negro whose major socialization experience occurred in the semicaste system of the South was less likely to assume an open opportunity structure and less likely to internalize high levels of aspiration than an individual socialized in the North, with its ethos of an open class system and the possibility of upward mobility for everyone. The migrants in our sample were, in fact, characterized by lower levels of goal-striving and less intense psychological stress associated with unfulfilled goals (Parker and Kleiner, 1966: 227-231).

These different rates of mental disorder for natives and migrants were inconsistent with the prevailing explanation of "culture shock," a phenomenon often assumed to accompany the migration experience. This concept has frequently been employed as a post hoc explanation of relatively high rates of mental disorder among migrants (Murphy, 1965). According to the culture shock idea, migrants who enter a social environment very different from their community of origin, may experience such psychological difficulties as severe role discontinuities, value conflicts, and social disorientation. Since our findings were not in accord with this reasoning, we decided to review the relevant literature more critically and carry out more extensive analyses of our own data.

With respect to mental illness and migration, some studies reported higher rates for migrants than for nonmigrants (or natives), other research showed no difference in illness rates between two such populations (Murphy, 1965), and our own data indicated higher rates for natives. In an early comprehensive review of relevant studies, Malzberg and Lee (1956: 43) concluded that most differences associated with migration "have been based on scant or otherwise inadequate data, and even the fact of higher incidence of mental disease among migrants is not firmly established, much less the theories as to cause." On the basis of more recent data about the white population of New York State, Malzberg (forthcoming) reported higher rates of mental illness for natives than for migrants when various status controls were employed.

Studies comparing illness rates for migrants moving into urban settings from rural areas versus other urban areas also seemed relevant. On the basis of the culture shock concept, migrants from rural communities should manifest higher illness rates, since the greater sociocultural differences between their point of origin and terminus of migration would imply more potential adjustment difficulties and psychological stress. However, Astrup and Ødegaard (1960) found higher illness rates among migrants to Oslo from other urban areas in Norway, than among those coming to Oslo from rural communities. Srole, et al. (1962: 253-281) found that migrants to New York City from other urban areas manifested more psychosomatic symptoms than those from rural settings.

Our own data failed to support the culture shock hypothesis on two bases: Negro migrants to Philadelphia from other urban areas had higher illness rates than rural migrants (Parker, Kleiner, and Needleman, forthcoming). Furthermore, if culture shock contributed at all to mental disorder in a given situation, its effects would presumably be most marked during the initial period of the migratory adjustment. However, recent migrants to Philadelphia (i.e., within the five-year period prior to data collection) had *lower* rates than migrants who had been living in that city for longer than five years. These findings suggested that culture shock either was not a useful explanatory concept or that it was associated with the development of mental disorder only in particular situations.

We searched for alternative lines of inquiry that might lead to a more comprehensive explanation of these rate differences. Our findings showed that stress resulting from goal striving was higher among the mentally ill than among individuals in the community sample, and also higher for natives than for migrants. In addition, high levels of goal striving and weak or ambivalent ethnic identification were associated with social mobility (in either direction), all of which related positively to mental disorder (Parker and Kleiner, 1964; Kleiner and Parker, forthcoming). Although these findings at first seemed unrelated, the theoretical concepts of the "marginal man," goal-striving, and ethnic identity suggested to us a unifying theme. This, in essence, provided the central focus of the present paper: would these theoretical ideas account for the relatively high rates of mental disorder in a northern Negro population[2] in terms of social mobility, high levels of goal-striving, and weak or ambivalent ethnic identification?

We hypothesized that northern Negroes would be characterized by greater social mobility than migrants, a prediction based on the assumption that the northern environment would provide more opportunities for mobility[3] than the traditional semicaste system of the South. Although upward mobility would be markedly restricted compared to that available to a white population, we felt that the North still offered relatively more occupational and educational mobility than the South. Southern migrants in the sample might have experienced some social mobility after coming to Philadelphia, but we assumed that for the majority of our sample, educational attainment and occupational limits had been established by late adolescence. For purposes of this study we defined social mobility in a given area in terms of an individual's achievement relative to that of his parents.[4]

Numerous studies have presented both a theoretical rationale and empirical data linking social mobility to various manifestations of interpersonal disturbance (Greenbaum and Pearlin, 1953: 480-491; Blau, 1956; Janowitz and Curtis, 1957; Lipset and Bendix, 1963). According to some authors, mobility disrupts existing relationships and adjustments to new ones, resulting in a relatively weak integration of the individual with his significant social reference groups. Others have postulated that extreme social mobility is often accompanied by discrepancies or inconsistencies between a person's various statuses (e.g., high education and low occupation), which cause role conflicts and interpersonal disturbances. A review of the research on the relationship of mobility to mental disorder, however, has been inconclusive (Kleiner and Parker, 1963). Our own work clearly indicated a relationship between mental disorder and the extremes of *both* upward and downward mobility.

Durkheim's writings on anomie and deviant behavior (1951) provided a further theoretical association among the major variables examined in this paper. Durkheim noted that suicide rates tended to increase during periods of economic depression and prosperity. The fact that suicide was associated with both types of situations suggested to him the operation of factors other than mere physical deprivation or poverty. Assuming that both depression and prosperity resulted in increased social mobility, Durkheim reasoned that these economic conditions were accompanied by disturbances in the "collective order" in which individuals were involved. The normative regulation by society (i.e., the controlling influence of psychologically significant groups to which individuals belong) was no longer effective in keeping one's aspirations within realistic limits.[5] The anomic situation was defined in terms of this regulatory force becoming relatively weak and ineffective. Under these conditions, individuals lost their social "anchors."

Durkheim saw social mobility in either direction as a phenomenon that disrupted the individual's integration with social groups that had previously provided him with consensus about appropriate levels of aspiration. During an economic depression, downwardly mobile individuals should realistically limit their goals, although in a "de-regulated" or anomic condition many people cannot reduce their aspirational levels realistically. The person experiencing upward mobility in the anomic situation also often fails to perceive realistic limits and continues to strive for unattainable goals. In both instances disappointment and a sense of failure result from "the futility of an endless pursuit." According to Durkheim, this situation provides the preconditions for anomic suicide.

We realized that there was considerable overlap among the variables stemming from level of aspiration theory and those implicit in Durkheim's approach. On the basis of concepts from both sources, we formulated and tested a number of hypotheses about the preconditions of mental disorder. This study showed that individuals in the community population characterized by high levels of either upward or downward mobility

(1) showed less integration with significant reference groups than those who were socially nonmobile;

(2) were less likely than the nonmobile to perceive realistic limitations to their ambitions;

(3) had higher status aspirations and ambitions than the nonmobile; and

(4) had lower self-esteem than the socially nonmobile population (Kleiner and Parker, forthcoming).

Our rationale for hypothesizing that the northern Negro group would have a weaker ethnic identification than their migrant counterparts is presented below. Frazier (1957) maintained that upper- and middle-class Negroes were ambivalent toward the Negro masses. He felt that they internalized many of the negative white middle-class attitudes toward lower-class Negroes, and that consequently they regarded their own ethnic group membership negatively. Lewin (1948) noted the same pattern among upwardly mobile Jews. Both represent instances of the classical "marginal man" pattern. Since northern Negroes in our study were presumably more socially mobile than migrants, we hypothesized that they were also more likely to manifest the characteristics described by Frazier.

In the present study, we shall first examine whether rates of social mobility are actually higher for northern Negroes than for southern migrants. Secondly, we shall see if northerners have higher levels of goal striving than migrants.[6] Finally, we shall try to ascertain whether the northern Negro population is also characterized by a more ambivalent or weaker identification with the larger Negro community.

Procedure

The data analyzed in this paper were collected as part of a larger project investigating the relationship between mental illness and such social-psychological factors as goal-striving stress, reference group behavior, and self-esteem (Parker and Kleiner, 1966). Information was obtained from two Negro samples:

(1) a mentally ill sample (n = 1423), representative of all in- and out-patient admissions to selected public and private psychiatric agencies during the period March 1, 1960 to May 15, 1961; and

(2) a representative community sample (n = 1489), drawn from the Philadelphia Negro community.

Respondents in both samples were Negroes, between twenty and sixty years of age, living in Philadelphia; both they and their parents had been born within the continental United States. The 206-item questionnaire, used for both samples, was presented to each respondent in a face-to-face interview.

The rate of mental disorder for a particular migratory status groups was derived by dividing the number of interviewed patients in that group by the number of community respondents in the corresponding group and multiplying the quotient by 100. This figure represented the number of interviewed ill individuals in a particular migratory status group per 100 interviewed community respondents.

A recurrent problem in interpreting questionnaire data involves establishing temporal relationships among the variables employed. It is usually assumed that the designated independent variable precedes the dependent variable temporally, even when an interaction between variables is postulated. In the present study, migratory status was taken as the independent variable, social mobility, goal-striving level, and ethnic identify as the intervening variables, and mental disorder as the dependent variable. These assumptions about the direction of causality or mutual influence were examined by means of a method we called the "yield procedure." This method is described below.

First, we computed rates of mental disorder for various subgroups within the community population (e.g., male-female, young-old, high-low status, migrants-northerners, etc.). These rates were considered as the *yields* of mental disorder for the various community population subgroups. We then compared high and low illness-yield subgroups to ascertain whether high yield respondents were also characterized by a response configuration similar to that obtained for the mentally ill population. Implicit in this procedure was the assumption that *antecedent* social-psychological factors differentiating community and mentally ill individuals would also discriminate to some degree between high and low yield *community* subgroups—presumably few of whom are mentally ill by our definition.

An example may further clarify our rationale for this procedure. If males were found to be overrepresented and females underrepresented in the mentally ill sample compared to their distribution in the community sample, males in the community sample would be designated a high yield subgroup and females a low yield subgroup. We would then determine whether the community sample males (the high yield respondents) were more likely than females to manifest various responses already known to characterize the mentally ill population. All the variables on which migrant and northern Negro samples were compared in the present study had already been shown to differentiate between an ill and a community population. On the basis of the reasoning underlying the yield procedure, comparisons on these variables were made between migrants (low yield subgroup) and northerners (high yield subgroup) within the community population.

At this point a few specific aspects of our questionnaire will be briefly described. One set of items, from which three measures were derived, was designated as the "self-anchored striving scale" (Cantril, 1965). Each respondent was presented with a diagram of a ten-step ladder, the top step labeled "the best possible way of life" and the bottom step "the worst possible way of life." After each respondent described what these two labels connoted for him, he was asked to select the step that best represented his current position. He then indicated the step-position that he hoped to reach in "a few years from now." The numerical difference between these two positions was designated as the "discrepancy on the striving-scale." Finally, the subject estimated the probability (ranging from one to ten chances out of ten) of reaching his aspired striving-scale step level.

Another item was designed to tap respondents' perception of the opportunity structure. Each respondent was asked the following question: "It has been said that if a man works hard, saves his money, and is ambitious, he will get ahead. How often do you think this really happens?" The four precoded responses ranged from "Very often" to "Hardly ever." A "Very often" response was taken to reflect a view of the opportunity structure as relatively open; a "Hardly ever" response was taken to mean that the subject saw the system as relatively closed.

Results

Data on rates of mental disorder for southern migrants and northerners are presented in Table 1. In analyses comparing the total groups, as well as in those controlling for age and sex, northern Negroes were characterized by significantly higher illness rates than southern migrants. Additional analyses were carried out to determine whether the social-structural and psychological factors shown to be associated with mental disorder also differentiated between these two migratory status groups.

Previously we noted that individuals who were mobile in both directions in the occupational and educational hierarchies had higher illness rates than their nonmobile counterparts. On this basis, we predicted that *both* upward and downward mobility would be more heavily concentrated among northerners than among southern migrants. This prediction was confirmed for occupational and educational mobility (see Table 2). These trends remained consistent when age and sex were controlled.

Since we had found previously that the socially mobile population tended to view the opportunity structure as relatively open compared to the socially nonmobile, we hypothesized that the northern Negro would see the opportunity structure as relatively open, while southern migrants would be more prone to view it as closed. This hypothesis was in fact supported by earlier analyses. However, in the present study we decided to examine a less global measure of perception of the opportunity structure—that is, to look at respondents' estimates of their probability of attaining occupational aspirations and goals as determined by the striving scale. It was predicted that for both goal areas northerners would have higher probability-of-success estimates than southern mi-

TABLE 1
RATES OF MENTAL DISORDER AND MIGRATORY STATUS

	Community Population (Percentages)	Mentally Ill Population (Percentages)	Rate[a]
MALE			
Southern Migrants	61	46	72[b]
Northerners	39	54	132
N	(600)	(576)	
FEMALE			
Southern Migrants	61	49	77[b]
Northerners	39	51	124
N	(885)	(845)	
YOUNG			
Southern Migrants	49	42	107[b]
Northerners	51	58	143
N	(798)	(1007)	
OLD			
Southern Migrants	75	63	51[b]
Northerners	25	37	89
N	(684)	(414)	
TOTAL			
Southern Migrants	61	48	75[b]
Northerners	39	52	128
N, Total	(1485)	(1421)	

[a] The number of mentally ill divided by the number in the community multiplied by 100.

[b] $p < .001$

TABLE 2
SOCIAL MOBILITY AND MIGRATORY STATUS

	Occupational Mobility[a] (Percentages)			
	Upward	Nonmobile	Downward	N
Southern Migrants	20	59	21	(827)
Northerners	36	39	25	(517)
	Educational Mobility[b] (Percentages)			
Southern Migrants	58	31	11	(901)
Northerners	60	25	15	(578)

[a] By chi-square, $p < .001$

[b] By chi-square, $p < .02$

grants. This prediction was confirmed (see Table 3). Differences continued to emerge where age and sex controls were introduced.

Since northerners viewed the opportunity structure as more open and were also more socially mobile, we predicted that they would be characterized by larger discrepancies between their aspirations and actual achievements. Data summarized in Table 4 shows that these predictions were confirmed with respect to aspiration-achievement discrepancies in the occupational and generalized striving-scale goal areas.

We also expected northerners to have higher occupational status reference groups than southern migrants, a characteristic previously found to be associated with social mobility. One questionnaire item presented respondents with three different paired combinations of occupations; each respondent indicated his preference for a white-collar, low-paying job versus a blue-collar, higher salaried job. It was assumed that when an individual selected the white-collar position, despite its lower financial remuneration, he was using a higher occupational reference group as a basis for his choice. Compared to southern migrants, natives and northern migrants clearly showed stronger preferences for the more prestigeful white-collar occupations (see Table 5). The differences were significant for all three paired combinations of occupations.

We have reported that compared to southern migrants, northerners had higher expectations of success, larger aspiration-achievement discrepancies, and higher occupational status reference groups. These characteristics have also been associated with social mobility. We also predicted

TABLE 3
SUBJECTIVE PROBABILITIES OF SUCCESS AND MIGRATORY STATUS

	For Occupational Aspirations[a] (Percentages)			
	Low	Medium	High	N
Southern Migrants	30	29	41	(789)
Northerners	24	33	43	(504)
	For Striving-Scale Aspirations[b] (Percentages)			
Southern Migrants	12	29	59	(895)
Northerners	4	31	65	(574)

[a] By chi-square, $p < .05$

[b] By chi-square, $p < .001$

TABLE 4
MEAN ASPIRATION-ACHIEVEMENT DISCREPANCIES AND MIGRATORY STATUS

	For Occupational Aspirations[a]		For Striving-Scale Aspirations[a]	
Southern Migrants	1.22		2.38	
Northern Migrants[b]	1.38	Mean for Combined Groups: 1.48	2.46	Mean for Combined Groups: 2.69
Natives[b]	1.50		2.74	

[a] Southern Migrants versus Natives, by t-test, $p < .001$

[b] These data were taken from previous publication (Parker and Kleiner, 1966: 270). For the present paper, no additional tests of significance were made between the southern migrant group and the combined northern migrant-native group.

TABLE 5
PREFERENCES FOR OCCUPATIONAL REFERENCE GROUPS, BY MIGRATORY STATUS

	Bricklayer vs. Teacher[a] (Percentages)		Machine Operator vs. Government Clerk[b] (Percentages)		Factory Worker vs. Sales Clerk[a] (Percentages)	
Southern Migrants	43	57	52	48	63	37
Natives[c]	31	69	42	58	54	46
Northern Migrants[c]	22	78	46	54	51	49

[a] By chi-square, $p < .001$

[b] By chi-square, $p < .01$

[c] These data were taken from a previous publication (Parker and Kleiner, 1966: 272). For the present paper, no additional tests of significance were made between the southern migrant group and the northern migrant-native group.

that northerners would be more dissatisfied than southern migrants with their achievements. It was assumed that individuals who placed themselves at lower positions on the striving scale were also more dissatisfied with their achievements than individuals who placed themselves at higher positions on this scale. This expectation was confirmed and reached statistically significance (see Table 6). These differences were also obtained with age and sex controls.

TABLE 6
SUBJECTIVE PERCEPTION OF CURRENT POSITION ON THE STRIVING SCALE, BY MIGRATORY STATUS

Perceived Striving-Scale Position	Southern Migrants[a] (Percentages)	Northerners[a] (Percentages)
Low	12	11
Medium	42	53
High	46	36
N	(945)	(615)

[a] By chi-square, $p < .001$

In summary, comparisons between northerners and southern migrants have shown that northerners are more heavily concentrated in the upwardly and downwardly mobile groups, and also that they are more prone to manifest the social-psychological goal-striving characteristics associated with mobility status and mental illness. Up to this point the discussion has been presented in the context of status striving within *vertical* hierarchies. Now we shall examine respondents' attitudes toward the larger Negro and white communities. This set of attitudes was previously referred to as "ethnic identity." Such attitudes may be conceptualized in terms of striving along a *horizontal* dimension, since they encompass the individual's view of his actual and desired position with respect to the two communities. In this paper we shall determine whether northerners, by virtue of their greater social mobility, have the same ethnic identity characteristics we have found to be associated with socially mobile groups.

A series of items in our questionnaire was designed to evaluate ethnic identity. Two of these items asked for respondents' reaction to the following situations: (a) "If you picked up a newspaper and saw the headline, 'Negro Receives Major Award,' which of the following would best describe your first reaction?" and (b) "If you picked up a newspaper and saw the headline, 'Negro Seized in Camden,' which of the following would best describe you first reaction?" The precoded responses to the first item ranged from "Very proud" to "Slightly annoyed," and to the second item from "Very uncomfortable" to "No feelings either way."

Since the northerners were found to be more socially mobile and thus would presumably be less strongly identified with the Negro community, we predicted that they would be "Slightly annoyed" or have "No feeling either way" about the positive headline. The direction of difference was as expected but not significant (see Table 7). On the

TABLE 7
REACTION TO POSITIVE AND NEGATIVE HEADLINES, BY MIGRATORY STATUS

	Southern Migrants (Percentages)	Northerners (Percentages)
Reaction to Positive Headline		
Very Proud	79	77
Fairly Proud	12	11
Slightly Proud	4	4
No Feelings Either Way	4	6
Slightly Annoyed	1	2
N	(955)	(618)
Reaction to Negative Headline[a]		
Very Uncomfortable	38	26
Fairly Uncomfortable	17	11
Slightly Uncomfortable	17	22
No Feelings Either Way	28	41
N	(949)	(615)

[a] By chi-square, $p < .001$

contrary, both northerners and southern migrants overwhelmingly manifested feelings of pride.

This measure taken alone would have indicated that our prediction was not confirmed. However, data relating to the *negative* headline yielded very different results. On this item we expected northerners to respond more often with "Slightly uncomfortable" and "No feelings either way," and southern migrants to say "Very uncomfortable" and "Fairly uncomfortable." The findings supported these expectations, and the differences were statistically significant (see Table 7). It is evident that both groups denied strong involvement in the negative situation, but the denial was significantly greater in the northern group. These differences were also obtained in analyses controlling for age and sex. We interpreted these findings as indicating greater conflict over ethnic identity among northerners than among southern migrants.

A third and perhaps more direct measure of a respondent's identification as a Negro was based on this item: "If you had a Negro friend who told you that he wanted to 'pass' for the advantages that it would give him, which of the following choices best describes how you would feel?" The precoded responses ranged from "Glad for him" to "No feelings either way." Consistent with our theoretical frame of reference, we expected that northerners would say either "Glad for him" or to have "Mixed feelings toward him," while southern migrants would tend to be "Angry with him." These predictions were confirmed, as shown by the data in Table 8, and the differences were statistically significant. The direction of the differences was the same with age and sex controls. The data from these three questions thus confirmed the view that, compared to southern migrants, northerners were less positively identified or integrated with the Negro community.

On the basis of this ambivalent or weak ethnic identity, we expected northerners to deny or deemphasize the importance of being Negro. One of the measures used to assess this notion was the following: "In your opinion, has being a Negro prevented you from getting the things you wanted?" There were four precoded responses ranging from "Yes, very much" to "No." We predicted that more northerners than southern migrants would respond, "No" or "Yes, slightly." These expectations were confirmed, and the differences were again significant (see Table 9). As with the previous measures, these differences were obtained regardless of age or sex.

TABLE 8

REACTION TO FRIEND WHO WANTS TO "PASS," BY MIGRATORY STATUS

	Southern Migrants[a] (Percentages)	Northerners[a] (Percentages)
Glad for Him	24	30
Mixed Feelings Toward Him	33	38
Angry with Him	27	19
No Feelings Either Way	16	13
N	(932)	(587)

[a] By chi-square, $p < .001$

TABLE 9

PERCEPTION OF RACE AS A BARRIER TO GOAL ACHIEVEMENT, BY MIGRATORY STATUS

	Southern Migrants[a] (Percentages)	Northerners[a] (Percentages)
Yes—Very Much	9	4
Yes—to Some Degree	33	27
Yes—Slightly	28	24
No—Not at All	29	41
N	(940)	(612)

[a] By chi-square, $p < .001$

In light of their ambivalent or weak identification with the Negro community and their denial of race as a barrier to achievement, we expected the northerners to be more "whiteward" mobile. A very direct reflection of this aspiration would be the racial composition of the neighborhood in which they preferred to live, given a choice. Table 10 summarizes respondents' neighborhood choices, controlling for their actual residential neighborhoods. It is interesting that the two groups did not differ in their descriptions of the racial composition of their actual neighborhoods.

For each type of actual neighborhood (see Table 10), significantly more northerners than southern migrants preferred to live in "Almost all white" and "Mixed-mostly white" neighborhoods. On the other hand, more southern migrants in every classification of actual neighborhood preferred to live in "Mixed-mostly Negro" or "All Negro" neighborhoods. These differences were also significant. Again age and sex did not change the response patterns appreciably.

In a further analysis of this issue, we compared the proportions of northerners and southern migrants who were satisfied with their actual neighborhoods. Satisfaction was inferred when an individual aspired to live in the same kind of neighborhood in which he currently resided. Comparisons were made for respondents living in the following kinds of neighborhoods: (a) "Almost all white" and "Mixed-mostly white;" (b) "Mixed-mostly Negro;" and (c) "All Negro." The proportion of northerners satisfied with their neighborhoods decreased in this order: 50, 29, and 22 percent, respectively. The proportion of southern migrants satisfied with their neighborhoods *increased* in this order: 19, 23, and 38 percent, respectively. These data clearly indicated that northerners were relatively more satisfied when living in predominantly white neighborhoods, and that southern migrants preferred living in predominantly Negro neighborhoods. These differences were statistically significant.

We also evaluated the reasons given by respondents for selecting various kinds of neighborhoods. We expected that northerners would be more concerned with physical and/or social class aspects of the neighborhood, with the issue of integration, and that they would deny more readily the relevance of race in selecting a neighborhood. On the other hand, we predicted that southern migrants would be more receptive to living with other Negroes and/or more hostile toward, or rejecting of, whites. The data went in the direction of our expectations (See Table 11), but reached significance only for two types of reasons: those reflecting (a) positive attitudes toward Negroes and/or negative attitudes toward whites; and (b) concern for "better neighborhoods" (physical or social class aspects). These differences were also obtained when age and sex were controlled.

TABLE 10

RACIAL COMPOSITION OF ACTUAL AND ASPIRED RESIDENTIAL NEIGHBORHOOD, BY MIGRATORY STATUS

	Actual Neighborhood							Total	
	Almost All White; Mixed–Mostly White		Mixed–Mostly Negro		All Negro				
	Southern Migrants	Northerners[a]	Southern Migrants	Northerners	Southern Migrants	Northerners[b]	Southern Migrants	Northerners[c]
	(Percentages)		(Percentages)		(Percentages)		(Percentages)	
Aspired Neighborhood								
Almost All White; Mixed–Mostly White	34	56	9	13	9	17	11	17
Half-and-Half	14	5	9	11	5	9	8	10
Mixed–Mostly Negro; All Negro	18	5	35	22	52	38	40	32
No Preference	34	33	47	44	33	36	41	41
N	(65)	(39)	(535)	(358)	(295)	(178)	(895)	(576)

[a] By chi-square, $p < .05$
[b] By chi-square, $p < .01$
[c] By chi-square, $p < .001$

TABLE 11
REASONS FOR ASPIRED NEIGHBORHOOD, BY MIGRATORY STATUS

	Southern Migrants (Percentages)	Northerners (Percentages)
Race Irrelevant	12	15
Positive Attitudes Toward Negroes; and/or Negative Attitudes Toward Whites	24	17[a]
Better Neighborhoods (Physical and Social Class Aspects)	31	36[a]
Negative Attitudes Toward Negroes; and/or Positive Attitudes Toward Whites	2	3
To Further Cause of Integration	13	15
Other	16	12
Can't Code	2	2
N	(906)	(583)

[a] Versus all other categories combined, by chi-square, $p < .01$.

Similar questionnaire items probed respondents' preferences regarding the racial composition of organizations they would want to join, and the reasons for their choices. The distribution of responses was similar to that obtained for "neighborhoods," and again age and sex did not alter these patterns appreciably.

In summary, we have shown that the Negroes' psychological orientation toward the Negro and white communities (i.e., horizontal mobility) is associated with vertical mobility status, mental illness, and migratory status.

It has been established that high goal-striving and weak ethnic identity coexist in the native northerners. Our theoretical position implicitly assumes that these two variables are directly related to each other, and we must establish whether or not this is so. As yet we have not probed this issue thoroughly, but we shall present some preliminary relevant analyses.

Previous analyses showed that, compared to southern migrants, northern Negroes stated higher probabilities of success on the striving scale measure, and were also more prone to deny that being a Negro constituted a barrier to achievement. To what extent were these factors interrelated? We found that increasingly higher probability of success estimates were directly associated with proneness to deny race as a barrier to achievement (see Table 12). This relationship was statistically significant.[7]

TABLE 12

PERCEPTION OF RACE AS A BARRIER TO GOAL ACHIEVEMENT, AND SUBJECTIVE PROBABILITY OF SUCCESS ON THE STRIVING SCALE

	Race Seen as Barrier[a]		
	Very Much; To Some Degree (Percentages)	Slightly; Not at All (Percentages)	N
Probability Estimate			
Low	55	45	(141)
Medium	42	58	(490)
High	33	67	(941)

[a] By chi-square, $p < .001$.

Another analysis related aspiration-achievement discrepancy scores on the striving scale to respondents' preferred residential neighborhoods. Results of this analysis are summarized in Table 13. For the total population, the aspiration-achievement discrepancy increased as preferred neighborhood moved from "All Negro" to "Almost all white." The relationship between these two measures was significant. These findings suggest that values with respect to associating with Negroes and whites relate to respondents' more general achievement striving.

Summary and Conclusions

Analyses discussed in this paper showed that northern Negroes, with higher rates of mental disorder than southern migrants, were also characterized by relatively higher levels of upward and downward social mobility, more intense goal-striving behavior, and weak or ambivalent attitudes toward their Negro group membership. These social-structural and psychological characteristics were previously shown to be associated with a known mentally ill population. The use of the "yield procedure," which involved demonstrating that these same characteristics prevailed in high illness-yield *community* population segments, increased the probability that such factors were not simply consequences of mental disorder, but rather dynamically involved in its development.

This conclusion was based not only on statistical findings which went in the expected direction, but also on the fact that these predictions derived from existing theory that linked seemingly disparate factors. The empirical findings that social mobility was associated with high rates of mental disorder, and also that high illness rates characterized northern

Negroes, indicated that the northern group was more socially mobile than a southern migrant group. The introduction of mobility as a variable suggested the relevance of Durkheim's theory of social mobility and anomie. This theory in turn suggested that goal-striving behavior and ethnic identity might relate to each other within a given individual and also be associated with mental disorder. Developing this strategy and carrying out the subsequent analyses impressed us with the value of integrating theory with the procedures of epidemiological research.

In reviewing the literature on migration and mental disorder we were struck not only by the lack of consistency in findings, but also by the fact that relatively few studies have utilized existing theoretical formulations in their research designs. This lack of systematic theory has resulted in frequent use of such post hoc concepts as "culture shock" to explain findings. We do not mean to imply that factors subsumed under this concept may not be very relevant to mental disorder and migration. However, the lack of studies done within comprehensive and unified theoretical frameworks, including concepts such as culture shock, has resulted largely in gathering discrete and unrelated bits of evidence. Until there are systematic investigations of how such gross structural factors as social class and migration are linked to intervening variables, we cannot adequately interpret the role played by more specific factors in the development of mental disorder. The crucial question is not *whether* migration is a factor in the etiology of mental disorder, but *why* it functions differently from one situation to another. Understanding something of the relationship between migration and mental disorder allows us to examine further how migration functionally articulates with other intervening social and psychological variables in different types of situations.

TABLE 13

MEAN ASPIRATION-ACHIEVEMENT DISCREPANCY ON THE STRIVING SCALE AND ASPIRED RESIDENTIAL NEIGHBORHOOD, BY MIGRATORY STATUS

Aspired Neighborhood	Southern Migrants	Northerners	Total Population[a]
Almost All White	1.30	1.91	1.48
Mixed–Mostly White	.81	1.57	1.91
Half-and-Half	1.11	1.30	1.19
Mixed–Mostly Negro	.87	.97	.92
All Negro	.73	.88	.76

[a] Since the distribution of respondents within each aspired neighborhood category was not appropriate for the Analysis of Variance procedure, the chi-square test was used instead. For the total population, by chi-square, $p < .001$.

The findings reported in this study invite speculation about the significance of the current Black Power movement in Negro life. The older and more traditional Negro leaders in civil rights organizations hoped that increasing numbers of Negroes would take advantage of opportunities to acquire additional education and better jobs. Integration with the larger dominant community would become facilitated as these Negroes "melted" into the larger middle-class "pot" via individual upward mobility. With the ascendance of Black Power, younger Negroes began to eye this traditional view critically. This type of solution would take too long to be effective, and grave doubts were raised about individual mobility as a viable solution for the masses of Negroes in our society. Critics also noted that upwardly mobile Negroes, who constituted the potential leadership for the larger Negro community, often acquired attitudes that ill-suited them for such leadership roles. As noted previously, Frazier (1957) maintained that middle-class, mobile Negroes were ambivalent toward the Negro masses. More recently, others (Lomax, 1960; Fuller, 1963) have accused the mobile, traditional Negro leaders of being "carbon copies" of the white middle class and of harboring covert contempt for the black masses. Those in the Black Power movement have accepted the idea that the fates of all Negroes are somehow linked, and that only by collective action can Negroes hope to improve their position in American society. The search for a positive group identity is seen in such slogans as "Black is Beautiful." It is not surprising that this point of view has been accompanied by opposition and polemics from both the white and black communities.

Our findings support the validity of the image held by these young critics. We found that upward mobility, coupled with high levels of goal-striving, were in fact associated with weak or ambivalent ethnic identity and mental disorder. A Negro leadership arising from the upwardly mobile segment of the population would more likely become lost to the larger Negro community, and might even be co-opted into a position of maintaining the status quo.

In this context it is also clear why Black Power has had, and continues to have, its *major* appeal among lower socioeconomic class Negroes with more positive, or less ambivalent, identity. Insofar as this movement is successful in creating a more positive identity for the Negro, it provides the potential for the development of a more effective leadership and better mental health.

NOTES

1. All subjects in this study were between twenty and sixty years of age. We classified a subject as a "native" or "migrant" according to his place of socialization (i.e., where he spent the majority of his first seventeen years of life).

2. The northern group includes both natives of Philadelphia and migrants to that city from other northern, predominately urban communities. In effect, this group includes all those socialized in the North, as this area is defined by the U.S. Census.

3. Respondents in the northern population showed more dispersion on measures of social status than those in the southern migrant population. This dispersion implies a greater likelihood of *both* upward and downward movement among the northern respondents.

4. A respondent's achievement in a given area was compared to that of whichever parent had attained the higher level. If his own level was higher than the parental level, he was classified as "upwardly mobile"; if his level was lower, he was called "downwardly mobile." If his achievement level was the same as the parental level, he was considered "nonmobile."

5. Merton (1957: 225-386), using more current sociological terminology, viewed such normative regulatory functions as deriving from significant reference groups which supply guidelines for one to judge his own performance and to set realistic aspirations. Group norms concerning appropriate levels of goal striving and probable rewards enable an individual to adjust to his place in relevant status hierarchies. Strong integration with significant reference groups facilitates the internalization of legitimate upper and lower limits to the range of goals potentially attainable to someone in his position.

6. This hypothesis will be approached in a manner different from that employed in the previously noted study (Kleiner and Parker, forthcoming). By this means, we shall both extend our knowledge of and confirm past findings about relative levels of goal striving in northern and migrant Negro populations.

7. This relationship holds for the migrant as well as the native population.

REFERENCES

ASTRUP, C. and ØDEGAARD (1960) "Internal migration and mental disease in Norway." Psychiatric Q. Supplement 34: 116-130.

BLAU, P. M. (1956) "Social mobility and interpersonal relations." American Sociological R. 21: 290-295.

CANTRIL, H. (1965) The Pattern of Human Concerns. New Brunswick: Rutgers Univ. Press.

DURKHEIM, E. (1951) Suicide (J. A. Spaulding and G. Simpson, translators). Glencoe, Ill.: The Free Press.

FRAZIER, E. F. (1957) Black Bourgeoisie. Glencoe, Ill.: The Free Press.

FULLER, H. W. (1963) "Rise of the Negro militant." The Nation 197 (September 14): 138-140.

GREENBAUM, J. and L. I. PEARLIN (1953) "Vertical mobility and prejudice: a socio-psychological analysis." Pp. 480-491 in R. Bendix and S. M. Lipset (eds.) Class, Status and Power: A Reader in Social Stratification. Glencoe, Ill.: The Free Press.

JANOWITZ, M. and R. CURTIS (1957) Sociological Consequences of Upward Mobility in a U.S. Metropolitan Community. Working Paper, The Fourth Working Conference on Social Stratification and Social Mobility, International Sociological Association, December.

KEPHART, W. M. and T. P. MONAHAN (1952) "Desertion and divorce in Philadelphia." American Sociological R. 17: 719-727.

KLEINER, R. J. and S. PARKER (forthcoming) "Social mobility, anomie, and mental disorder." In S. C. Plog, R. B. Edgerton and W. C. Beckwith (eds.) Changing Perspectives in Mental Illness. New York: Holt, Rinehart & Winston.

——— and S. PARKER (1963) "Goal-striving, social status, and mental disorder." American Sociological R. 28: 189-203.

LEWIN, K. (1948) Resolving Social Conflicts. New York: Harper.

——— T. DEMBO, L. FESTINGER and P. S. SEARS (1944) "Level of aspiration." Pp. 333-378 in J. McV. Hunt (ed.) Personality and the Behavior Disorders: A Handbook Based on Experimental and Clinical Research. New York: The Ronald Press.

LIPSET, S. M. and R. BENDIX (1963) Social Mobility in Industrial Society. Berkeley: Univ. of Calif. Press.

LOMAX, L. E. (1960) "The Negro revolt against Negro leaders." Harper's: 220: 41-48.

MALZBERG, B. (forthcoming) "Migration and rates of mental disorder." In S. C. Plog, R. B. Edgerton and W. C. Beckwith (eds.) Changing Perspectives in Mental Illness. New York: Holt, Rinehart & Winston.

——— and E. S. LEE (1956) Migration and Mental Disease: A study of First Admissions to Hospitals for Mental Disease, New York, 1939-1941. New York: Social Science Research Council.

MERTON, R. K. (1957) Social Theory and Social Structure. Glencoe, Ill.: The Free Press.

MURPHY, H. B. M. (1965) "Migration and the major mental disorders: a reappraisal." Pp. 5-29 in M. B. Kantor (ed.) Mobility and Mental Health. Springfield, Ill.: Charles C Thomas.

PARKER, S., R. J. KLEINER and B. NEEDLEMAN (forthcoming) "Migration and mental illness: some reconsiderations and suggestions for further analysis." Social Science and Medicine.

——— and R. J. KLEINER (1966) Mental Illness in the Urban Negro Community. Glencoe, Ill.: The Free Press.

——— and R. J. KLEINER (1964) "Status position, mobility and ethnic identification of the Negro." J. of Social Issues 20: 85-102.

SAVITZ, L. (1960) Delinquency and Migration. Philadelphia: Commission on Human Relations.

SROLE, L., T. S. LANGER, S. T. MICHAEL, M. K. OPLER and T. A. C. RENNIE (1962) Mental Health in the Metropolis: The Midtown Manhattan Study. New York: McGraw-Hill.

TUCKMAN, J. and R. J. KLEINER (1962) "Discrepancy between aspiration and achievement as a predictor or schizophrenia." Behavioral Science 7: 443-447.

PART IV

PROGRAM PLANNING and RESEARCH

Chapter **15**

A Simulation Model of Urbanization Processes

ROBERT C. HANSON, WILLIAM N. McPHEE, ROBERT J. POTTER, OZZIE G. SIMMONS and JULES J. WANDERER

Our objective is to report on a computer simulation model which reproduces the life experiences of rural migrants as they learn to adjust to the city environment. The working model is the outcome of several years of empirical work and theoretical reformulations; the history of the development of the model is a story in itself, not presented here.[1] We wish to describe the model as a "final" product, to show the simplicity of its assumptions but also to indicate something of the complexity of the consequences of these assumptions as they are put into operation in a computer.

We believe that the major contribution of the model is its ability to simulate the dynamic interaction of newcomers with the social structures of the city. It is in their new roles as particular neighborhood residents, job holders, consumers, and leisure participants that newcomers begin to

---

Author's Note: *This is a revised version of a paper read at the annual meeting of the American Sociological Association, San Francisco, August 1967. The research is supported by Grant No. MN-90208 from the National Institute of Mental Health. The project is one of several being carried on in the Program on Social Processes, Institute of Behavioral Science, University of Colorado. We wish to acknowledge the assistance of Marjorie C. Schultz and Dan Anderson, expert programmers for the project.*

integrate into unfamiliar social structures and a different style of life. It is the changes in these roles and the consequent changes in associated attributes that reflect the progress of a migrant toward integrated adjustment or rootless maladjustment to the city. To be able to reproduce the processes through which men acquire new roles and attributes is to have in hand a means for the systematic exploration of a theory of urban socialization.

The basic assumptions of the model can be stated as follows:

(1) Men search their environment for opportunities to achieve goals arising out of their current living situation and recent past experience (e.g., a man may be searching for a larger apartment, a better-paying job, a new car, membership in a social club).

(2) The city provides particular opportunities through its "opportunity structure" composed of four major "market sectors," (i.e., the housing, employment, consumer goods, and leisure activity sectors).

(3) Opportunities are advertised through three types of communication structures (i.e., impersonal bureaucratic, natural proximity, and social net structures), and men search for opportunities within these same structures.

(4) As a consequence of the interaction of advertised opportunities and searching men, a man may accept opportunities which change his current living situation and goal orientation (e.g., he may have acquired a new job and its associated attributes).

(5) The changed living situation and goal orientation of the man effects changes in his search for opportunities next time.

What follows in this paper is an attempt to spell out how these assumptions can be represented in a computer: (a) how opportunities and the opportunity structure of a city can be represented; (b) how the interaction of opportunities and men can be represented within communication structures; and (c) how the future search behavior of a man can be estimated from his current living situation, the latter conceived as a list of current attributes of the migrant and his family.

The Concept and Representation of Opportunities

Formally, an opportunity is a set which can be occupied or not.[2] An opportunity may be represented as a box (a set) as shown in Figure 1. A point inside is an occupant (element) of that set; a point outside is a

occupied opportunity

unoccupied opportunity

element in the universe of possible occupants

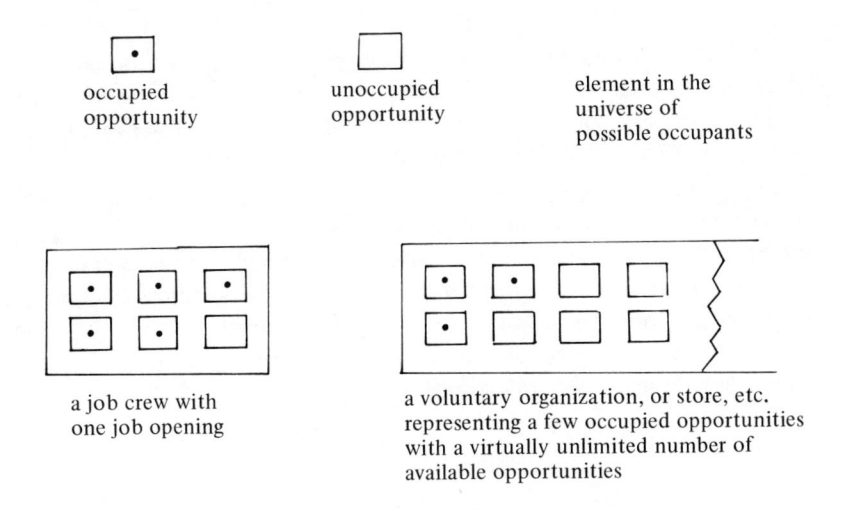

a job crew with one job opening

a voluntary organization, or store, etc. representing a few occupied opportunities with a virtually unlimited number of available opportunities

Figure 1. OPPORTUNITIES AS SETS

person not occupying any of these opportunities (an element in the universe, but not in these sets); and a vacant box is an opportunity without an occupant (a vacant house, an unfilled job).

Opportunities are usually valued, but need not be. A vacant house, a job on the market, unpurchased durable consumer goods in a store, a seat in a bar are examples of valued opportunities. Although we do not do so in this model, we could also treat a bed in a hospital or a cell in a jail as opportunities to be occupied, whether sought or not. Some opportunities, like jobs, are scarce; others, such as a picnic spot in the mountains, a seat in a church, a particular make and style of automobile, and so on are practically always available.

A cluster of opportunities may also be a set, such as all the jobs in one work crew, all the houses in one neighborhood, all the cars in one car lot, and the like. The larger cluster of similar opportunities may be represented as a box containing the smaller boxes representing occupied and unoccupied opportunities. For example, in Figure 3, the numbered boxes represent typical Denver neighborhoods with different characteristics, each with ten houses or apartments either occupied or vacant. Each of these housing opportunities carries with it a set of attributes: a certain number of rooms, a monthly rental or mortgage payment, and a degree of trouble risk from living in a safe or slum-like neighborhood.

[380] BEHAVIOR IN NEW ENVIRONMENTS

Each box containing a cluster of opportunities may also be conceived of as an organization subunit: a church, a neighborhood, the work crew of a construction company, the appliance division of a department store, and so on. Then lines connecting the boxes represent various types of linkages among the organizational units, such as nearness in physical space or nearness in theoretical similarity. The result of the construction of such linkages is a lattice which can be converted to a graph, and such graphs can be represented in a computer. In the model the linkage lines become the lines of communication flow for formal advertising or informal communication about unoccupied opportunities, and are also the paths which men follow in search of opportunities.

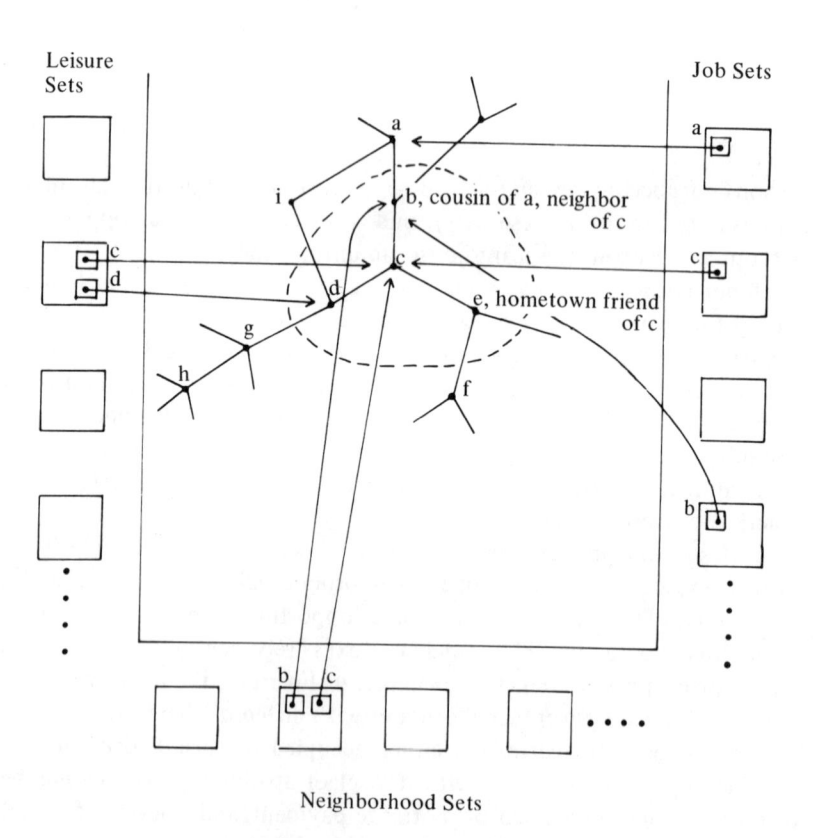

Figure 2. **A PORTION OF THE SOCIAL NET SHOWING C'S FIRST ORDER ASSOCIATES**

We have constructed lattices representing four major market sectors in the city—housing, employment, consumer goods, and leisure—each offering opportunities to newcomers. Each opportunity carries a set of attributes with it so that when an unoccupied opportunity is filled by a man, he acquires the associated attributes as a consequence.

As illustrated in Figure 3, the opportunity structure of a city may be represented in a computer by lattices converted into graphs, but we have not yet described the structures within which men and opportunities interact and how the dynamics of this interaction occurs within three types of communication structures, which we label the social net, the natural proximity structure, and the impersonal bureaucratic structure.

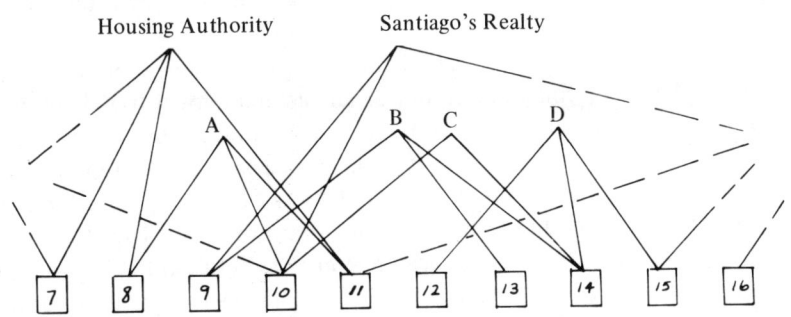

Each numbered box represents a neighborhood with occupied and unoccupied housing opportunities.

Graph Structures Derived From the Lattice

The lines represent communication linkages among organizational units. The natural proximity graph is constructed by connecting the member sets of nodes A, B, C, and D of the lattice. The impersonal bureaucratic graph is constructed by connecting the member sets of the nodes "Housing Authority" and "Santiago's Realty."

Figure 3. THE LATTICE STRUCTURE OF A PORTION OF THE HOUSING SECTOR

Communication Structures

The Social Net

Information about available opportunities may come to one's attention, or opportunities may be searched for, through one's close friends and relatives, and, at second and higher removes, through their friends and relatives.

Figure 2 illustrates a portion of the social net represented as a graph. Suppose person a occupies one of the jobs on a construction crew. Vacancies arise from time to time, and a can tell his cousin b, who works in a different job in another part of town, about them. Then suppose b has a neighbor friend c, not known directly by a. The information about the job vacancy can be passed from b to c at the second remove, in two steps from a.

Dotted lines enclose the close friends of $c-b$, the neighbor friend; d, hunting and fishing companion; and e, an old hometown friend—within one remove of c, whom he can contact, or be contacted by, in one step. Others, a, i, g, and f, can contact him in two steps; h and unrepresented others can contact him in three steps, and so on. Thus, if c were looking for a job, he could inquire only among his first order associates (a dense, short-range pattern), or he could search along a path of extended personal relations to some person several removes from his own social net location (a long-range, rifle type pattern).

The Natural Proximity Structure

The top part of Figure 3 illustrates a portion of the lattice structure of the housing sector, including both natural proximity and impersonal bureaucratic communication structures. The lower left illustration shows the natural proximity structure represented as a graph; the lower right illustration shows a portion of the impersonal bureaucratic structure represented as a graph. Let us examine the natural proximity structure first. The numbers in the boxes represent different types of relatively homogeneous neighborhoods in which housing opportunities of a similar type are available. The nodes A, B, C, and D connect those neighborhoods which have one or more features in common, e.g., a similar rent range, an emphasis on furnished apartments, a safe versus a slum-type neighborhood, and so on. Suppose a person in neighborhood 8 were looking for a larger unfurnished apartment for his family. Instead of, or in addition to, inquiring about housing opportunities in the social net, he could walk or drive through neighborhoods looking for vacancy signs and contacting landlords directly. This "stumbling" type of search proceeds along the lines of the natural proximity structure. Even a blind search is structured by such aspects of the city environment as the physical concentration of various types of housing, or jobs, or stores. In addition, the stumbling search is biased because of the migrant's own attributes, e.g., the size of his family, his job skill level, and his current income. In this illustration, we may presume that the apartment hunter from neighborhood 8, engaged in a stumbling type of search for housing, would first look for vacancies in his own neighborhood, and then proceed either to neighborhood 10 or 11 which are nearby and which offer essentially the same kind of housing, e.g., large unfurnished apartments for rent. He would be unlikely to stumble directly over to neighborhood 15 in a different part of town which may feature small furnished apartments for rent.

Although less likely, information about housing vacancies may also be advertised to residents of a particular neighborhood along the lines of the natural proximity structure, without a search being initiated. For example, landlords may post vacancy notices on bulletin boards in public places in nearby or similar type neighborhoods.

The Impersonal Bureaucratic Structure

A still different kind of communication structure is that represented by mass media (such as the ad sections and the entertainment page of the newspaper) and bureaucratic organizations (such as unions, public and private employment agencies, the civil service, real estate agencies, and the public housing authority). The prime characteristic of these bureaucratic resources is that they bring together in one location information about a variety of opportunities. Thus a searcher who utilizes the impersonal bureaucratic communication structure gains access to whole sets of opportunities. In the housing illustration in Figure 3, for example, the apartment hunter in neighborhood 8 can, in one step, go directly to the housing authority which lists public housing vacancies in neighborhoods 7, 8, and 11. Similarly, if a bureaucratic organization decides to advertise its opportunities, the information becomes immediately available to the occupants of many different sets.

The representation of communication structures within which opportunities are advertised, and within which men search, may be summed up as follows. Each market sector is represented as a lattice composed of a natural proximity and an impersonal bureaucratic structure, each of which is represented as a graph in a computer. Men are located as members of sets within these graphs and "hear about" and "search for" opportunities along linkage lines connecting the sets, starting from their origin location. In addition, men may spread information to their friends in the social net about opportunities known to them in their origin sets in each of the four sectors, and men may inquire in their social net for information on all four types of opportunities. The social net is also represented in the computer as a graph with the linkage lines connecting friends in a communication network. There remains to be described our representation of the advertising of opportunities and the search behavior of men.

Circulating Information About Opportunities

The sequence of model operations each month is quite simple. First, in a given sector, the unoccupied opportunities are processed one at a time. In effect, the processing determines how densely or how widespread the information about an opportunity has been communicated along the linkage lines throughout the three communication structures. Then each man is processed, one step at a time, in each communication structure. His movement from set to set along the linkage lines determines what opportunities he will "hear about" and thus have a chance to accept.

Two parameters determine the extent to which information about an opportunity is communicated in each of the three structures. They are p and r. p stands for "persistence," or the probability that an opportunity will be communicated from its origin location or present set to another connected set. r stands for "range" or "rifle path" and determines whether information about the vacancy moves from its present location forward to another set (a "rifle" type communication path), or, alternatively, whether the information starts back at the preceding origin and moves to another set from there (a dense, "spoke" or fan type of communication path; this latter type of movement of information is the complement of r, namely, s).

Another parameter, c, stands for the "criteria" offered and demanded by the opportunity before it may be accepted. For example, a job requires a certain skill level, offers a certain pay, and has a degree of risk of unemployment attached to it. These are the attributes of the opportunity. They are set at the initialization of the model and may be systematically varied for experimental runs, as can the p's and r's of opportunities. Further model experimentation will come through our manipulation of exogenous factors affecting the type and number of opportunities available. We can explore, for example, what happens when a low cost housing area is torn down and many house seekers flood the market. Or,

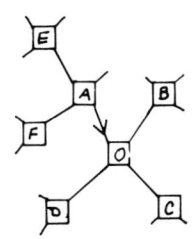

Step 1. The p of an opportunity is realized and the information about that opportunity has reached A.

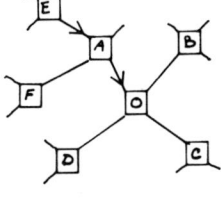

Step 2. p is realized within the r range. Information about the opportunity has moved from A to E.

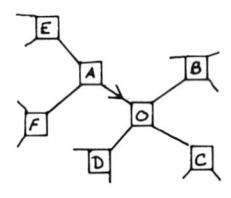

Step 1. The p of an opportunity at 0 is realized and the information about that opportunity has reached A.

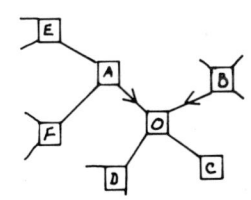

Step 2. p is realized within the s range. Information about the opportunity has moved from the preceding origin set, 0, to B.

An opportunity is "advertised" when its p value is realized, either in its r or s range. The top illustration shows the movement of information about an opportunity from its origin set 0, to a connected set A, on the first step, and from A to E on the second step, p being realized in the r range. The bottom illustration shows the movement of information when p is realized within the s range at the second step. Notices, represented by the arrows, trace the movement of information from the origin sets and later, in the search process, direct men to the origin sets. After all opportunities have been advertised in the three communication structures, each man searches in each structure for vacant opportunities, from set to set, his movement depending on realization of his p value and whether the realization falls within the r or s range. The movement represents, for example, inquiring among friends, reading ads, or wandering from place to place. But when a man arrives at a notice (arrow), he is directed, on his next realized step, toward the source of information about the opportunity. When a man reaches the location of an unfilled opportunity, he will immediately accept it if the c's of the opportunity "fit" the c's of the man. Otherwise he continues searching until his q is realized.

Figure 4. THE SIMULATION OF ADVERTISING AND SEARCHING PROCESSES

we can explore what would happen if all job vacancies were offered with a guaranteed annual wage as compared with the normal situation. It is obvious that certain kinds of practical administration problems can be systematically studied through manipulation of beginning states and exogenous inputs from the sectors representing the market sectors of the city. We can imagine Denver businessmen and city administrators in particular as the manipulators of the p's, r's, and c's of various kinds and numbers of opportunities.

How can the circulation of information be represented in a computer? For an illustration, we will briefly describe the movement of information about opportunities in the natural proximity structure. The dynamics of information flow in the other structures and all sectors is essentially the same except for some qualifications pointed out below.

Each opportunity has its natural origin in its own organization unit, e.g., a vacant apartment in the Flat Street neighborhood, an unfilled job in the Ace Company construction crew, a used car in Sam's Used Car lot, a seat at Casey's bar. For a particular run or set of runs, the p's and r's of opportunities have been assigned by the experimenter. A random number is generated for each unfilled opportunity, processed in random order, one step at a time. As illustrated in Figure 4, at the first step, if the generated value falls within the p range, information about the opportunity travels from its origin set O along a natural communication link to one of the organizational units (e.g., sets A, B, C, or D) to which it is linked (each unit having an equal chance of being selected). Information about the opportunity becomes "known" at that set it has reached (call it set A), and a "flag" or "notice" is set at A directing the channel of information flow (for a searcher) from A to O, where the opportunity is located.

If the generated value falls within the q range, i.e., the complement of p, information about that opportunity does not get circulated this month, or moves no further than the set it had reached when the q was realized. After all opportunities have been processed one step, those whose p's were realized at the first step are processed again. If the p value is realized again within the r range, then information about the opportunity moves forward from set A along a natural communication line to one of the organizational unit sets connected to set A, (call it set E) and a flag is marked directing information flow (for a searcher) from E to A. Information about the opportunity now has been advertised to both E and A from its source or origin O. When p's are realized within the r range, a long ranging communication channel becomes defined for that opportunity. The flow of communication for a man later searching the structure is from E to A to O, set by flags or "notices."

At the second step, if the p value is realized again but within the s range (s being the complement of r) then information about the opportunity travels out from the preceding origin, set O, to one of the sets connected to it (B, C, or D), with equal chance except for the set from which the information has just come, i.e., set A. Assume the information this time went from O to B. Then a notice is set directing information flow for a searcher from B to O. When p's are realized within the s range, short-ranged but dense or concentrated communication channels are defined for that opportunity. In this case, the flow of information for a man searching the structure is from A to O if he "comes to" A or B to O if he "comes to" B, as set by notices.

The processing of opportunities is repeated until all q's are realized. Information about an opportunity is known (i.e., has been "advertised") at each set it has reached thus defining the extent of the circulation of information about that opportunity in the natural proximity structure of that market sector this month.

If the p of an opportunity is realized in the impersonal structure, then information about it is formally advertised this month in all sets connected to that bureaucratic agency. If q is realized, the opportunity does not get advertised by that bureaucratic agency this month.

In the social net, the circulation of information is as has been described for the natural proximity structure with the exception that a man carries the information about unfilled opportunities to the social net from the man's origin in the sector sets. Once the information is in the net, its further circulation is dependent on further realization of its p, and the spread of the information is determined by which range values (r or s) are realized.

Men Search the Communication Structures

Once the information flow notices in each of the three communication structures in a given sector have been set, the men, in random order, search the structures, one step at a time, first in the social net, then in the natural proximity structure, and finally in the impersonal structure. Empirically based multiple regression equations set the p's (persistence), r's (range), and c's (criteria) for each man. These parameters determine the character of each man's search in each of the three structures in each of the four sectors.

First, however, it is assumed that each man is aware of the vacant opportunities in his own origin set in the natural proximity structure. For example, if a house with more rooms is vacant in his own neighborhood, then he and the other men in the neighborhood have the first opportunity to rent that house. Thus, regardless of p or r values, each man, in random order, checks the vacancies in his sector origin set. A match is made between the attributes (c's) of the opportunity and the criteria (c's) specified for the man. If a fit occurs, the man accepts the opportunity; immediately that vacancy is marked filled in all structures. Note, however, that notices about that vacancy still exist in the communication path to a heard-about vacancy only to discover that it has already been filled.

When the vacant opportunities within origin sets have been checked out, each man, in random order, has a chance to move one step in the social net, depending on the intensity of his drive for a particular goal (such as for better housing) and his proneness to utilize this type of communication structure. This is represented by his persistence value (p) set by a regression equation for him for this communication structure in this sector this month. When his p value is realized, he moves one step to another set where he may or may not come across information about an opportunity. If the source of the information about the opportunity is at that set, he immediately gets the chance to accept it. If information about an opportunity is there, but just a notice and not the source, then on the next step, if his p is realized, he follows the notice path toward the source of information about the opportunity, one step at a time, until he reaches the source, or until his q is realized. Until his q is realized, his search path is determined by the r or s realizations each time he gets another chance to move through his realization of p.

After each man has had a chance to move one step in the social net, the processing is repeated in the natural proximity structure and then in the impersonal bureaucratic structure. The values of p's and r's are different, again dependent on the regression equations for him for these structures, and men move to sets representing organizational units rather than personal associates. But the processing steps, the p and r realization procedure, the following of notice flags to the source of information about opportunities, and the matching of the c's of men and opportunities when the source of information is reached, are the same.

The processing continues until q's have been realized for all men (or until an arbitrary number of moves have occurred with nothing happening, as when a man with a very high p for housing may keep searching for a house with a very low rent, but there are none that low). In the processing, men will have accepted opportunities which give them new attributes, e.g., a better paying job, a less crowded house, a new car and a new debt, membership in a church or social club, and so on. The new set of attributes affects the future values of $p, r,$ and c that will be estimated for the man in each of the three structures and each of the four sectors for the next month.

An Auxiliary Process: Exogenous Events and Their Consequences

In our model we also activate rarely occurring events, something like acts of God, which change men's attributes. We treat illnesses, jail and legal trouble, changes in size of family and family composition, and changes in social net size and composition as incidents which are determined from experience rate tables or are predicted from regression equations. Although we cannot describe our treatment in detail here, the essence is as follows. The probability of an illness or legal trouble incident is estimated by a regression equation. Then a random number is generated. If the generated number falls within the probability range, the consequence may be loss of work and/or a monthly cost for a particular duration of time. These consequences are determined by generating another random number with which to enter an empirically-based experience table. The row of the table hit by the number specifies the consequences of the event. The consequence of a marriage, birth, separation, or dependents' arrival and departure (these events being determined by realization of a probability established by a rate table, for those eligible) is a change in the size of the family and possibly a change in the family cycle variable. Since both of these major variables enter other predictive equations, the consequences of such changes are far reaching. Similarly, if a friend is gained or lost, the incident affects the social net size and net composition variables such as the proportion of urban peer associates versus hometown-kin associates and the relative number of deviant associates. These major variables also enter the predictive regression equations affecting future behavior of the man.

The Estimation of p's, r's, and c's of Men

As has been described elsewhere (see Hanson, Simmons and McPhee, 1968), we have used factor analysis to help us build major variable indexes which enter multiple regression equations used in predicting the p's, r's, and c's and the various incident probabilities for each man each month. While these major variables cannot be described in detail, a brief mention of the types of variables we are using seems necessary. First, one group of measures may be characterized as "on arrival constants and variables"

[390] BEHAVIOR IN NEW ENVIRONMENTS

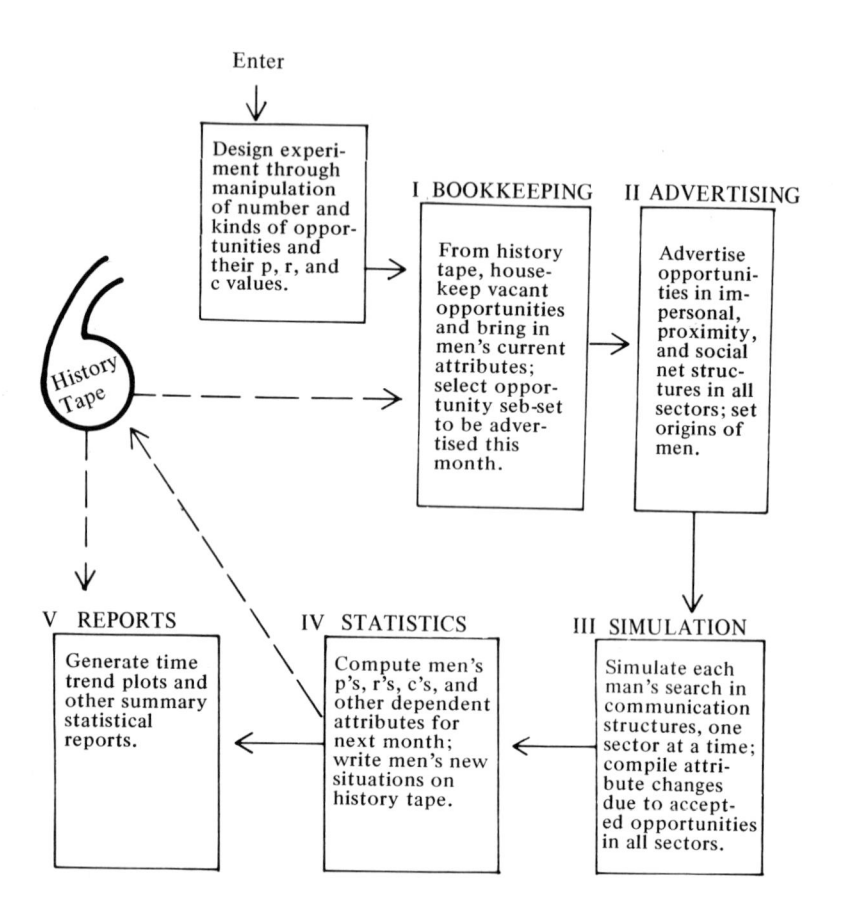

Enter ↓

Design experiment through manipulation of number and kinds of opportunities and their p, r, and c values.

I BOOKKEEPING
From history tape, housekeep vacant opportunities and bring in men's current attributes; select opportunity sub-set to be advertised this month.

II ADVERTISING
Advertise opportunities in impersonal, proximity, and social net structures in all sectors; set origins of men.

History Tape

V REPORTS
Generate time trend plots and other summary statistical reports.

IV STATISTICS
Compute men's p's, r's, c's, and other dependent attributes for next month; write men's new situations on history tape.

III SIMULATION
Simulate each man's search in communication structures, one sector at a time; compile attribute changes due to accepted opportunities in all sectors.

Figure 5. TIME SEQUENCE OF MODEL OPERATIONS FOR THIS MONTH'S SIMULATION

including common measures such as Socioeconomic Status, Age, Size of Family, and personality constants such as IQ, Self-Esteem, and Prudence. Second, there is a large group of "current situation variables" such as degree of employment, degree of settled housing, total property value, earned income, current payments, organizational participation, current satisfaction variables, and so on. Third, another group of variables reflect recent past behavior such as "propensity to get drunk," and "rootless situation." The latter includes recent unemployment, change of residence, and turnover in friendships. Fourth, there are the dependent variables, the p's, r's, and c's in each of the three communication structures in each of the four sectors. All of the preceding variables may play a predictor role in an equation estimating next month's lagged dependent variables. Finally, there are the dependent incident variables such as illness, legal trouble, and family and net changes which are predicted only and do not enter the regression analysis as major predictor variables.

For the regression analysis, all of the variables have been standardized from .00 to .99; they have been complemented, if necessary, to conform to the data showing their relation to adjustment. The regression equations thus contain only those predictor variables that have positive coefficients and are theoretically justifiable. We believe such steps allow us to interpret the partial regression coefficients as causally meaningful values.

Summary

Figure 5 provides an abstract view of the sequence of model operations each month. For any particular run or set of runs designed to explore systematically the implications of the model, the experiment can be formulated by manipulating sector graph structures and the number, kinds, and p's, r's, and c's of available opportunities at initial input. Each month, thereafter, the processing proceeds through five major operations. Part I prepares the sector graph structures for this month's run by removing the filled opportunities and listing the vacant opportunities still unfilled from the last time (general housekeeping). A subset of vacant opportunities is selected for advertising this time, and the current origin locations and p, r, and c values of men are brought in from the history tape. Part II takes the p's, r's, and c's of the vacant opportunities and circulates information about them throughout the impersonal, proximity, and social net structures in the four market sectors. Origins of men in the sets of the structures are determined by the location of opportunities they presently hold, or by their origin of search last time. After the advertising is completed, Part III simulates men's search behavior in the three communication structures, one sector at a time. When the searching and matching stops, current data on men's attribute changes due to acceptance of opportunities are compiled. Part IV takes the current situation of the man and computes his p, r, and c values, and other dependent attribute values, for next month. The current situation attributes of each man are written on the history tape, month after month, providing an historical record of changes over time. Finally, from this month's experience and from the history tape, Part ·V generates periodic summary statistics and time trend plots for the migrants as a whole and for various subgroups under consideration in the experiment.

At the present time, adjusting or "tuning" the model until its outputs reasonably reproduce the empirical data remains to be done. We set out to simulate complicated social processes in a computer and we know that this can be accomplished. But the systematic exploration of the model and consequent assessment of its theoretical and practical contributions must wait until a later date.

NOTES

1. Papers and dissertations currently available from the project are listed as the references at the end of the paper. The "role path" paper (Hanson and Simmons, 1968a) describes the data collection procedure; the "quantitative analyses" and "time trends" (Hanson and Simmons, 1968b) papers illustrate data analysis procedures; the "world of work" paper (Simmons, Hanson and Potter, 1969) presents case study materials in the light of model concepts; and the "experience paths" paper (Hanson and Simmons, 1969, herein) provides an empirically based interpretation of the process of adjustment of rural migrants to the city.

2. This section follows closely a portion of a mineographed paper by W. N. McPhee on opportunity structures, available from the Institute of Behavioral Science, University of Colorado.

REFERENCES

HANSON, R. C. and O. G. SIMMONS (1970) (forthcoming) "Differential experience paths of rural migrants to the city." Herein.
--- (1968a) "The role path: a concept and procedure for studying migration to urban communities." Human Organization 27 (Summer): 152-158.
--- (1968b) "Time trend analyses of the urban experiences of Spanish American migrants." Boulder: Institute of Behavioral Science, prepared for the Conference on Adaption to Change, Foundation's Fund for Research in Psychiatry, Puerto Rico. (Unpub. mimeo)
HANSON, R.C., O. G. SIMMONS and W. N. McPHEE (1968) "Quantitative analyses of the urban experiences of Spanish American migrants." Pp. 65-83 in J. Helm (ed.) Spanish-Speaking people in the United States. Seattle: Proceedings of the 1968 Annual Spring Meeting of the American Ethnological Society, distributed by the University of Washington Press.
KURTZ, N. R. (1966) "Gatekeepers in the process of acculturation." Unpub. Ph.D. dissertation, University of Colorado.
RENDON, G. (1968) "Prediction of adjustment outcomes of rural migrants to the city." Unpub. Ph.D. dissertation, University of Colorado.
SIMMONS, O. G., R. C. HANSON and R. J. POTTER (1969) (forthcoming) "The rural migrant in the urban world of work." Pp. 21-31 in the Proceedings of the XI Interamerican Congress of Psychology, Mexico City, 1967.

Chapter **16**

Policies for Planning Rural-Urban Migration: *Urban Villages Reconsidered*

RICHARD L. MEIER

The "urban village" is a concept born of experience and despair. The despair has been voiced by all urban planners faced with the floods of immigrants yet to come, more in Asia than elsewhere, when they recognize how standard procedures for planned accommodation of growth will be overwhelmed in a number of regions crucial to economic development. The experience was derived from a number of accidents of history and a few limited experiments which together suggest a strategy. The natural processes of urban development, especially the steps whereby an immigrant community learns how to make a living and integrate itself into the larger urban scene, can be accelerated by occasional intervention of planners. A number of impediments to this development process can now be recognized; they are all large in scale, so the plans for the mass production of urban settlement opportunities need not be reviewed block by block but as programs for community building. Viewed from the offices of the regional planners, the urban village is a "training camp" for people of one ethnic origin (or some complementary grouping of them) that produces recruits for the productive activities of the metropolis. However, as is the case for many a camp, it is expected to crystallize into a self-organizing permanent settlement, surrounded by many other such settlements, and increasingly a participant in urban culture.

Partial precedents can be found that go back for centuries, but the generalization as a policy of urban regional planning seems not to have been formulated before 1956 (Meier, 1956). Since that time a number of new experiments were undertaken, mainly in Latin America, Turkey, and Israel, utilizing various stimuli for self-help in newly settled "squatter" communities. Now a distillate of that experience is available (Turner, 1966). The experts have come down firmly on the side of "encouragement of popular initiative through the government servicing of local resources." Highest priority should be given to (a) the assignment of a homestead to each new settler, (b) access to schools and market from this plot, and (c) an opportunity to add modern utilities by connecting in the main grid at a later date.

A number of administrative actions of unexpected complexity follow from this simple set of requirements. A guarantee of security of tenure to the occupant of a plot, for example, requires that the land must have been previously acquired, surveyed, subdivided, and the boundaries marked. The terms of lease or ownership must be simple, well advertised, and available to all. The community itself must be encouraged to incorporate legally so that formal responsible relationships are established with the central city and the province. These provisions greatly encourage private investment in dwellings and enterprises.

Access to central metropolitan markets, services, and employment is planned in advance. Earnings are brought into the settlement by such proximity. Also, many of the most competent households are encouraged to remain, providing community leaders and educators for the self-improvement process. In contrast, we have learned that isolated new urban settlements often experience a stunted development, misusing land and human resources, and are then unable to solve collective problems.

The supply of new plots must be maintained at a high enough level to prevent the bidding up of land prices in the communities already settled. Escalation of land prices is usually accompanied by uneconomic crowding in the poorer parts of the city. This means that monopolistic behavior, especially the withholding of urban fringe land from the market, may require stern action on the part of the government.

The public health features of the area to be settled need reconnaissance and later some supervision. This will include the supply of water, the handling of sewage, the solid waste disposal, the drainage sytem, and occasionally the control of disease-bearing organisms.

Training people to build with new kinds of materials will be particularly important since traditional construction may be prevented by the short supply of timber. The main contribution of self-help housing lies not in the accumulation of "sweat equity" as capital, because it is worth only $.30-$1.00 (U.S.) per day and hardly pays for the extra expenditure of human energy in the form of food, but arises from the network of cooperative relationships created in the community to get these jobs done. These same relationships can then be utilized in organizing new cooperative arrangements for meeting collective needs. Foresight is required at this stage because the methods of layout and construction that are most natural to the settlers make subsequent connections to metropolitan service grids uneconomic or physically impossible.

Further rationalization of the settlement process involves steps like immediately introducing electricity for public lighting purposes, which greatly enhances educational possibilities. Street surfacing allows internal vehicular access and therefore greater social organization. The provision of credit through cooperatives, for which some of the seed capital comes from the outside, will speed up the transition to an urban community with a finished physical appearance. Any interference at this level will often need to be more thorough than appears on the surface, because the poor are exploited by moneylenders and racketeers who resist rationalization in many devious ways.

The programs to date have been small scale, with very few cities in the world responsible for coordinating the construction of as many as 10,000 dwellings on a self-help basis. Now it is evident that schemes must be worked out which will handle tens of millions. We must therefore find a way of getting into the mass production of communities. The agency responsible must add refinements to the planning as it learns, but a number of these can already be proposed.

A curious byproduct to these programs makes the expenditure of scarce planning effort on them far more defensible than mght be judged from the size of the budget or the criticality to immediate political issues. Perhaps the best way to make this point is to assert that "the self-help, self-organizing community is a happy community." It is a community that presents fewer insoluble political and administrative problems to the central government. Community development workers and anthropologists have said so repeatedly, but their samples were small and their criteria somewhat ambiguous. However, more thorough studies have shown that when social participation is high, regardless of the outcome of that participation, the reports of the individuals say they are personally "very happy" or "pretty happy."[1] Negative feelings bear no relation to participation. Therefore, arguing from what is known in general about the politics of communities the trouble involved in creating new urban villages indirectly should save trouble in the relatively short run by enhancing political stabilization. Participant self-help is expected to be one of the cheapest possible ways of producing happiness in cities!

[398] BEHAVIOR IN NEW ENVIRONMENTS

The Urban Village as Programmed Instruction

Viewed from above, in the perspective available to most planners, an urban village is an institution for the development of human resources. It is therefore regarded as a special educational unit, one that transmits concepts and insights to its members in such a way that a clear image of the whole system for organizing the metropolis is formed.

The task is that of transforming rural immigrants, the majority of them 19 to 24 years of age and the remainder distributed above and below these middle two quartiles in a highly asymmetric pattern, into competent urbanites, capable of making their own way and contributing to the surplus required for the support of the less fortunate.

The recruits for early settlement are certain to be very poor, but the predominant majority will be literate. In the Indian states on the Bay of Bengal today about thirty to eighty percent of the children of school age are enrolled, the highest proportion being in the general vicinity of Calcutta and Madras. Perhaps half will have advanced to the stage where they could undertake the keeping of records and the maintenance of simple accounts. A majority of them will stay less than two years before returning to their home village or their town. During such a visit to the city a great deal can be learned about the opportunities and hazards offered by urban living. These personal findings are brought back home, seriously discussed in the village, and many items are added to the repertory of expectations carried with the newer migrants when they make their attempt to find a niche in the big city. Thus members of the urban village, residing in the enclave of a number of villages in the hinterland, transmit information about employment opportunities, family planning, mass media, politics, labor unions, and a number of other phenomena strange to rural village life. Quite a few immigrants who failed to get established the first time will try again and again, giving up around the age of 30 if they have not made it by then. The backflow to the rural areas is however not all made up of failures since a number of vacated posts in the community must be filled and the ambitious families conspire to bring about the recall of the hereditary successors living in the city who will prefer a secure position in the village to an uncertain one in the urban area. Some of the backflow will be made up of persons who have succeeded in the city and have gone back to the rural area to buy property.

Villages which have established contact with an urban settlement will show discernible advances in sophistication within a decade, nevertheless, in comparison the urban village visualized here is a "pressure cooker" which must produce a *significant shift in behavior* within a matter of months and go on from there to create a competent citizenry over the years. The "program" for the instruction of urban villagers is outlined in Figure 1. It contains a *main sequence* for the transformation of the rural immigrants that remain. Inevitably a smaller *fast stream*, made up of energetic *achievement-oriented* individuals, will short-cut and move on to the cosmopolitan core of the city where less care is taken to cushion the shocks of getting acquainted with urban life. A *slow stream*, made up of tradition-oriented individuals, kin to the other two and supported by them, will also need to be accommodated.

One way of illustrating what the programmed instruction model might be designed to do is to follow a typical person through the migration and instructional process from infancy to the burning ghat. By "typical" here is meant that at each choice point he will follow the center of the mainstream (the distribution mode anticipated at that time). See pp. 400-1.

The standard projection of income improvement incorporated in this career allows the family to achieve the minimum adequate standard of living about 20 years after the head of the household arrives in the city for permanent residence. It implies an increase in personal productivity of three to four percent per year over a period of three decades, followed by levelling and decline. Performance in the modern industrial sector of Calcutta-Howrah, comprising about 300,000 employees in 1966, is higher, while the antiquated jute textile industry (almost as large) and the food processors do more poorly. Therefore these expectations seem quite reasonable.

An urban village evolving in this fashion will accumulate three kinds of people over time (See Figure 1). Perhaps the most important of these is the skilled, externally employed, community-oriented type who will retire and grow old in the community. Another is the less stable, semiskilled person working in light industries, mostly close to home, who gets only peripherally involved in the development of the community (as in the case described above) and retains strong family ties with the hinterland as well as other locales in the metropolis. Another category finds urban life too demanding and is satisfied with simple service jobs, intensive gardening, or food production in the reserved spaces surrounding the urban village.

Normally it should be planned that a community of about 30,000 persons will require five to ten years of growth to occupy an area of about 800 acres. These dimensions reflect a size and density needed to impart urban character to the settlement, particularly access to metropolitan educational services and employment opportunities, at the same time that self-sufficiency in perishable foods (likely to remain the unrationed portion of the diet) can be achieved. Rapid settlement of the immigrant

THE LIFE AND TIMES OF SRI Y

1967 — born in a village of 3,000 persons
       Hindi-speaking parents, earnings Rs. 15000/yr.
       third child, second son
       not a Brahmin

1969 — another sibling (dies in infancy)

1972 — another sibling (survives)

1972 — enters school, one of 45 students
       male teacher
       one text book available

1976 — goes to city (for funeral or wedding)

1980 — drops out of school, family crisis

1982 — returns to complete 8th year of schooling
       obtains casual construction work, Rs 3/day

1984 — obtains bicycle (Rs. 100)
       visits metropolis and returns

1986 — contract worker in construction, Rs. 5/day
       returns home

1987 — finds opening in 5-year old urban village
       works as food distributor, income variable

1988 — marriage arranged
       girl from neighboring village back home

1989 — constructs a shelter on own plot

1990 — nuclear family household formed, one child born
        builds a satisfactory house
        achieves steady work in food processing, Rs. 6/day
        assists migration of siblings and cousins
        participates in community activities
        adds two more surviving children
        raises income to Rs. 11/day

DECADE

2000 — becomes established member of community
        eldest son completes 11 years of school
        expands house to 300 sq. ft. floor area
        often entertains relatives and keeps "boarders"
        family income to Rs. 20/day (children contribute)

DECADE

2010 — two children move into central city area
        household is receiver of new immigrants (relatives)
        one parent becomes minor official - voluntary post
        marriage is effected (traditional rites)
        earnings reach Rs. 20/day on his own

DECADE

2020 — debility and more frequent sickness set in
        head of household dies (2024)
        widow transfers lease to relatives
        offspring in urban village reaches Rs. 15/day pay
        offspring in central city average Rs. 20/day salary

DECADE

2030 — widow dies, estate exceeds Rs. 1000
        offspring are acquiring seniority in bureaucracies

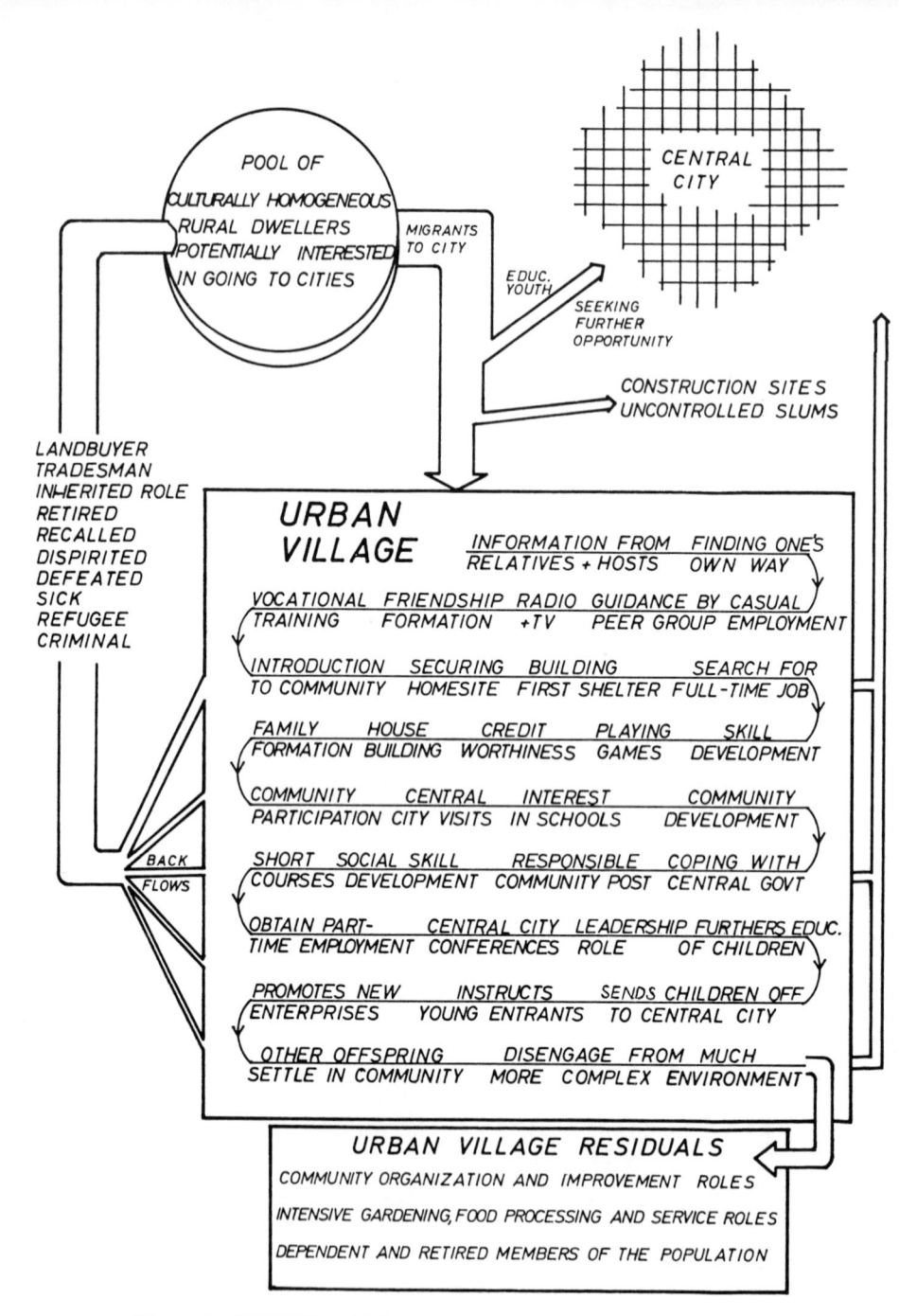

Figure 1. THE URBAN ENVIRONMENT AS A PROGRAMMED
INSTRUCTIONAL SYSTEM

(sequence for creating productive citizens)

SOURCE: Richard L. Meier, "Material Resources," in *Mankind 2000,* Robert Jungk and Johan Galtung, eds., Oslo: Universitetsforlaget; London: Allen & Unwin, 1969, P. 108. Reprinted with permission of the publishers.

population means that the urban village will be afflicted with an extremely abnormal age distribution so typical of new towns and newly created suburbs. As a consequence schools may have to expand to two or three times the normal size, only to contract a few years later to a third of the previous enrollment. It appears to be easier to put up with the dislocations due to unique age structure, taking advantage of them sometimes in community participation schemes, rather than elaborate a complex plan for restoring the age structure to a more typical composition. One doesn't know, until the events have transpired, what the *age structure of human resources* in the community will be. Therefore, much of the initiative remains in the hands of the community leaders and those in the central city who are assigned to the task of managing the "growing communities."

This is written as an interim report in a series of studies on large scale urbanism to which I have been contributing over the past dozen years. They are more fundamental in most cases than the points made here. It is useful, therefore, to cite the work upon which this study is based.

(1) R. L. Meier, *Science and Economic Development* (Cambridge, MIT Press, 1956, 1966).

(2) Roy Turner, ed. *India's Urban Future* (Berkeley, University of California, 1963).

(3) C. A. Doxiadis, J. Papaioannou, et al., "The City of the Future," a series of presentations in *Ekistics*, particularly July 1965 and July 1967, most of which have been available in mimeographed reports.

(4) R. L. Meier, *Developmental Planning* (New York; McGraw-Hill, 1965).

(5) R. L. Meier, *Studies on the Future of Cities in Asia*, Center for Planning and Development Research, Institute of Urban and Regional Development, University of California, Berkeley, July, 1966, 110 pages, mimeographed.

(6) R. L. Meier, *Resource Conserving Urbanism for South Asia* (Ann Arbor: School of Natural Resources, University of Michigan, January, 1968. Available from University Microfilm-Xerox, Ann Arbor.)

(7) R. L. Meier, "Material Resources." In R. Jungk and Johan Galtung (eds.) *Mankind 2000* (London: Allen & Unwin, and Oslo: Universitetsforlaget, 1969.)

(8) R. L. Meier, "Technologies for Asian Urbanization." *Economic and Political Weekly*, IV, 1969, Nos. 28-30, 1-7.

NOTE

1. See Phillips (1967). The conclusion holds even more strongly as education increases. See also Bradburn and Caplovitz (1965).

REFERENCES

BRADBURN, N. M. and D. CAPLOVITZ (1965) Reports on Happiness. Chicago: Aldine.

PHILLIPS, D. L. (1967) "Social participation and happiness." American Journal of Sociology 72, No. 5: 475-488.

MEIER, R. L. (1956) Science and Economic Development: New Patterns of Living. Cambridge: Technology Press.

TURNER, J. F. C. (1966) "Uncontrolled urban settlement: problems and policies" (Proceedings of the Inter-Regional Seminar on Development Policies and Planning in Relation to Urbanization, held by the United Nations Bureau of Technical Assistance Operations in Pittsburgh, October and November). Reprinted in Ekistics, February, 1967.

Chapter 17

Alaskan Natives in a Transitional Setting

**ROBERT L. LEON
and HARRY W. MARTIN**

The Seattle Orientation Center was established in 1963 by the Employment Assistance Branch of the Bureau of Indian Affairs to acquaint Alaskan natives with the problems of city living as they migrate from villages in Alaska to urban centers in the south forty-eight states. Under programs of the Employment Assistance Branch, Indians and Alaskan natives who so choose can migrate to selected large cities where the Branch maintains field offices for either immediate direct employment or for vocational training and subsequent employment. The Bureau provides transportation for Indians and their families, vocational training in the cities for those who elect it, job placement and economic assistance until the relocatee becomes self-sufficient.

During the course of several years of operation, staffs in the field office had observed that Alaskan natives, principally Eskimos, had greater difficulty in adjusting to city living than did Indians coming from the south forty-eight. This is not to say that the latter do not frequently

---

Author's Note: *We wish to express our appreciation to all those persons in the Employment Assistance Branch who helped make available the information on which this report is based. Particular gratitude is due to the staff members of the Seattle Center and the field offices who completed the forms in the midst of all their duties.*

experience problems or do not return to the reservations. However, the problems of adjustment were more marked among the Alaskan natives and more of them returned to Alaska.

Eskimos in particular seem to be "lost" in the cities. They were less able to adjust to employment and vocational training and more frequently failed at these. They more frequently tended to remain isolated. They displayed more anxiety. They had greater problems with alcohol, and those with families showed more signs of family disorganization.

For these reasons the Employment Assistance Branch conceived the idea of establishing an orientation program which would expose the Alaskan natives to the complexity of urban living and teach them some of the skills to survive in the urban settings before they arrived at their destination city. Since Seattle, Washington, was the nearest urban center to Alaska and the first stop in the south forty-eight en route to major destination cities from Alaska, the center was placed in Seattle.

Once the decision was made to establish the center, we were asked to act as consultants to outline the actual program for the center. At the same time the center director was selected. The director participated in developing the program and was given a special training program in group and individual counseling.

The site selected was a new motel near the University of Washington campus across from a shopping center and with access to bus routes to downtown Seattle. There were several reasons for selecting this site. The facilities were modern and pleasant and provided an example of some of the better apartment living available in the cities. The motel provided small apartments where the Alaskan natives could live close to one another while still relating to other guests of the motel. The motel also provided office space and group discussion rooms for use of staff and clients. The adjacent shopping center was available for experiences in banking, eating in restaurants and shopping in supermarkets. The food purchased could be prepared in the motel kitchens with guidance from staff. Proximity to the University of Washington enabled the staff to obtain professional consultation.

The program provided an opportunity to learn the skills of urban living and at the same time, within the context of a therapeutic community utilizing principally group counseling techniques, provided for expression of the anxiety related to relocation and new learning and expression of the grief associated with leaving one's home and loved ones. Each morning the clients met with the center director or a member of the staff to discuss first their experiences in coming to Seattle and later their experiences in adjusting to the city and learning new skills. The sessions were designed to encourage expression of feelings around these issues. Programs for the remainder of the day contained one or more specific experiences in urban living. Clients were given money which was used to buy groceries in the supermarket. Another day they might open an account in the local bank. Accompanied by staff, clients would board a bus for a destination in the city, and the following day they would take a bus trip to the same or different destination on their own. Another day might be occupied by an auto trip to the city in which clients guided the staff driver to the destination by reading a map. Clients went through the entire procedure of applying for a job at a local industrial plant. They were also encouraged to develop projects on their own.

The program was designed to approach both the cognitive and affective problems of rural migrants to an urban area. The cognitive aspects of the program were most easily understood and implemented by those untrained in mental health or behavioral sciences. The methods designed to deal with affective problems were not easily understood and were frequently subverted by untrained personnel.

The material which follows is selected to illustrate an interaction between migrant and those who are seeking to help him in his adjustment to an urban setting. The illustrations are presented not to show new methods—which they are not—but to show how it is comparatively simple to adapt group work methods and methods and techniques of the therapeutic community to deal with problems of rural-urban migrants.

The material also illustrates what can easily happen when a program for migrants falls to untrained people who are potentially destructive in their interactions with others.

The following, taken directly from an audio-tape, is an example of an effective group counseling session around the problem of drinking: "Clients in this group were outgoing, generally high school level, and had met several times with a warm relationship developing among each other and with the counselor. There were two new people in this group. Although there had been no complaints from the neighbors about their drinking and no night calls to staff members, their morning appearance indicated much drinking. One or two of the clients had also complained about the drinking habits of at least three of the members.

Counselor: You know — I think we do have one more field to cover and that is drinking. Now again, you know — as long as you drink and you don't cause a disturbance — and you don't get sick — and you're wide awake the next day — and it doesn't make you lose work — and you don't spend too much money, then it's strictly your business. But, look at all those things you have to overcome before it's strictly your business.

Client: All depends on how you handle your drinking. You can drink and get drunk and you can drink sociably too.

Counselor: How about telling me the difference —

Client: Well, if you're a steady drinker, you drink a lot — and you just can't stay away from it — If you're a sociable drinker, you just go out and get a couple — or have a couple in your apartment — and that's about it. If you're a heavy drinker, that's where trouble starts.

Client: Yes — some people go out to get drunk — just to get drunk — especially quite a few people have troubles I know that they'll go out and just get plain drunk —

Counselor: Uh-huh — You mean they leave home with the idea of getting drunk?

Client: They do. The next morning they find out it doesn't help them at all. Especially if they had an argument with their wives. She'll be more than madder —

(Laughter)

Counselor: Well, how do you — how do you fellows feel about drinking? Do you drink to get drunk?

Client: I did a couple of times.

Client: Hate the hangovers.

Clients: Yeah — sometimes.

Counselor: (to client) How do you handle your drinking?

Client: To be sociable.

Client: Only when I know there's not going to be any trouble — just to have a good time — if it's perfectly safe, then I drink —

Counselor: How can you tell when this is going to be? — the circumstances?

Client: Well, I don't really know about this place. But in Sitke, there are places where you can have parties and nobody bothers you. Away from town where we have cabins and pianos and —

Client: As long as we're drinking at home —

Client: And where you know the people — you know —

Client: You know the people that way you know it's pretty safe. But places like this — you're new around here. You know a few people here and there — You have a party and you never —

Client: If there's trouble, you can sense it — if you watch the people — you can tell if they're out to get you.

Client: Do you have any suggestions?
Counselor: Well, I was thinking. The last time I really suggested this, a boy lost his wallet. I will tell you what the police told us — if you will go to taverns in your own neighborhood —
Client: We located one not too far from the Center.
Counselor: Is this a fairly decent one?

(Silence)

Counselor: Or is it rough?
Several Clients: No... No... No...
Client: It's real nice — have music. Quiet.
Counselor: You see as long as that man wants to stay in business — most of the people coming there live in his vicinity and are fairly well behaved I'm sure. They're not going to knock you in the head and get your wallet — But, I think any place in this neighborhood might be a good place so long as you don't get drunk. Once you get drunk, any place can be a real hazard I believe. Don't you agree?
Client: Uh huh.
Counselor: I suggest you go up to the University — the first beer tavern there when you leave the campus (discuss it — location — near campus, etc.) — must be a good place — So, try these out and you can let me know. Try them out. I think when you go you want to think about how you look. I might add, all you fellows always look nice. I don't have to tell you how to dress — But I think if you've been out all day, be sure you have on a fresh shirt — you know, little things that make the difference. And go sober."

Here the center director not only encouraged expressions of feelings about drinking but also encouraged socially acceptable ways of drinking.

Although drinking is a serious problem with American Indians and Alaskan natives in socially disorganizing situations, drinking was not a serious problem at the Seattle Orientation Center. Only rarely did a client get drunk enough to create a disturbance or land in jail.

At one point, however, drinking did become a serious problem at the center. This occurred when an unusually large number of clients had been sent to the center and, therefore, staff from other offices was sent to the center to assist the director. Some of the detailed staff people were quite authoritarian and repressive in their approach and constantly expected clients to misbehave and in subtle and not so subtle ways made this expectation known to the clients. As these people began working with clients, acting out increased. Clients overslept, got drunk, created disturbances and generally made life difficult for the staff.

"Mr. B. saw his group of young married men and a few single men as 'boys.' He saw them as irresponsible people who had to be awakened every morning. At first, I insisted that Mr. B not round up his group every morning for the first meeting. I pointed out that routinely clients come to meetings without being called. This did no good. *Every* morning, Mr. B with much anxiety, rounded up his people. Sometimes he went for them and sometimes he sent for them. Regardless of how he rounded them up, he was never in his meeting on time. Most mornings, he and Mr. H and Mrs. R could be seen as late as 9:15 in the office wondering where they were going to meet and what they were going to talk about. You are reminded that each client and each staff member had a schedule giving the names of the persons who were to be in the group and where they were to meet. This was also posted on the bulletin board. In our staff meeting, I had listed topics to be discussed in group meetings and suggested always that these topics not be introduced or that they be set aside if clients wished to discuss other things. I pointed out over and over again that the first meeting in the morning was hopefully not an information giving session, but one in which clients could feel free to discuss their feelings and their attitudes about their experiences at SOC, about leaving home, and about what they expected to find at destination. Some marital discord developed between two of the couples and each time, Mr. B. made no effort whatever to assist, but instead turned these problems over to me. In one case the young man left the center in the afternoon and did not return until the next morning. That evening his wife called two staff members for a total of three or four telephone calls. The morning he returned he left again and did not return that day until about midnight. In the other instance, the young man was disinterested. He'd go on tours and appear to make a point of not even listening or looking at what the employer might point out or at what was going on if it happened to be recreation. He missed many, many of Mr. B's sessions. Because I was hearing some complaints and because the clients were balking at our activities, I made the suggestion many mornings that the counselors ask the clients specifically for comments about our activities and for suggested changes. Mr. B's group never made a single suggestion. They did, however, let him know over and over again that they would like to go on to their destinations.

"To summarize, Mr. B saw them as boys. He seldom kept them longer than thirty minutes in a meeting. He received no complaints or suggestions. Members of his group wanted to leave SOC early. They missed sessions or were late. They were rounded up in the morning and they were drinking excessively at night."

A more subtle influence of counselor's attitude is demonstrated in the following excerpt from a report:

"When Mr. A., for example, meets with a group, he is their buddy – he is their equal. He not only says this, but he spells it out, that 'Here I am an Indian. I can make it, and you can make it, too, if you want to.' He had met with the group and cut the meeting short, and I met with them two or three days later. Some were leaving, and I felt that because of the size of the group and our lack of staff, we perhaps neglected to make them face the reality of moving into a new community. I met some angry protests when I started talking about some of the problems such as homesickness, dislike of the training or school, dislike of the jobs, and that sort of thing. One young man said that Mr. A. had talked to them about what to expect and why was I such a pessimist? Mr. A., on the other hand, extremely dislikes working with Indians who goof up and who are failures in his eyes."

In this example the counselor's attitude toward the clients was influenced by his own anxiety about being an Indian. This attitude blocked clients' expression of feelings in the group, and they were not able to deal with anxieties about what they would find at their destination cities.

The migrant is anxious, depressed and certainly regressed in his behavior. He is far more susceptible than usual to environmental forces about him. These forces are potentially destructive in that they can trigger behavior which calls for the retaliation by the community and thus begins a vicious circle leading to personal or family disorganization. But this need not be. The state of transition can be a stimulus for personality growth if a way of dealing with disruptive affects, emotional support and opportunity for new learning are provided. Therefore it is important not only to study the problems of migration, but also to promote a large measure of behavioral sciences input into programs for migrants.

This paper first of all illustrates principles, methods, and techniques of a program to aid rural migrants to urban areas. It is a specialized program but with modification could be applied to most rural-urban migrants. Although the elements of the program are not new and have been well known to behavioral scientists for quite some time, this is the first time to our knowledge that these program methods have been applied strictly to migrants. The examples are used to illustrate how methods developed by psychiatry for use in the therapeutic community can be applied in a different setting. The Seattle Orientation Center has shown that mental health does have a valid role in the vast problem of rural-urban migration. The examples of how the program can be subverted by an authoritarian repressive approach were given to demonstrate that the problem of migrants should not be left to untrained people who have no knowledge of human behavior and methods of modifying human behavior. Appended to this paper is a brief follow-up of some of the Indians and Eskimos who have come through the orientation center. This is not a scientific sampling nor a controlled study since we were forced to rely on reports submitted by various staff members of the field offices in the destination cities. It does contain, however, some interesting information regarding high and low risk groups.

NOTES

1. The median divides the distribution in half; one-half the cases fall below and one-half above these points.
2. The range is the difference between the maximum and minimum lengths of stay; here, this is indicated by giving these two points.
3. Education of one individual was not available.
4. The rating procedure is described in greater detail on attachment I of this document.
5. Five of the six has a score of 9.
6. Three were rated 8, and nine were rated 9.
7. It could be argued that persons who have broken contact with field offices or moved elsewhere are returning home by stages. This may or may not be the case. The data we have do not permit answers in this regard.
8. That is, they were not instructed to either include or exclude such information.

ATTACHMENT I

Procedure for Rating Adustment of SOC Clients at Destination FEAO'S

(1) *Good*: The information supplied by FEAO'S shows no evidence of adjustment problems; or the information indicated some problems of adjustment at outset but these appear to be either diminishing or have been overcome, i.e., doing well at time this information was provided. May or may not be at destination.

(2) *Fair*: Record indicated problems of adjustment (social, personal, work); may be inability to cooperate, improper use of alcohol, poor work or school performance, tardiness or absenteeism from work or school; however client remains at FEAO and continues to try. May be described generally satisfactory but has dicernible problems. May or may not be at destination.

(3) *Poor:* Record indicates various personal, social and work problems; little or no evidence of improvement; may or may not have departed FEAO.

(4) *Not Rateable*: Unable to rate because of insufficient information.

APPENDIX *

An Analysis of Clients Served by the Seattle
Orientation Center from its Opening Through
December, 1964.

The Seattle Orientation Center (SOC) supplied information on each client—e.g., age, tribe, education, etc., and brief description of each client's behavior while at the Center, and made a prediction on each client's adjustment at destination points. Forms were forwarded to destination Field Employment Assistance Offices (FEAO's) where a brief description of each person's adjustment at destination points, the number of days each client had spent at these points, and the present whereabouts of each client were supplied. The request for descriptions of behavior from both points—SOC and FEAO—gave no specification as to what should be included or excluded in the descriptions.

Information was requested on all clients served by the Center from its opening in mid-1963 to December 31, 1964, excluding those persons going to New York for training at the RCA Institute. A total of 182 forms were received. One of these arrived after tabulations were complete and is not included in this report.

Of these 181 cases, seventeen returned to Alaska without proceeding to their originally selected destination cities; the remaining 164 cases contained seventy-nine persons (48.2 percent) relocating for employment, and eighty-five (51.8 percent) relocating for training. Table 1 gives the number of days spent at SOC and at destination cities in median[1] days and the range[2] of days by destination points. The median lengths of stay at the Center varied from 18.0 days for the Cleveland group to 25.0 days for the group going to San Francisco. Those people returning to Alaska had a median stay at the center of 16.3 days.

---

*This appendix is an analysis of data supplied by the Seattle Orientation Center and the several field offices to which clients of the Center went following their orientation. The report cannot be considered an evaluation of the influence the Center had upon its clients with regard to how they adjust at their training or employment destinations. It is, in part, a step in that direction. For example, if similar information were available on Alaskans who proceeded directly to their destination points without stopping at the Seattle Center, certain evaluative comparisons could be made.

TABLE 1

MEDIAN NUMBER AND RANGE OF DAYS SPENT AT SOC AND DESTINATION POINTS, BY THE CITY OF DESTINATION

Destination	N	Length of Stay At SOC MDN Days	Range	Time At Destination MDN Days	Range
Chicago	33	20.0	5-55	174.5	15-438
Los Angeles	31	19.2	4-45	152.5	21-434
San Francisco	29	25.0	3-47	152.5	18-454
Cleveland	19	18.0	4-39	131.8	21-457
Oakland	19	21.0	7-48	152.5	7-458
San Jose	12	23.0	13-48	199.0	10-436
Denver	11	20.3	12-27	129.0	9-358
Dallas	10	24.5	17-53	108.0	10-364
Return to Alaska	17	16.3	2-124	---	---
Total	181	20.1	2-124	137.0	7-458

Table 2 displays the total number of units going to each field office, and gives the percentage of persons in each of these groups by sex and marital status. The Chicago group contained the largest proportion of single men (75.8 percent), and Denver the smallest (36.4 percent); Denver received a large proportion of single women (45.5 percent), and Chicago the least (12.1 percent). The Los Angeles group contained the largest population of families (30.7 percent), and Cleveland the smallest (5.3 percent). The group returned to Alaska contained 64.7 percent single men, 29.4 percent families, and 5.9 percent single women.

TABLE 2

NUMBER OF PERSONS TO EACH FIELD OFFICE BY PERCENT IN EACH SEX-MARITAL STATUS CATEGORY

Total Received	Chi 33	L.A. 31	S.F. 29	Ok 19	Cl 19	S.J. 12	Den 11	Dal 10	Total 164
Single Men	75.8%	35.5	62.1	63.2	57.9	58.3	36.4	40.0	56.9
Single Women	12.1[a]	38.7	17.2	15.8	36.8	16.7	45.5	40.0	19.3
Married Men (families)	12.1	25.8	30.7	21.0	5.3	25.0	18.1	20.0	23.2

[a] Contains 1 divorce.

Training Status and Destination

There were rather marked differences in the percentages of persons going to the various field offices by relocation status, that is, whether for training or employment. For example, 100 percent of the persons who went to Dallas were trainees, and trainees accounted for more than 80 percent of the Cleveland group; in contrast, about 70 percent of the Oakland and San Jose groups were on employment status. As can be seen in the table below, the California field offices attracted a significantly larger proportion of the 184 relocatees seeking employment while the other field offices attracted a larger proportion of trainees.

	Employment	Training	
California FEAO's	(69.5%) 55	(42.4%) 36	91
Other FEAO's	(30.4%) 24	(57.6%) 49	73

$x^2 = 11.11, P < .001$

Age and Education

The median age of the 180[3] persons served by the Center was 22.2 years; and the median years of schooling was 9.1. The plus (+) and minus (-) symbols were used to represent each person with respect to his age and education relative to these medians. This produced four age-education patterns and a distribution over them as follows:

Pattern	Total N	Percent in Pattern	
+ +	29	16.1	over both medians
+ -	55	30.6	over age median, under education median
- +	54	30.6	under age median, over education median
- -	42	23.3	under both medians

Thus, persons going to the Center from Alaska are about equally likely to either be over the median age and under the median education level or the younger better schooled persons. Persons over both medians were least likely to be represented. It should also be noted here that the - + pattern contains twenty-six of the forty-two women relocatees.

Predicting and Rating Adjustment

The Center's staff predicted each client's chances of making a good adjustment on relocation; these predictions were grouped as follows: 1 = yes, good; 2 = yes, with some reservations; and 3 = yes, with stronger reservations. The adjustment of the clients was rated as follows: Three raters, working independently, read the descriptions of each client's adjustment supplied by the field offices.[4] Each rater assigned the following scores to each case: 1 = good, 2 = average, 3 = poor. We were able to rate 151 of the 164 cases in this way. The thirteen cases not rated contained insufficient information for rating; including these thirteen cases, perfect agreement on ratings occurred in 70.7 percent of the cases. The numerical ratings of three raters were summed to give a range of scores from 3 to 9.

Among the 151 persons rated on adjustment, 63.6 percent received a high prediction and 57.5 percent were rated high (3-4) on adjustment. At the opposite end of the scale, 13.9 percent received low predictions and 23.8 percent low adjustment ratings. In the middle ranges, predictions of 2 and adjustment scores of 5-7, the prediction percentages of 22.5 exceeded that of the rating percentage (18.5 percent). These percentages reflect a fairly close correspondence between the predictions and behavior ratings.

A somewhat more precise way of showing the relationship between the predictions and adjustment ratings is as follows:

Adjustment	Predictions High 1	Medium 2	Low 3
High (3-4)	69.8%	47.1%	19.0%
Medium (5-7)	17.7	20.6	19.0
Low (8-9)	12.5	32.3	61.9

$$x^2 = 21.91, 4df, P < .001$$
$$C = .36$$

Here we see that 70 percent (69.8) of the persons receiving a high prediction from the Center were rated equally high on adjustment, and only 12 percent of those with a high prediction were rated low. At the other extreme, 62 percent (61.9) of the persons receiving a weak or low prediction of success were rated low, and less than 20 percent (19.0) of this group received a high adjustment rating. The x^2 of 21.91 and the coefficient of contingency of .36 show that the predictions are significantly associated to a moderately high degree with adjustment outcome.

Age-Education Pattern and Predictions: The relationship between clients' age and education, and the judgments of the Seattle Staff regarding an individual's chances of making a good adjustment may be shown as follows:

Age-Education Patterns	Predictions in percent 1	2-3	Total II
- +	84.9	15.1	53
+ +	78.6	21.4	28
+ -	52.1	47.9	48
- -	37.1	62.9	35
			164

$x^2 = 26.86$, 3df, $P < .001$

These percentage distributions indicate that the predictions are related to the age and education of clients. Among persons over the median education level, the younger receive a larger proportionate share of the high predictions; among those with less than the median education, older persons get a higher proportion. The most dramatic differences occur between the - + and - - patterns: eighty-five out of 100 persons among those under the median age and over the median education level get a high prediction while only thirty-seven of 100 under both medians do as well.

The following table, with medium and low prediction and adjustment ratings combined, compares predictions with adjustment by age-education patterns. By following the row of percentages for each pattern, the discrepancy between predictions and adjustment ratings can be determined in percentages. The smallest discrepancies occur for the + - pattern. The difference between the two percentages is 2.3. The greatest discrepancy

Age-Education Pattern	N	High Pred.	Adj.	Medium & Low Pred.	Adj.
- +	50	84.0	74.0	16.0	28.0
+ +	24	79.2	58.3	20.8	41.7
+ -	44	50.0	47.7	50.0	52.3
- -	33	39.4	49.5	60.6	51.5
	151				

occurs for the + + pattern, 20.9 percent; the discrepancy for the - - and - + patterns is 9 and 10 percent, respectively. Thus, there seems to be a tendency to overpredict for the better educated and underpredict for persons under the median education level. This underprediction occurs primarily among younger persons. The bet on education appears well placed for the younger, i.e. - +, group. Age appears to weaken the positive influence of education in the + + group.

It is necessary, of course, to exercise caution in drawing conclusions from the data. The extent to which the staff takes age and education specifically into account in making their predictions is unknown. There is also a question of the validity of the adjustment ratings, that is, how accurately they reflect the behavioral adjustment of these clients. This too, is unknown. The present procedures for making such determinations is relatively simple to use and inexpensive. It needs, however, validation and refinement.

Age-Education Patterns and Adjustment: The table that follows shows the percentage distribution of clients within each age-education pattern by adjustment rating scores. These scores in this table show the percentage of cases in each pattern receiving perfect rater agreement that adjustment was either good (score 3), or poor (score 9), or between these extremes (scores 4-8).

Adjustment Scores	- + N = 50	+ + N = 24	- - N = 33	+ - N = 44
3	68.0%	50.0%	42.4%	36.4%
4-8	24.0	29.2	30.3	45.5
9	8.0	20.8	27.3	18.1

$$x^2 = 13.12, \ 6df, \ P<.05$$

To the extent that the ratings possess validity, these distributions show a difference in the adjustment of clients according to their age and education, with education being more important than age. The x^2 value and its associated probability indicates that the distribution of ratings among these patterns is greater than expected on chance. If these percentages are taken as empirical probabilities, the chances for an exceptionally high adjustment are 68 out of 100 for the - + pattern while the + - pattern is only 36 out of 100. At the opposite end of the rating scale, it is the - - pattern with the greatest chance of poor adjustment, i.e., 27 out of 100.

Table 3 gives three orders of information by age-education patterns: (a) the percent of persons on employment and on training status, (b) their location at the time of the study, and (c) the percent in each pattern who had and had not encountered problems in the use of alcohol.

TABLE 3

RELOCATION STATUS, PRESENT LOCATION, AND DIFFICULTIES WITH USE OF ALCOHOL BY AGE-EDUCATION PATTERNS

	Age-Education Patterns			
	+ + 28	+ - 48	- + 53	- - 35
A. Relocation Status	%	%	%	%
1. Trainees	60.7	14.6	92.5	34.3
2. Employment	39.3	85.4	7.5	65.7
B. Present Location				
1. At Destination	60.7	72.9	90.6	74.3
2. Left Destination	39.3	27.1	9.4	25.7
	$x^2 = 10.26$, 3df, $P < .02$[a]			
C. Problems with alcohol				
1. No evidence of	85.7	56.2	54.3	60.0
2. Probable or certain	14.3	43.8	5.7	40.0
	$x^2 = 24.89$, 3df, $P < .001$[a]			

[a] x^2 values calculated from frequencies.

Relocation Status: Among the 164 persons proceeding to their destination cities, eighty-five or 51.8 percent were trainees. Part A of Table 3 shows that better than 90 percent of the younger persons above the median education level relocated for training. This is encouraging and speaks well for the guidance given these persons as they apply for relocation. Just under two-thirds of the older persons above the education median also went into training. Only one-third of the younger group under the education median were in training, and less than one-fifth of the older persons under the median education level were trainees.

As can be seen in the table immediately below, a significantly larger proportion of trainees received a high adjustment rating (score 3): 63 percent of the trainees were rated this high while only 37 percent of the persons on employment received a rating of 3. In the latter group, the percentage rated poor (score 9) was almost five times larger than that of trainees (6 to 29 percent).

	Total N = 180	To Alaska N = 16[a]
+ +	16.1%	6.3
+ –	30.6	43.8
– +	30.0	6.2
– –	23.3	43.7

[a] Education not available on one case.

Since we have seen that the younger and better schooled got a greater share of the high adjustment ratings and since the vast majority of them were in training, these results are not surprising. Examining the data in this way did, however, pose a question: Is the risk of making an unsatisfactory adjustment less for training status regardless of one's age and education?

To explore this question with the present data, it was necessary to collapse the adjustment ratings into two categories: 3 to 6, and 7 to 9. In the - + pattern, there were only four persons on employment status and all of these fell into the 3 to 6 category and 85 percent of the trainees were rated this high. All but seven persons in the + - group were on employment status, 70 percent of this group were rated in the 3 to 6 category of adjustment, and 65 percent of those in the + - pattern were so rated. None of these differences proved to be significant.

A different picture emerged, however, for the + + and the - - groups. Among the former (+ +), six of the nine persons on employment (66.6 percent) were rated in the 7-9 adjustment range.[5] All (100 percent) of the fifteen + + trainees were rated 6 or below; actually, ten of the fifteen were rated 3. These differences were significant at the .01 level. In the - - pattern, twenty-two were on employment, and eleven were trainees; 55 percent of the former were rated 7 or better,[6] and 100 percent of the eleven trainees were in the 3 to 8 adjustment range; eight of the eleven were rated 3, a perfect high score. These differences by employment and training status were significant at less than .01. These findings suggest that + - and - + persons are likely to adjust about as well on employment status as on trainee status. On the other hand, it appears that + + and - - persons are considerably more likely to make a better adjustment on training than on employment status. Should these findings hold up upon further investigation, they could have important implications. Also, if these findings should remain firm, there is the problem of explaining why this should be so. The answer may be that persons in these age-education groups seeking training are more stable than their counterparts seeking employment and/or training status itself may be more conducive to adjustment than is employment status.

Present Location: At the time of the study, thirty-eight of the 164 persons (23.2 percent) were no longer at their destination points or their whereabouts were unknown.[7] The number and percentage of the persons no longer at their destination points were:

>Twenty-three to Alaska ....60.0 percent
>Ten were elsewhere in the U.S. ....26.0 percent
>Five had broken FEAO contract ....14.0 percent

As Part B of Table 3 indicates, the persons most likely to have left their destination city or to have lost contact with field offices were those over the median age and education (+ +) levels; those least likely to have done so were the - + persons. About 40 percent of the former and 10 percent of the latter had departed or broken contact with field offices. Persons most given to returning to Alaska were older men, i.e., men over the median age, and these were most heavily represented in the + - group. Among the thirteen persons in the group (+ -) no longer at their destinations, eleven had returned to Alaska; only six of the + + group no longer at their destinations were reported to have returned home. Younger persons (- + and - -) were more likely to have broken contact or gone elsewhere in the country: among the thirteen persons in these two groups no longer at their destinations, only six were indicated as having returned home.

Persons who had departed their destinations by the time the study was done were much more likely to be on employment rather than training status. Among the thirty-eight persons in this groups, thirty were on employment status and accounted for 38 percent of the total in this status. The other eight were trainees who accounted for less than 10 percent of the total trainees. These differences were statistically significant (x^2=17.21, d.f.1, P <.001).

Problems with Alcohol: In preparing their descriptions of the adjustment of clients, the field offices were free to include or exclude references to problems resulting from the use of alcohol.[8] The descriptions provided made specific reference to this problem in thirty-nine instances and in three cases the problem was suspected; the forty-two cases amount to slightly more than one-quarter of the total group. The records indicate that eleven of the 164 had experienced some problems with liquor while at the Center; seven of these persons repeated this pattern at their destination and are included in the forty-two above. As Part C of Table 3 shows, problems with alcohol differ significantly among the four age-education patterns. It can be seen in Part C that the less well educated are more susceptible to this hazard. The + - and - - pattern percentages are three to six times greater than the + + and - + patterns.

Persons Not Proceeding to Their Destinations

Seventeen or approximately 9 percent of the persons coming to the Center (exclusive of the RCA trainees) did not proceed to their destinations; they either chose to return home or were returned by the Center. This group contained eleven (65 percent) single men, six (30 percent) married men, and one (6 percent) single woman. Are these persons equally distributed over the four age-education patterns? The following table gives the percentage of the total group and the percentage of turn-arounds in each of the patterns:

Relocation Status	3	Adjustment Scores 4-8	9
Training	63.3%	30.4%	6.3%
Employment	37.5	33.3	29.2

$$x^2 = 16.43, 2df, P < .001$$

It can be seen that + - and - - persons were most heavily represented among the turn-arounds. These two patterns accounted for 54 percent of the total but contained more than 85 percent of those returning home from the Center. Nine (50+ percent) of these seventeen experienced some problem with alcohol while at the Center. The records indicate that only eleven (less than 7 percent) of the 164 going to their destinations exhibited such behavior while at the Center. Also, it can be seen from Part C of Table 3 that the + - and - - groups were the persons most likely to have problems in this regard after arriving at the field offices. And, as will be recalled, these patterns received smaller proportions of high predictions of success and high adjustment ratings. All told, it appears that the percentage loss by turn-arounds is relatively modest, and that those persons who return home at this point are among those with greater risk of poor adjustment at destination points.

Chapter **18**

To Be or Not to Be Political: *A Dilemma of Puerto Rican Migrant Associations*

LLOYD H. ROGLER

For many years social scientists have recognized the importance of voluntary organizations in the life of immigrants. Their research documents the variety of functions of migrant organizations — from preserving some continuity in the members' social experience to serving as collective mechanisms for adaptation to new demands.[1] To perform such functions, however, the organizations must be relatively stable. This has not been the case of organizations formed among the 4,500 Puerto Rican migrants living in Maplewood, a middle-sized city on the eastern seaboard.[2]

Over the years the Puerto Ricans in Maplewood have tried repeatedly to organize groups to serve their ethnic interest, but many, perhaps most, of the groups came to an end before they were even given a name and they are now recalled simply as meetings held here or there by interested persons. One such organization held one meeting and another three before collapsing, reportedly because of conflict among the members. Since 1955,

---

Author's Note: *This investigation has been aided by a grant from the Foundations' Fund for Reserach in Psychiatry. The research has also been supported by an award from the Ford Faculty Research Fund given to me by Yale's Concilium on International Studies. I am grateful to Janet Turk for her editorial assistance on this paper.*

about fifteen associations developed to the point of being named. At present, there are only two active Puerto Rican associations in Maplewood. Thus, the thrust to become organized has been almost entirely offset by the inability of the groups to endure.

This paper reports preliminary ideas and observations relevant to the chronic instability of migrant associations. It demonstrates that Puerto Ricans find formidable, indeed, the problems the new social setting imposes upon their efforts to develop organizations. In Maplewood, pressures stemming from the local political system create an unsettling but persistent dilemma of organization. Three alternatives present themselves to the organizations — to become affiliated with a political party, to remain politically independent but support as a group the candidate or party with the most to offer, or to remain completely uninvolved in political activity — but whatever choice is made, dissension is bound to arise among the members. Each alternative evokes opposing views, and without a firm consensus, which is difficult to achieve and maintain, the issue remains a continuing threat to the unity of the group. This dilemma has roots which extend deep into the migrants' experience in the new and unfamiliar social setting to which they have moved.

The observations in this paper derive from intensive interviews with Puerto Ricans and North Americans who have been involved in the development of such groups. Included were lawyers, the mayor of the city, coordinators of neighborhood school programs, antipoverty workers and officials, and Puerto Rican leaders. Also, participant observation was used as a method, group meetings were tape-recorded, and field notes taken. One Puerto Rican group, the Hispanic Confederation of Maplewood, was selected for detailed case study, and its meetings observed from the group's inception to thirty-one months later in order to identify the evolving phases of its life, as related to its broader setting. After the Confederation had been in existence for nineteen months, forty-two persons who had attended at least one meeting were interviewed individually for about three hours with a preformulated schedule. Also, semistructured interviews were conducted with the leaders of the Confederation. Most members of the group are married adults who were born in the rural areas of their homeland; they have lived almost a third of their lives in the United States where they are at the bottom rung of the socioeconomic class hierarchy of the city.[3]

The role of politics was first observed early in the life of the Hispanic Confederation when, after considerable discussion, the members decided that the group would not engage in political activities. This decision was referred to as a "premise" of the Confederation and, together with the goals of working for the betterment of the migrant population, comprised part of the modest legacy with which the group began. However, the policy of avoiding political activities evoked strong feelings from the members and was not easily maintained. Six months after the group's inception a bitter argument broke out over the question of providing transportation to register Puerto Ricans as voters. Some members thought this was one of the group's civic responsibilities. They believed that it could be done without influencing the registrant if political topics were avoided during the process of registration. The opponents, however, countered that to help persons register as voters, if not an outright political action, came dangerously close to being one; also, there were doubts that members would have sufficient self-control to suppress their own political opinions at such a time. Though proscribed by policy, politics was still viewed by some members as an insidious threat to the organization.

In the ninth month of the Confederation's life when dissatisfaction in the group had grown, the threat materialized. Although the Confederation was an action group according to its goals, there was little continuity in the pursuit of objectives. The members became locked into repeated discussions of their own personal, troublesome experiences. From meeting to meeting arguments exploded, and word soon spread among the Puerto Ricans of Maplewood that the Confederation was like "... a bunch of squabbling women." Interest in the organization began to fade.

In response to this crisis, the president of the Confederation consulted with *the* Puerto Rican political boss of Maplewood, Vicente de Serrano, about the possibility of converting the group into a political organization. (Ironically, at the early meetings it was the president's voice that had been the strongest against political activities.) After a meeting with the mayor of the city, arranged by Serrano, the president went to the Confederation to persuade the members that the group should turn to politics. He was accompanied by two outsiders, sent by Serrano, who also spoke persuasively, but the members were mindful of the history of failures in which earlier groups had been split into politically opposing camps. They remembered other organizations delivered into the hands of those who were interested in the welfare of the Puerto Ricans only during election campaigns. After a series of stormy sessions, the group rejected the proposal to become political. Subsequently, the Confederation withstood successive political pressures, acquired a new president, and continued as an independent civic organization, with a reaffirmed policy of avoidance of political activity.

In the larger Puerto Rican community, however, opinions on this issue divide themselves into the three alternatives of the dilemma affecting groups. Representing the first alternative are those who play an active role in politics, particularly in the highly organized Democratic Party which has overwhelming control in Maplewood; they are the strongest and most convinced advocates of politicizing migrant groups. Chief among these persons is Vicente de Serrano, the Puerto Rican political boss of Maplewood. For several years, Serrano has been an important link between the mayor's office and the Puerto Rican migrants of the city. Among compatriots, he has been the most active in registering voters for the party and encouraging them to vote. As an interpreter at the local court, translating for Puerto Ricans who do not know English, Serrano enjoys the benefits of political patronage; meanwhile, he consolidates his power among migrants brought to court who feel obligated to him for his help. To enlarge his grip on the migrant community, Serrano has several Puerto Rican aides who assist in his political activities and, in turn, receive the favors of the political machine. One important duty of the aides is to participate in newly formed migrant associations, preferably as officers. At election time, the aides are allowed to give brief talks at political rallies for the Democratic Party, but Serrano maintains firm control over his ubiquitous role as master of ceremonies, introducing visiting dignitaries and political candidates. Thus, Serrano's role at political rallies is symbolic of his broader role in the community, the powerful intermediary between the migrant and City Hall. To maintain this position, he finds it important either to bring nascent migrant groups under his control, as he tried to do with the Confederation, or to see that groups threatening to establish independent channels to City Hall be eliminated.

Serrano's position in the system makes of him a defender of the efficacy of politics. He believes politics is a democratizing force, bringing together persons of unequal background to work together. In so doing, he believes, politics introduces a cohesive force into the life of a group, enlarges the opportunities for members to participate, and suffuses the group with the democratic ethos of the country to which the persons have migrated. Politics also mitigates a primary source of instability in Puerto Rican social or civic groups — jealousy among members who have unequal levels of education. Serrano told us:

> ...those that make a little bit more money or live a little bit higher [better] —he [the person of higher status] does not want to mix with the other because he thinks that he is better. That is why [social] groups and associations do not succeed. That is why

politics is the only way, because in politics you have to treat everyone the same way. Politics accepts anyone, drunks, bums, . . .

Serrano went on to say:

Yes, it [the group] has to be [political]. Everything is moved by the political machine . . . Today most of the people are in government jobs, the judges, the mayors — all these jobs are politically based. That is why we [the Puerto Ricans] need politics . . . [Politics] is the only thing. Social activities don't mean anything in this city. They don't care how many social activities you have, how many dances you have. They [the Puerto Ricans] can't get a job through that. They get a job through politics.

Serrano believes also that through politics the identity of the ethnic group is established and recognized by those who have political influence. By means of political rallies, he told us, a minority group presents itself to influential persons in the city in a show of unity, as an alive and vital group:

We have a [political] rally so that we can be recognized as a group. We use the club [the Pan-American Association] so that we, the Puerto Ricans, can be recognized as a group as are the Italians, the Greeks, and so on. Why do you think the colored people are in a position today to demand? Through politics! That's why! Otherwise they would be nobodies!

Serrano sees the politicians as omnipotent, in contrast to the powerless Puerto Ricans. The tactic of the minority then, he feels, must consist of involvement with the political elite, demonstration of the group's vitality, and support in political activity in order to gain benefits and avoid being ignored.

Echoes of Serrano's political views can be heard from his political aide, Mr. Romero Ponce, who enjoys the benefits of political patronage as a refuse collector for the city.

. . .I say that everything has to be linked with politics. If one does not embrace politics, one gets nowhere . . . .Because these people [the politicians] are the ones who control the city and the state. We [the Puerto Ricans] don't control anyone. They can throw us out of here whenever they feel like it and send us somewhere else. If you are here, you have to be with them. . . . That's the way it has to be.

The appeals of Vincente de Serrano and Romero Ponce are most persuasive to the migrant who is at an elementary level of assimilation. Knowing little or no English, restricted to menial work at the bottom rung of the occupational ladder, and subject to discrimination because of his race, ethnicity, and language, the unassimilated Puerto Rican is a marginal man striving to understand the new environment in which a welter of public agencies—public welfare, the antipoverty agency, the employment service, the department of education—surround him as a broad system with which he must contend. His perception of the system is indicated by the questions asked of local officials invited to address the groups about the services the agencies provide. At one meeting of the Hispanic Confederation, the members of the audience brought up unpleasant incidents experienced in their contacts with the agency. The official was expected to know the details of the incident—why the client was made to wait for several hours, why a particular agency clerk was impolite or insulting—and to explain the agency policy that was responsible for such treatment. To the audience it came as a surprise that the official could not answer the questions.

The gaps in the communicative and control structure of agencies, the social cleavages and rivalries which are endemic to large organizations, as well as the differences in the orientation of their staffs are at best dimly perceived by the migrant who, though a marginal outsider, must cope with the formality attending the agency services he receives.

The Puerto Rican migrants expected agency officials to conform with agency policy. At a meeting of the Confederation when an employee of the local antipoverty agency expressed disagreement with policies of his own agency, the members of the group were surprised and confused. After the meeting, in casual give-and-take conversation among the members who lingered, one commented that the antipoverty worker did not really disagree with his own organization but had been instructed by his superiors to oppose agency policy at the meeting just to confuse the members of the group. To the migrant there is bureaucratic design behind the most casual, even gratuitous, comments of agency officials.

The advocate of the importance of political activity in migrant associations links his appeals to this image. The system, he argues, has been created by politicians and stands at their command; politics, as it were, provides the cohesive force. The message is not lost on the unassimilated migrant. If memories of the homeland *patron*, who looked after his moral and personal welfare, still linger, he decides it is now the political figure who has parallel resources in the form of agencies affecting the grass-root level of his life. "If you don't have a godparent, you don't get baptized." Thus, arguments in favor of associations being political are coherent, credible, and meaningful, for they have roots in the migrant's experience. Nonetheless, there is strong opposition to such an idea. Politics also connotes shady deals, schemes, and intrigues designed for personal gain at the expense of the migrant. To Antonio Tejada, who has been a leader in Puerto Rican associations, politics is anathema:

> ...If you want to see the Puerto Ricans united, don't talk to them about politics! The Puerto Rican who comes here, comes to work, not to be in politics; he comes to see his friends, get a job, and to live religiously. Those who come here with the idea of politics are in a minority.

Tejada resigned as president from the now defunct Pan-American Association because the other officers of the group wanted it to become political:

> I did not want them to include me in politically oriented activities because here the politician says he has the Puerto Rican between his thumb and forefinger.

Tejada feels politicians cynically accept the support of the Puerto Ricans without assuming any obligation in return to help the migrant. He thinks this is because the migrant population is small. Contrary to expectation, Tejada's unfavorable attitude towards politicians does not deny the importance of political activities. He suggests:

> If you make an organization so that all Puerto Ricans can belong to it, then we don't need to be political. If we are united, some two or three thousand strong, we can fight with a united vote for the betterment of the community. With good leaders, they [the politicians] will come to us and offer us what we want without our going to them.

Unlike Vicente de Serrano who believes that political groups have the advantage of developing solidarity by striking a common democratic balance among members of unequal education and income, Tejada emphasizes the need for educated leaders:

> I have more confidence in a person who has education than in one who is a politician of any kind. A person with a good education can guide and teach those who are in need of learning something, but political groups have leaders without the qualities of leaders — a leader only because he knows the mayor and others! This is a leader?

Thus, Tejada recommends that the group should have educated leaders, be organizationally outside the party structure, but engage in political activity according to what would most benefit the Puerto Ricans. This position represents the second alternative in the dilemma of politics and is referred to in the migrant vernacular as being for *candidatura* (for candidates) or as "following the course of liberal politics." It advocates transitory political attachments depending upon the parties or candidates who promise to serve the Puerto Rican. It assumes that members of the organization will agree as to the best candidates or party and then, disciplined and unified, obey the group decision, although party lines may have to be crossed to support individual candidates. The political preference of the member, if different from that of the group, would have to be suppressed in conformance to group decisions. Then, to provide active political support, a fine-grained allocation of effort would be instituted in the organization. It is difficult to explain this attitude at group meetings, for it straddles the fence in the dilemma of to be *or* not to be political and invokes the group to be *and* not to be political. The organizational tasks imposed upon such a group would be extraordinary indeed, and efforts to develop groups of this kind have not succeeded.

The third and final alternative invokes the group to be neither linked to a political party nor involved, as a separate body, in actively supporting parties or candidates; all political activities must be clearly and unambiguously ruled out. This is the alternative the members of the Hispanic Confederation chose.

After the Hispanic Confederation weathered a series of political assaults and the dissident members who advocated politicizing the group either dropped out or were silenced by the group, it bypassed political intermediaries and established direct contact with the mayor of the city. This was a unique and bold innovation in the history of the city's nonpolitical Puerto Rican groups. In a face-to-face conference, a committee from the Confederation presented to the mayor a number of proposals to benefit the Puerto Rican population of the city. The mayor was impressed by their requests and instructed officials in the antipoverty program, in the urban redevelopment agency, and in the educational system to cooperate with the Confederation. Subsequently, an office was established in the antipoverty agency to serve the needs of the Puerto Rican community. The mayor thought the Confederation's actions were unusual. He said:

> The ... interesting thing ... about the Latin American group [the Confederation] is that they scorn the power brokers who attempt to represent them politically and instead they prefer to negotiate with City Hall on the basis of what they feel the community should do for these people, not in terms of jobs, political jobs, patronage, but in terms of assisting these people in adjusting to urban life. This is a totally different approach, as you know, from the old concept. When the Irish first came to America they were greeted at the docks in Boston with their citizenship papers in one hand and a certificate to go to work for the city in the other hand. This was a political venture and nothing else. But today in the Latin American Confederation they are here for different purposes.

As "chief magnate" of the city, how does the mayor view the dilemma that politics imposes upon the migrant group? To answer this question we directed the Mayor's attention to the decision the Confederation faced as to whether or not to turn political, and we asked if the Confederation would have been more effective as a political group. He responded:

> It doesn't really make any difference. I've watched many ethnic groups go through this convulsion. The Negroes have tried to organize on a political basis and the Jews tried fifteen years ago when I first began to run for mayor. Actually what happens is that the average voter is far more sophisticated than anybody realizes. They understand what candidates stand for and what their programs mean. Frankly speaking, I have never had the problem of communication with any of these groups. I think if we organize on the basis of what is needed in the community the political benefits become obvious.

The mayor's statement coincides with his prior reaction to the proposals submitted to him by the Confederation. He was surprised that the Confederation had bypassed political intermediaries and negotiated with him directly. Correlatively, the members of the Confederation were surprised that, having bypassed such intermediaries, they met with a favorable response from the mayor without political favors being promised or exchanged.

There are present in Maplewood minority pressure groups which operate outside the conventional political party structure. One Negro group has assumed a militant stance in relation to City Hall and the service agencies of the city and has, in its negotiations, achieved considerable success; although this Negro group is located in the largest Puerto Rican neighborhood of the city, the Puerto Rican migrants had not benefited from the example. For this reason, channels for exercising influence, aside from the political as represented in the person of Vicente de Serrano, remained unestablished. At present it is too early to determine if the Confederation's actions will set a precedent for subsequent migrant associations or what its influence will be. Should its actions become widely known and accepted, a necessary step will have been taken in the development of a model of an independent action group as a part of migrant culture. The pervasive confusion introduced into the group effort by the dilemma under discussion will then be mitigated.

Yet the central question of how the Puerto Rican is to participate in the higher levels of decision-making still remains unanswered. The political boss, his aides, and his associates are embedded in the political party structure. To maintain their position as intermediaries they cannot take the mayor's catholic view of organizing "... on the basis of what is needed in the community ...." They must induce the migrants to continue to believe that partisan politics is the primary means of acquiring group influence. The experiences of the migrants in the new social setting, particularly their efforts to understand the array of city agencies and services that affect their lives, support the appeals of the politicians, and it is with the advocate of politics that persons forming groups associate daily, for they are all part of the web of social relations which binds the migrant community. The effective world of the Puerto Rican extends horizontally into his own ethnic group at the lowest level of the city's power structure; it does not rise vertically. No informal web of stable interpersonal relations of friends, neighbors, and associates links him to the focal points of decision-making.

In brief, there are many precedents of persons who got jobs or had complaints remedied by means of political contacts, but *within* the life space of the migrant there is no precedent for a successful *independent* Puerto Rican group. By deciding not to be political, the migrant group denies itself access to the most evident instrumentality for upward influence.

As first-generation migrants with limited formal education, the Puerto Ricans have had few experiences from which they can draw in developing effective action groups. The array of voluntary associations which proliferate in middle-class American life does not have a counterpart in their experience either on the Island or the mainland. In the absence of such learning experiences, the fund of skills and attitudes needed to maintain organizations remains undeveloped, thus making the groups vulnerable to the delemma and the problems of choosing an alternative—to be linked to a political party, to be independent of a party but engaged in political negotiations, or not to be political either in organization or in activities. Group consensus on this point is difficult to achieve and to maintain. The stable operation of social forces impinging upon the Puerto Rican serves, in an ironic manner, to perpetuate his unorganized position and to deny him a voice he very much wants.

NOTES

1. Among these are Fallers (1967); Little (1965); Mangin (1965); and Glaser and Sills (1966).
2. The names of the city, the persons, and the Puerto Rican groups are fictitious.
3. All the members of the Confederation migrated to the United States, ninety percent of them from Puerto Rico. Two-thirds were born in the rural areas of their original homeland. The majority migrated directly to the state in which Maplewood is located a mean number of eleven years ago. They attended school a median of eight and one-half years. The median age is 36; two-thirds are male; eighty-six percent are married. Spanish is still their first language. According to Hollingshead's Two Factor Index of Social Position (1957), which is based upon occupation and education, seventy-four percent of the members are in Class V, the lowest class in a fivefold system.

REFERENCES

FALLERS, L. A. [ed.] (1967) Immigrants and Associations. The Hague: Mouton.
GLASER, W. H. and D. L. SILLS [eds.] (1966) The Government of Associations. Totowa, N.J.: Bedminster Press.
HOLLINGSHEAD, A. B. (1957) Two Factor Index of Social Position. New Haven: Privately Printed (1965 Yale Station, New Haven, Conn.).
LITTLE, K. (1965) West African Urbanization: A Study of Voluntary Associations in Social Change. Cambridge: Cambridge University Press.
MANGIN, W. P. (1965) "The role of regional associations in the adaptation of rural migrants to cities in Peru." In D. B. Heath and R. N. Adams (eds.) Contemporary Cultures and Societies of Latin America. New York: Random House.

Chapter **19**

Preventive Planning and Strategies of Intervention:
An Overview

EUGENE B. BRODY

Intervention aims at promoting the economic integration, optimal function and personal stability, in short, the adaptation of the migrant. Conversely it aims at avoiding unemployment, family fragmentation, mental illness or other forms of behavioral deviance, and the development of destructive tension between migrant and host population. Whether or not the promotion of early acculturation and eventual cultural assimilation is a legitimate goal of intervention still seems debatable.

Intervention may be aimed at the individual migrant, the donor society, or the host society during any phase of the migratory process including the transitional. With respect to the donor society goals may range from preparation of the potential migrant for his move to diminishing push-factors which encourage migration, or in other ways encouraging the migrants to remain in their home communities. With respect to the host society customary goals include reducing the likelihood of prejudice-based hostility or discrimination, the development of integrating mechanisms and institutions, and the creation of transitory environments or other means for accomplishing a smooth intersystemic move.

With respect to the migrant himself, or his family, the goals generally include attention to those factors which differentiate him from members of the host community or interfere with his adjustment, and to those

---

Editor's Note: *Much of the material of this chapter is taken from the daily conference discussions. Thus, in many respects it reflects the thinking and experience of all conference participants.*

which might increase the likelihood of his making a satisfactory transition. Examples of these are literacy-training, orientation and anticipatory information about the new community, the acquisition of marketable skills and language training. (The most visible differences between a migrant and the new community, i.e., skin color, are not susceptible to change.) An important subgoal is the stimulation of motivation to begin educational and other programs. This may require a fundamental change in value systems.

Public Policy Level Intervention

Migration may be discouraged in extreme cases by the use of armed force at international boundaries, or in cordons around cities. Less extreme but more common are restrictive laws such as the United States McKerran Act.

The needs which motivate migration may be taken care of in the donor society by such public measures as building new factories, providing new jobs or educational opportunities. Granting subsidies to create new industries and a new labor market will reduce the number of rootless people in the host as well as the donor society. Factory work may produce literacy and serve rural workers as a partial substitute for going to school. Jobs make it possible to buy television and transistor radio sets which further stimulate modernization in the potential donor system.

Modernization, in general, including advances in agricultural productivity may decrease migration. This can help prevent the depopulation of donor areas and reduce the number of emotionally vulnerable individuals and fragmented families. Reduction in the volume of migration can relieve the pressures upon host system municipal relief rolls and the demand for municipal services to persons who are unable to contribute to the tax base. Premigratory educational programs or policies aimed at diminishing the flow of illiterates may contribute to the same end.

At the level of public policy (covert as well as overt), a government may promote migration as a substitute for birth control. This may involve lowering the costs of transportation (which in turn can facilitate circular migration patterns and the maintenance of continuity with the donor system) and alerting employers in the host area. This loss may on one hand reduce economic tensions among the migrants, but on the other increase the hostility of competing higher-priced workers in the host society. Public policy decisions about low-cost housing may promote ethnic enclave or ghetto formation which promotes tension among the migrants and backlash in the surrounding community; or policy decisions against such housing may promote diffusion of migrants throughout the host community, which may deprive them of needed transitional period support. Policy decisions may also lead to the building of new towns to house newcomers. Similar considerations attend decisions about the development and distribution of health, education and welfare services and the placement of new industry.

Migration is influenced not only by governmental policy but by the needs of the industrial-agricultural complex. Thus manufacturers or ranchers may in some periods facilitate the (illegal as well as legal) entry of migrant workers into an area and at other times discontinue their support. In this sense migrants may be regarded as a type of import. A series of undesirable interactions with the host area may be predicted from this. Some of these are based on the nature of those who come in, and their loneliness and frustration since they are more likely to come without families and no expectation of eventual assimilation. Some are based on the likelihood of resentment among native workers, and the lack of concern by employers once the migrant's economic usefulness is passed. At the same time the donor areas may profit from the income which these workers send back home.

The deliberate use of acculturative mechanisms such as military service, or employment in government projects (cf. WPA, CCC of the F. D. Roosevelt era), may be effective in the social, economic and cultural integration of migrants. Placing television sets in public places makes information more available, facilitates learning to read and stimulates acculturation. The establishment of educational facilities up to the community college level will contribute to the development of a pool of human capital in both donor and host areas; the community college may have special value in moving people out of the ethnic enclave. The host system has a normative language and behavior structure against which it measures the behavior of the migrant. A problem in intervention is how to get away from the built-in systems of punishment in school for children who don't "talk right" even though their primary language is called "English."

The timing of special programs should be determined by the S-shaped learning curve which includes a "familiarization" function. Thus, it has been suggested that adult education programs for recent migrants should not be five, but fourteen or even twenty-four months. In the Israeli, Soviet, Brazilian and other armies, educational programs are used to promote acculturation and to develop general cognitive skills and coping capacities.

Some of the educational needs of migrants have been conceptualized in terms of their learning to use "the urban language," which is regarded by some as part of a "modernization" process. Curricular planning for schools serving migrant areas poses particular problems. What, for example, are the relative advantages and disadvantages of teaching or requiring Spanish as a major primary school language for Puerto Rican children in New York City or Mexican Americans in San Antonio?

Individual Intervention

Job placement services, scheduled meetings with local personnel officers, newsletters to home villages (as in the case of Southwestern Alaska natives), meeting services at airports and bus stations, the construction of Welcome Centers or Relocation Centers (as in Anchorage and Seattle) are examples of initial or "gateway" services aimed at individual migrants themselves. The same may be said for adult education programs, language training and the like. Recruiting for literacy programs may be a problem. Store front missions have been used to promote literacy with a drivers license as a reward. Another approach has been to have people in coffee shops watching to see who can't read the menu.

A "gatekeeping" agency might collect information about the basic personal and socioeconomic characteristics of migrants and predict on the basis of such information what will happen to them; at the same time it could supply counsellors whose intervention would be based on the predictions. This might be especially important for "stumblers" and "strugglers" who can go either way. Some "losers" seem condemned from the beginning by potentially treatable poor health.

A comprehensive "gateway center" arriving at an assessment of the liabilities and assets of migrants as they enter the host system would facilitate their effective acquisition of needed skills and job placement. It may be useful in this respect to think of constructing a *modal profile* of a migrant, a profile of the host system, and a prediction about fit in terms of outcome. The construction of "profiles" could lead, however, to prematurely crystallized concepts of migrants or environments without general applicability. On the other hand, if done at the donor site, perhaps by the U.S. Employment Service or some similar agency, this kind of systematic assessment might prevent certain adaptive crises from developing later on.

Intervention with Groups

The question of how to bring migrants into the decision-making process—especially insofar as it concerns their own fate—is an important one. The development of self-aware migrant groups with political leverage or the transformation of an agglomeration of individuals into a collectivity committed to action on the basis of a shared value system is an aspect of this.

An important approach to this problem has been through teaching patterns of formal organization as 4-H Clubs in Appalachia. Many migrant groups, e.g., from Puerto Rico, seem unable, however, to develop a working group history. They, at least the Puerto Ricans and Mexicans, do not function well, according to sociological observers, as organized groups in terms of conducting meetings, note keeping and producing messages aimed at power groups. Some of this may stem from earlier histories of relating to a patron who acted as a recorder and transmitter of appeals. Other factors inhibiting effective organization and group function are fear of exploitation by whoever motivates the group's formation and deep-seated distrust of being used for political ends. This last seems to arise from patterns of political practice in the donor societies.

There are certain principles of organization embedded within cultures. It takes time to develop particular group norms, e.g., that personal revelations are not sanctioned in a parliamentary group. Groups also soon realize that just talking about or reiterating the presence of problems is a dead end. They must move on toward some kind of action. In the process they must overcome the suspicion that someone is getting personal advantage from the group. This is a problem because such groups (e.g., of Puerto Ricans in the continental United States) usually are initiated by a white North American. Here intervention must deal with a problem in culture, values and time orientation. Intervention aimed more specifically at problems of emotional adjustment may include group therapy meetings, the development of migrant social clubs of people from the same locality and individual counselling.

Specific intervention possibilities include: (a) teaching principles of organization, parliamentary procedure, and the use of social power; and (b) promoting the development of group norms.

Intervention research could be concerned with how knowledge is related to social demands. How may people be reached through appropriate communication networks, e.g., through taverns, neighborhood centers? What are the communication systems of greatest relevance, e.g. how can local political bosses be useful?

Migration Counsellors

Who should do the individual intervening and what should be their training and background? Are social workers and other similar personnel capable without added education? Should the counsellor be from the same area or of the same ethnic group as the counselee? How do they fit in with voluntary associations of migrants? Some experience suggest a variable outcome with placing migrants close to an already settled family from their area.

Circular Migration

In Brazil, and probably in other countries, most returnees to home areas are those who have become disillusioned or they are women and minors whose husbands or parents have died or deserted them. Services have been developed to provide transportation back to home areas.

Ethnic Enclave Development

Individual behavior may be determined by the shape of ethnic enclaves. Does one of irregular form with pseudopods extending into the dominant community increase the speed of acculturation and adjustment by providing opportunities for maximum contact? Or does it impede adjustment by extending boundaries at which there exists a potential for conflict?

Some investigation has been done on the nature and location of artificial bridging environments. Thus a new colony of an old ghetto may be set up ten minutes away on a subway. This can reduce the tension which appears when two different ghettos expand against each other, which happens with successive waves of migration from different countries or from different U.S. rural areas into the city.

Questions relevant to ethnic-enclave intervention include the following:

(1) How can artificial bridging environments best be constructed?
(2) What is the optimum shape of a bridging environment or a ghetto? What are the advantages of self-contained, round or square (unambiguous, with fewer conflictful possibilities) versus more ambiguous pseudopodial (with more proximity possibilities and perhaps a broader range of available options) boundaries?
(3) What is the best approach to the development of educational facilities, especially community colleges to promote outmobility from the ghetto?
(4) What is the best approach to development of hand industries as well as manufacturing in the ghetto itself?
(5) What is the best approach to development of available service and other jobs in the cosmopolitan urban environment to promote movement out of the ghetto?

Involuntary Migration

Involuntary migration is often a reaction to natural or political catastrophe or man-made emergencies such as flooding to create lakes in the Sudan. Here intervention must begin early because these are uprooted people who haven't the time to make decisions. They have few options. Should they go alone or with families? To what degree should they be able to depend on bureaucratic institutions? People who come into contact with paternalistic bureaucracies sometimes want to perpetuate their dependence. Should they begin language lessons or job training before their arrival in the host system?

What kind of preparation must be made in receiving countries? When large numbers of refugees must be handled by camps how should these camps be operated? What can be done to arrest deterioration in the camps?

Are there special problems around high status people who lose their status supports as they become refugees? How can counsellors or refugee-helpers be recruited and trained?

There is presently no record system which follows people through the refugee system. Can one be devised to which they, themselves, can contribute?

How can communication be established with the people left behind? What should be done about loneliness fostered by rapid spatial mobility? How might communication be facilitated between organizations which receive refugees? Should there be a central clearing house?

All of these questions suggest forms of intervention at both public and private levels which may prevent later problems in adaptation.

APPENDIX

Chapter **20**

Resources for Refugees

Compiled and Annotated by
HENRY P. DAVID

Agencies and organizations working with refugees may be divided into three major categories: Intergovernmental, U.S. Governmental, and Non-Governmental (voluntary). Voluntary organizations are often subdivided into denominational and non-denominational agencies. They tend to work more closely with individual refugees and families, focusing on preparations for migration, transportation, and follow-up services in receiving countries.

The following list of organizations with brief reports on their activities is based on information gathered for the 1968 World Report, prepared annually by the U.S. Committee for Refugees, 20 West 40th Street, New York, New York, 10018.

Further details can be obtained by writing directly to the agencies listed. More extensive reports on major international nongovernmental organizations were published in the July-August 1967 issue of *Migration News*, available from the International Catholic Migration Commission, 65 rue de Lausanne, 1202 Geneva, Switzerland.

Author's Note: *Much of the material contained in this Appendix was gathered by the U.S. Committee for Refugees and is printed here with their permission.*

INTERGOVERNMENTAL AGENCIES

Office of the United Nations High Commissioner for Refugees
United Nations Plaza
New York, New York 10017

The work of the Office of the United Nations High Commissioner for Refugees is of a social and humanitarian nature. It provides international protection to refugees and seeks permanent solutions to their problems, either through voluntary repatriation or, if this is not possible, through assimilation within new national communities. The scope of the High Commissioner's mandate is global. Only those refugees who are the direct concern of other UN offices (such as the Palestine refugees who are the concern of UNRWA) and those who have the nationality of the country granting them asylum or equivalent status (for example, refugees in Vietnam) are not the High Commissioner's responsibility. The UNHCR's role is to promote, organize, coordinate and supervise international action on behalf of those refugees who are its concern. It acts upon the request of governments and works in close cooperation with UN Specialized Agencies and other UN programs, regional organizations such as the Council of Europe, the Organization of African Unity, and the Organization of American States, as well as many Voluntary Agencies and nongovernmental organizations.

At the beginning of 1967, an estimated 2,350,000 refugees were subject to UNHCR protection and/or assistance activities, in some 50 countries throughout the world. The concentration of refugee situations of concern to UNHCR has moved from Europe to Africa and Asia where vast new refugee problems have arisen during the last few years.

All UNHCR material assistance programs are financed through voluntary contributions, primarily from governments. In spite of modest yearly target figures intended to meet the minimum requirements agreed upon by the UNHCR Executive Committee, adequate financing of these programs has been one of the High Commissioner's constant preoccupations. The target figure agreed for the 1967 program, for example, was just under $4,850,000 but, by May 1967, was still $1,500,000 short. More than half of the 1967 program ($2,600,000) consisted of assistance projects for refugees in Africa.

United Nations Relief and Works Agency for Palestine Refugees
United Nations Plaza
New York, New York 10017

UNRWA is a subsidiary organ of the United Nations established by the General Assembly to aid the Palestine refugees. In cooperation with the "host" Governments, it carries out its task of providing relief (basic food rations, shelter, supplementary feeding and welfare to refugees in special need) and health and education services for more than 1,000,000 refugees.

The Palestine refugees are the Arabs who, as a result of the Arab-Israeli conflict in Palestine during 1948, left their homes in the territory which is now Israel and took refuge in the neighboring areas.

At the beginning of 1967, there were 1,330,077 refugees registered with UNRWA, of whom 859,966 were eligible for all UNRWA services, including food rations. *THESE FIGURES DO NOT INCLUDE, OF COURSE, THE ADDITIONAL REFUGEES OF THE 1967 CONFLICT.*

UNRWA headquarters are in Beirut and Lebanon. The Agency operates in Lebanon, Syrian Arab Republic, Jordan and in the Gaza Strip and West Bank areas occupied by the military forces of Israel with liaison offices at the Palais des Nations, Geneva, Switzerland and the United Nations, New York.

Intergovernmental Committee for European Migration
370 Lexington Avenue
New York, New York 10017

The movement of European refugees to countries offering them opportunity and security is a major function of ICEM. To ensure the efficient resettlement of 35,000 to 40,000 uprooted people each year, ICEM works closely with the United Nations High Commissioner for Refugees, the United States Government through the U.S. Escape Program, as well as the other governments, international organizations and voluntary agencies. The current Director is an American, John F. Thomas.

From 1952, when the agency began operations, to December 31, 1967, ICEM moved a total of 1,527,340 persons of whom 704,866 were refugees.

For the period January 1 to December 31, 1967, 34,936 refugees were resettled by ICEM. The main countries of resettlement were: Israel (10,002), USA (9,201), Australia (8,772), Canada (3,035), Europe (2,920), South Africa (580), etc.

The ICEM 1968 Refugee Program approved by the Council in November 1967 foresees the emigration of 32,800 refugees at an estimated cost of $6,395,700.

U.S. GOVERNMENT

Department of Health, Education and Welfare
Welfare Administration
Washington, D. C. 20201

Total registrations at the Cuban Refugee Program's Miami, Florida Center reached 269,095 persons as of December 15, 1967. U.S. chartered flights operate twice daily, five days a week between Varadero, Cuba and Miami, Florida. The relative-to-relative airlift from its start December 1, 1965, through December 8, 1967, brought 90,413 Cuban refugees. Total arrivals in that period were 97,327; resettlements 75,063, or seventy-seven percent of arrivals. Almost 4,000 refugees a month arrive in the United States via this means.

Established by Presidential directive early in 1961, the Program's mission has been to relieve the burden imposed on Miami-Dade County, Florida by the sudden exodus of thousands of destitute Cubans fleeing their island home. While resettlement is the main objective of the Federal operation, emergency financial assistance is available to eligible refugees in Miami as well as surplus food distribution, medical care, and—for certain airlift arrivals—housing.

Federal funds underwrite health and educational activities in Miami-Dade County to meet problems created by the large refugee population.

As of December 15, 1967, over sixty-three percent of total registration (or 169,971 refugees) had been resettled to homes and job opportunities in more than 2,200 communities and in every state of the nation. This vast and continuing resettlement effort is carried out by four national voluntary resettlement agencies with offices at the Cuban Refugee Center in Miami.

Department of State
Office of Refugee and Migration Affairs
Washington, D. C. 20520

Several programs are administrated directly by the State Department through its Office of Refugee and Migration Affairs. The United States

Refugee Program (USRP) operates in Europe and the Near East where, through contracts with voluntary agencies, it assists directly in the integration and resettlement of refugees from East European countries. USRP assisted over 10,000 in 1967.

Through the Far East Refugee Program (FERP), ORM provides assistance to refugees from Communist China in Hong Kong and Macao. FERP projects include educational facilities, medical services, vocational training, resettlement assistance and other projects which help the refugees to become self-sustaining.

Through the Department of State, the United States supports the Office of the United Nations High Commissioner for Refugees (UNHCR). The United States is a member of the Executive Committee of the UNHCR and is the principal contributor both in cash and food commodoties.

The Department of State also provides United States support to the Intergovernmental Committee for European Migration (ICEM) of which the U.S. has been a member since its inception in 1951. In 1967, the United States contribution was $3,350,000.

Many African refugees, primarily young people, have left southern Africa in search of educational opportunities denied them in their homelands. Since 1961, the U.S. Government has maintained a scholarship program to provide higher education for these refugees. More than 400 students from a number of southern African countries are currently studying in the U.S. under this program, for which the U.S. Government allocated approximately $2,300,000 in 1967.

Over the years, the United States has been the principal contributor to the United Nations Relief and Works Agency for Palestine Refugees (UNRWA). Since the establishment of UNRWA in 1950, the United States has contributed nearly seventy percent of the funds received by UNRWA from governments to meet its overall costs in behalf of the Palestine refugees. The U.S. contribution to UNRWA's regular program for 1967 was $22,200,000 in cash and food. In addition, a special U.S. contribution of $2,000,000 was made to UNRWA in 1967 to help fund UNRWA's emergency relief program following the Middle East hostilities.

The Agency for International Development (AID), of the Department of State has arranged for other major contributions to refugee programs in Asia and Africa. For example, during 1967, AID provided about $20,000,000 in Public Law 480 foods to refugees under the Food for Freedom Program and about $22,000,000 to refugees in Vietnam and Laos through regular AID programs. Other general AID programs provide substantial peripheral benefits to refugees.

VOLUNTARY AGENCIES

American Council for Emigres in the Professions, Inc.
345 East 46th Street
New York, New York 10017

The American Council for Emigres in the Professions, Inc., specializes in counseling and placement services to refugee professionals. It also retains persons whose foreign academic background is not immediately adaptable to the requirements of employment on a professional level in the United States.

In addition to the approximately 2,000 academically trained emigres currently in its active files, ACEP registered 1,086 new refugee professionals during the fiscal year ending May 1, 1967. The number of registrations shows an increasing trend in the current fiscal year. New registrants were trained originally in some professional field including architecture, engineering, medicine, social, political and natural sciences, education as well as in the arts.

In recognition of the fact that there is urgent need for trained personnel in several professional fields, ACEP carried out special retraining projects geared to meet the requirements for these employment opportunities. For example, the Lawyer-Librarian Retraining Project was initiated nine years ago. One hundred four applicants received their Master's Degrees in Library Science and found immediate employment.

The American Council for Judaism Philanthropic Fund
201 East 57th Street
New York, New York 10022

Last year the Philanthropic Fund assisted in the resettlement of 2,176 refugees of Jewish faith through offices in Paris, Rome, Vienna, Stockholm and New York. During 1966, the Fund's two primary concerns were: (a) North Africans of Jewish faith seeking to achieve final resettlement in France; (b) East European refugees aided both in their initial asylum in Western Europe and in resettlement in the United States. Within a year, the Fund brought its five hundredth refugee immigrant to this country.

American Council for Nationalities Service
20 West 40th Street
New York, New York 10018

ACNS and its member agencies are primarily concerned with services to immigrants and refugees after they arrive in American communities.

ACNS also helps in the resettlement of refugees and is particularly involved today with refugees from Cuba and Hong Kong. ACNS cooperates actively with the several resettlement agencies, aiding in finding jobs and housing for Cubans. ACNS, through its San Francisco International Institute, has sent a staff member to Hong Kong to provide preimmigration planning for the Chinese.

Member agencies of ACNS, which number thirty-six in eighteen different states, also seek to stimulate and maintain a spirit of "welcome" and acceptance for the foreign-born. To this end, member agencies have organized nondenominational resettlement committees composed of representative elements in the community.

American Council for Voluntary Agencies for Foreign Service, Inc.
200 Park Avenue South
New York, New York 10003

The American Council of Voluntary Agencies for Foreign Service was established in 1943 to provide a means for consultation, coordination and planning and to assure the maximum effective use of contributions by the American community for the assistance of people overseas. Through the Council, forty-four member American voluntary agencies engaged in programs of active service overseas now coordinate their plans and activities both at home and abroad, not only among themselves but also with nonmember agencies and governmental, intergovernmental and international organizations. Since 1955 the Council has operated the Technical Assistance Information Clearing House under contract with the United States Agency for International Development.

American Friends Service Committee, Inc.
160 North Fifteenth Street
Philadelphia, Pennsylvania 19102

The AFSC's major programs to aid refugees include:

Algeria: AFSC has established a series of community development programs in rural villages composed of former refugees and displaced Algerians which includes training in manual skills, health education, agricultural extension work and assistance in the organization of community projects. Two hundred fifty thousand returning refugees have merged with the rural population and benefit from this program.

Hong Kong: A program has been established within the largest resettlement housing project, offering such services as a cooperative nursery school, a mothers' and fathers' club, counseling and recreational groups for youth and vocational training. Community development projects have been established in fishing villages inhabited by refugees hear Hong Kong. Ten thousand refugees were directly involved in AFSC programs while some 55,000 were reached indirectly.

Austria: Some 100 refugees are involved in the final stages of phasing out a settlement program for Hungarian refugees.

Central Africa: Material aids are supplied to approximately 50,000 Angolan, Congolese, and Sudanese refugees in the Congo and Uganda.

Vietnam: Relief work in Quang Ngai, South Vietnam, consists of a child day-care center, physical and occupational therapy, and prosthetics services for amputees.

American Foundation for Overseas Blind
22 West 17th Street
New York, New York 10011

As the world's only specialized agency dealing exclusively with the problems of all blind people regardless of nationality, race, color or creed, AFOB provides service to thousands of refugees who are among the world's fourteen million blind people. Programs in agricultural training, industrial training, rehabilitation and education reach a large refugee population, especially in the Middle East and Southeast Asia. AFOB cooperates with agencies of the United States Government and the United Nations in providing these services.

American Fund for Czechoslovak Refugees, Inc.
1775 Broadway, Room 430
New York, New York 10019

There has been substantial increase in the number of refugees escaping from Czechoslovakia into Germany, Austria and Italy and other West European countries during the last several years. The AFCR program is one of continuing assistance in resettlement, counseling, local integration, material and medical help, education, recreation and rehabilitation of refugee children. Newer refugees are resettled as soon as possible in the United States or other free-world countries.

American Immigration and Citizenship Conference
509 Madison Avenue
New York, New York 10022

The AICC, as an association of voluntary agencies, does not conduct programs of direct relief but its work is of importance to refugees in that it bears upon their admission to this country and integration once they are here. The AICC is a coordinating agency for over ninety nonprofit and nonpolitical agencies interested in promoting a nondiscriminatory immigration policy. Standing committees initiate and carry out studies on immigration legislation, immigrant integration, citizenship preparation, international migration affairs and immigration research. AICC acts as a clearing house for information, stimulates conferences on immigration and refugees and provides the means for joint action by its member agencies.

The American-Korean Foundation
345 East 46th Street
New York, New York 10017

Thousands of Korea's 4,000,000 refugees are still living in makeshift hovels and without the skills to earn an adequate livelihood. To help overcome this situation, the AKF is providing earth block "do-it-yourself" houses, vocational training opportunities, scholarships and school buildings; it is conducting a bench-terracing program to reclaim thousands of acres of unused land for food production.

AMA Volunteer Physicians for Vietnam
American Medical Association
535 North Dearborn Street
Chicago, Illinois, 60610

This program was initiated in 1965 and is financed by contract with the United States Agency for International Development (USAID).

Personnel: Thirty-two volunteer physicians. In 1967, approximately 200 volunteer doctors served in overlapping two-month tours. Thus far 236 physicians have participated.

American Middle East Rehabilitation, Inc.
777 United Nations Plaza, Suite 7E
New York, New York 10017

AMER's aims and objectives are three-fold: (a) to help provide vocational training for Palestine Arab refugee youths; (b) to provide emergency relief when needed in the Middle East and (c) to promote and help support projects designed to help raise the standards of living in the Middle East.

AMER solicits cash donations and "gifts-in-kind" (chiefly medical supplies) the proceeds of which are used to provide scholarships chiefly in UNRWA's vocational training schools and, also at the Jordan River Project of the Arab Development Society. As of November, 1967 AMER had provided 432 of these scholarships to worthy Palestine Arab youth.

American ORT Federation
222 Park Avenue South
New York, New York 10003

ORT (Organization for Rehabilitation through Training) services to refugees are directed toward the problem of their economic integration and the education of high-school-age members of refugee families. The program includes vocational and technical training for adults, vocational training for youth, apprentice placement for on-the-job training and aid in finding employment. The programs are primarily for Jewish refugees.

Asia: Newly arrived refugees in Israel received training and schooling at various levels at thirty-three localities. The Teheran ORT Center trained Tibetan youngsters to become trade school teachers for Tibetan refugees in India.

Europe: Many special courses were organized at the ORT centers in Paris, Lyon, Strasbourg, Toulouse and Marseilles. About 4,000 Jewish refugees from North Africa received assistance to these programs. ORT centers in Rome and Genoa continued the dual operation of language training and trade instruction to refugees enroute to English-speaking areas of ultimate settlement.

North America: The ORT vocational school in New York, which is entirely devoted to refugee aid, enrolled about 720 persons last year.

American National Red Cross
17th and D Streets, N.W.
Washington, D. C. 20006

The American National Red Cross provides assistance in certain refugee situations in cooperation with other National Red Cross Societies, the International Committee of the Red Cross and the League of Red Cross Societies. The latter two international organizations work closely with the Office of the United Nations High Commissioner for Refugees and often serve as the operations service for that organization.

The American Red Cross meets refugee needs by providing funds, supplies and assistance of qualified specialized staff. Other services including reuniting refugees with their relatives in this country and maintaining an international foreign location inquiry service to help refugees locate missing family members.

Brethren Service Commission
1451 Dundee Avenue
Elgin, Illinois 60120

BSC, cooperating with Church World Services (CWS) and the Mennonite Central Committee, serves refugees in South Vietnam through the Vietnam Christian Service program. BSC also operates service centers across the United States where material aid, clothing, blankets and medicines are processed for Church World Service, Inter-Church Medical Assistance and Lutheran World Relief. BSC works with CWS in resettling refugees in the United States including those from Cuba.

Catholic Relief Services
United States Catholic Conference
350 Fifth Avenue
New York, New York 10001

Catholic Relief Services-USCC maintains relief, social and economic development projects and various health, education, and welfare programs in more than eighty countries throughout Asia, Africa and Latin America. These programs include food, clothing and medicine distributions to the needy—regardless of race, religion or color—as well as vocational training, rural education, urban renewal projects, the provisions of emergency supplies and other services.

During the 1966 fiscal year, CRS sent relief supplies overseas (including U.S. government-donated foodstuffs) valued at the highest total in the agency's history—$135,867,910.

South Vietnam: CRS continues to place major emphasis on assistance to war victims in Vietnam. CRS supplied food, medicines and clothing to 200,000 refugees awaiting settlement. Where the need existed for refugee shelters or camps, CRS constructed temporary shelters for 100,000 refugees.

India: CRS is providing food and clothing to 21,000 Tibetan refugees as part of its continuing program in India.

United Arab Republic: CRS is providing food and clothing to 7,532 Palestinian refugees through units of the Ministry of Social Affairs.

Christian Children's Fund
Richmond, Virginia 23204

The Christian Children's Fund provides a wide range of aid—financial, medical, educational, food, clothing, care, adoptions—for both refugee and nonrefugee children all over the world. Its principal refugee relief projects last year included:

Israel: Educational and emergency aid to children of Arab refugees near Bethany.

Hong Kong: Assistance to several roof-top schools for refugee children, providing books, clothes, emergency medical aid, and one meal a day.

India: Maintains refugee children from Tibet in orphanages close to the border areas.

Korea: Financial aid to orphanages caring for children of North Korean refugees, either abandoned or separated from their families, as well as assistance to many displaced and deprived widowed families through Family Helper Projects.

Vietnam: Orphanages for children of parents killed in the war receive financial aid. Family Helper Projects for children of refugee families are established and maintained.

Church World Service
475 Riverside Drive
New York, New York 10027

Church World Service, a program of the National Council of Churches, is the coordinating and administrative facility for major Protestant refugee and relief agencies in the United States.

Church World Service works in close fraternal relationship with the Division of Inter-Church Aid, Refugees' and World Service of the World Council of Churches whose headquarters are in Geneva, Switzerland.

Refugee relief and rehabilitation have been included in CWS programs of assistance in some fifty countries during the last twenty years. In 1966 this program was supported by cash disbursements of $11,355,810 and material aid distributions valued at $36,865,455.

Notable refugee assistance programs of CWS include:

Congo: Food, clothing, blankets, emergency medical treatment and self-help through the plowing of fields and distribution of tools and seeds, all through the Congo Protestant Relief Agency. Refugees are from the Sudan and Angola and are a result of continuing internal political strife.

Burundi: Food, blankets, clothing, soap, medicine, and vitamins provided for refugees from the Congo during their brief asylum.

Tanzania: Material aid supplied through Lutheran World Federation; support for self-help projects using indigenous staff. Refugees are from Rwanda, Mozambique and other countries.

Hong Kong: Hot lunches provided through the Children's Meals Society; family planning, TB and dental clinics within the Hong Kong Christian Welfare and Relief Council.

India-Pakistan: Resettlement and temporary material aid to refugees created by border disputes.

Middle East: A wide variety of long-range self-help and rehabilitation projects for Arab refugees are conducted under the auspices of the Near East Council of Churches Committee for Refugee Work.

Cooperative for American Relief Everywhere, Inc. (CARE)
660 First Avenue
New York, New York 10016

In its daily operations, CARE does not serve refugees as a special group but directs its aid to them as part of its assistance to all groups of needy.

In consistent efforts to provide refugees with the means of reestablishing themselves economically, CARE has given and continues to give substantial support to resettlement and job-training projects in South Korea, Israel, Hong Kong, Jordon, the Gaza Strip and other regions.

Community Development Foundation
Boston Post Road
Norwalk, Connecticut 06852

Community Development Foundation, having worked in the past with refugees in Greece, Korea, and France, is currently laying the groundwork for a future program with Arab refugees in the Middle East and actively involved with program services for refugees in Vietnam.

Vietnam: The nature of CDF refugee service is a training program preparing Vietnamese personnel to initiate and implement self-help activities in order to improve living conditions in temporary camps and to prepare for economic and social needs of resettlement.

The Thomas A. Dooley Foundation
442 Post Street
San Francisco, California 94102

The Dooley Foundation provides technical and material assistance to the developing nations of Asia in the fields of medicine, health education and community development. Assistance programs are presently carried out in India, Nepal, Vietnam and Laos.

Foster Parents' Plan, Inc.
352 Park Avenue South
New York, New York 10010

Foster Parents' Plan, Inc. a nonsectarian agency, has a twofold purpose—the relief and rehabilitation of destitute children in war-ravaged and underdeveloped countries and the preservation and rehabilitation of the family through help to one of its members. This is done by means of financial "adoptions." A United States or Canadian individual or group "adopts" an overseas child. The Foster Parent pays $15 a month for a minimum of one year. Benefits to the child and his family include monthly cash grant, new clothing, household equipment and medical care, plus counseling and guidance by local social workers. Emphasis is on the education of children.

Of the 47,300 children currently enrolled, 20,000 live in Vietnam, Korea and Hong Kong. Many are refugees. Foster Parents' Plan has a long-range program for all children it its care. A Foster Child remains with PLAN until he becomes 18 so long as the family continues to need assistance and the child remains in a suitable form of education.

International Rescue Committee
386 Park Avenue South
New York, New York 10016

Formed in 1933 by a group of Americans to assist persons fleeing from the first waves of Nazi persecution. Specifically directed toward saving leaders who some day might return to their countries.

U.S. programs include Cuban refugees, resettlement of refugees from Eastern Europe and Hong Kong, and a special project for Haitian refugees. European and Chinese resettlement caseload in 1966: 2,000 plus.

Programs in Asia, Europe, Canada, and Latin America.

International Social Service American Branch, Inc.
345 East 46th Street
New York, New York 10017

International Social Service American Branch is part of a worldwide network of individualized services to families and children whose problems require help in more than one country or involve nationals of more than one country. It aids thousands of refugees with problems of divided families, lost or missing parents, international marriage, lost or abandoned children.

Established in 1921, ISS world headquarters are in Geneva, Switzerland. ISS American Branch and International Social Service have programs in Argentina, Venezuela, Greece, France, Switzerland, Hong Kong, and Vietnam.

International Voluntary Services, Inc.
1555 Connecticut Avenue, N.W.
Washington, D. C. 20036

International Voluntary Services, Inc. is a private, nonprofit corporation chartered under the laws of the District of Columbia. It was incorporated in July, 1953, is nonsectarian, nonpolitical and is governed by a 21-member Board of Directors representing a cross-section of American academic and professional life. It was organized for the purpose of providing technical assistance to people of developing countries with an objective of developing the human resources through direct involvement of people of the host country in planning and implementation of programs. IVS participates in a supporting role.

Since its inception, IVS has fielded teams of young men and women to fourteen countries. Current programs operate in Vietnam and Laos. In Vietnam, 159 volunteers from ten countries work in agriculture, education and community development. In Laos, 106 volunteers from five countries work in education and rural development.

Joint Distribution Committee, Inc.
60 East 42nd Street
New York, New York 10017

In 1966 the Joint Distribution Committee assisted 36,000 refugees, repatriates, etc. Of these, 9,200 were citizens in a country of asylum; 26,800 did not have the protection of citizenship and were scattered over a number of countries. In addition, there are thousands of former refugees living among the "settled" populations of their new homelands who still do not have citizenship status. The JDC helped these refugees with food, clothing, shelter, medical assistance, vocational training, low cost loans and school and guidance counseling. JDC also maintained a community assistance program which provided funds and technical assistance to synagogues, schools and community centers.

In 1966 JDC focused particular attention on the following refugee areas:

France experienced an influx of 29,100 refugees during 1966 mostly from Tunisia, Algeria, Morocco, Egypt and Eastern Europe. Smaller groups of refugees also made their way to Italy, Austria and Belgium.

JDC is still assisting some 5,000 residual refugees and displaced persons from World War II. Most of them (3,100) are in Australia. There are also 350 in Sweden and a lesser number in Belgium, Portugal, and Spain. JDC aid is also going to approximately 1,000 refugees from the Hitler era in such widely scattered areas as Tangier, Haiti, the Philippines, Portugal, and Spain.

In 1966 the health, welfare, rehabilitation and cultural programs on behalf of some 400,000 refugees and other needy Jews in all parts of the world were sustained by a budget of $22,594,800, provided chiefly by the campaigns of the United Jewish Appeal, a contribution by the United States of 14,710,000 pounds of Food for Peace supplies, valued at close to $900,000, and other material aid.

Lutheran World Federation
U.S.A. National Committee
315 Park Avenue South
New York, New York 10010

During 1967, the American Lutheran Church and the Lutheran Church in America contributed through Lutheran World Action about twenty-two percent of the cost of the refugee services of the LWF Department of World Service, budgeted annually at over $2,750,000. This does not include the $10,000,000 worth of clothing, food and medicine shipped annually by Lutheran World Relief.

Australia: Resettlement of immigrants from Europe through travel loans; housing, trade, craft and farming loans (under agreement with ICEM); counseling, welfare relief, assistance to the immigration chaplaincy and referral service.

Hong Kong: In addition to distribution of material aid from Lutheran World Relief and other relief sources, the LWF-WS administers a large program of self-help and rehabilitation, education, youth work and vocational training (800 students); and medical and health services, primarily for Chinese refugees.

India: Resettlement aid to several thousand refugees from East Pakistan and Tibet including housing, farming, production cooperatives and food-for-work projects such as building schools and digging wells and reservoirs.

Jordan: Distribution of food and clothing to refugees and needy persons both East and West of the Jordan River. LWF-WS operates work projects, a vocational training center (120 students), a handicraft center for thirty-six blind boys, a nurses' school (60 students), a hospital (subsidized by UNRWA) and thirteen medical clinics (five mobile).

Syria: LWF-WS operates five medical clinics in the Damascus area and contributes to the nursing school of the Palestine Arab Refugee Institute.

Tanzania: At least 33,000 refugees in Tanzania have received help from LWF-WS. This includes 18,000 Mozambiquans, 12,300 Rwandese (mostly Watutsi), 2,500 Kikuyus and 300 Congolese. A tripartite agreement between the United Nations High Commissioner for Refugees, the Tanzania Government and the LWF made possible the creation of an agricultural community in the Mwesi Highlands of Tanzania for 3,000 refugees from Rwanda.

Zambia: LWF provides resettlement aid in Zambia as in Tanzania. In cooperation with UNHCR and the Zambia Government, the LWF established the "Zambia Christian Refugee Service" early in 1967 to express the ecumenical concern of Protestant Christian groups. Three comprehensive rural settlements are now under development with 4,000 Mozambiquan and 1,500 Angolan refugees who fled civil upheaval in their homelands.

Vietnam: LWF provides services in Vietnam through Vietnam Christian Service (See Church World Service).

**Mennonite Central Committee
21 South 12th Street
Akron, Pennsylvania 17501**

The Mennonite Central Committee, the official agency of the North American Mennonites, conducted refugee relief operations in Asia and Africa. Last year, the Committee assisted refugees in India, Hong Kong, Jordan, Vietnam, Burundi and the Congo. The relief program in these areas consisted of:

India: A health program, self-help project and educational program in a government-assigned refugee colony for East Pakistanis in Calcutta.

Hong Kong: Educational and family-child assistance to 450 children and general distribution of clothing.

Vietnam: Assistance to refugees from North Vietnam with donations of food, clothing and medicines. Medical treatment at Nha Trang was expanded. A doctor and nurse were sent to Pleiku to initiate another medical project.

Congo: Shipments of food and clothing distributed through the Congo Protestant Relief Agency for refugees from Angola.

Burundi: Food, clothing and medical supplies for refugees from Rwanda and the Congo. This work was conducted in cooperation with Church World Service and the World Relief Commission.

Jordan: Seventy-six thousand pounds of blankets and clothing were distributed to refugees on the East Bank. On the West Bank ten tons of flour, three and one-half tons of meat, three tons of bedding and clothing and six tons of soap were distributed.

**Polish-American Immigration and Relief Committee
156 Fifth Avenue
New York, New York, 10010**

PAIRC conducts a continuing program for Polish post-war refugees in Western Europe as well as newly arrived escapees. The scope of the Committee's work covers immigration and integration help for new refugees, registration and documentation for "old" refugees under special immigration schemes, provision of sponsorships for regular immigration cases, financial assistance to needy refugees and cooperation in UNHCR integration programs.

The Salvation Army
120-130 West 14th Street
New York, New York 10011

The Salvation Army is serving refugees in several countries as part of its overall social service programs in those areas. Elsewhere, refugees are aided from time to time on an individual basis.

In Hong Kong, The Salvation Army operates nurseries, kindergartens and primary schools which offer education to 5,400 children. Free libraries are attended by 700 to 1,000 children daily. At Salvation Army vocational training centers, more than 300 persons annually learn a trade that will enable them to support themselves. It also operates a shelter for street sleepers, a children's convalescent home, medical clinics, and gives direct relief to the needy, especially the aged and ill.

In Calcutta, The Salvation Army Social Service Center, with the financial backing of OXFAM (Oxford Famine Relief), carries on a daily "Meals on Wheels" service, which provides meals and dried foodstuffs for approximately 1,000 poor people, of whom it is estimated about 125 are displaced persons. Of the 125, there about 25 lepers. They and their dependents receive food each day.

In Bombay, there is a similar program. Every day (Sunday included) since July, 1964, an average of 800 needy people have been fed. The only qualification is need.

The Salvation Army is also aiding Cuban refugees in Florida. In Key West, it is the agency responsible for distributing government commodities to refugees and also assists with emergency relief. In Miami, Cuban refugees make up the largest part of the Salvation Army's emergency family relief program which involves food, assistance with rent, utilities, clothing, etc.

Save the Children Federation, Inc.
Boston Post Road
Norwalk, Connecticut 06852

Save the Children Federation has worked in the past with refugees in Greece, Korea, and France and is currently active in a program for refugee children and their families in South Vietnam.

In 1966, Save the Children Federation established field headquarters in Bin Dinh Province in South Vietnam and began a Sponsorship program to help refugee children and their families. American families, schools, business firms or clubs enrolled as Sponsors exchange correspondence with Vietnamese families. They support a program which provides funds and counseling to help families add to their resources for better child care, increase their ability to be self-sustaining, and join with others in self-help projects to establish services and facilities desired by the community.

Spanish Refugee Aid, Inc.
80 East 11th Street
New York, New York 10003

Spanish Refugee Aid, organized in 1953 to aid the Spanish Republican refugees who had fled to France after the Civil War, faces the dual problem of sharply rising costs in France and the need of an increase in aid as the refugees grow older. Of the 100,000 still in France, 10,000 need help currently, 1,549 persons over the age of 60, 1,300 invalids, 282 persons with tuberculosis and 42 blind men and women are on SRA lists. SRA gives general financial help to the neediest. It sponsors an adoption program (278 refugees have been adopted), a scholarship program (69 students are being helped through school) and a special fund-raising campaign for the hospitalized refugees.

At the Foyer Pablo Casals, a center for old people in Montauban, France that was organized by SRA, 259 refugees receive monthly food packages and clothing as against 243 last year. SRA maintains three offices in France which are used as distribution centers, and staff workers make regular visits to homes. The problem of the aging refugees, living on small pensions ranging from $30 to $40 a month, is SRA's greatest concern.

Tolstoy Foundation, Inc.
250 West 57th Street
New York, New York 10019

The Tolstoy Foundation is in continuous contact with some 20,000 refugees and escapees registered with its offices overseas and with some 2,000 cases in the U.S.A.

While the majority of refugees assisted by the Tolstoy Foundation are Russians, the Foundation also assists Yugoslavs, Bulgarians, and Rumanians who sought asylum abroad, and a group of Tibetan students from among the refugees in India.

The Foundation continues to give counseling assistance in emigration in Belgium and Latin America. It also assists in local integration through the establishment of small business enterprises, workshops, etc. in conjunction with programs sponsored by the United Nations High Commissioner for Refugees (UNHCR).

Cultural activities, an important part of the Foundation's program, are centered in the L.C. Stevens Library in Munich and at the Tolstoy Foundation Center in Valley Cottage, New York.

The Tolstoy Foundation and its branches assume the direct responsibility for supervision of homes for aged refugees in Latin America, Western Europe and in Rockland County, New York with some 1,000 residents.

A Tolstoy Foundation Nursing Home at Tolstoy Center in Valley Cottage, New York is projected for opening in 1969 with a capacity of eighty beds.

Assistance to children is centered mainly in West Germany, with valuable cooperation of British Voluntary Agencies and Swiss Foreign Aid, and continued summer camp programs for children supported by the Tolstoy Foundation in the United States and in Belgium.

United Church Board for World Ministries
475 Riverside Drive
New York, New York 10027

The Division of World Service of the Board of World Ministries of the United Church of Christ last year administered programs and allocations for relief and rehabilitation in some fifty countries to the extent of about $1,500,000 with half that much again in gifts-in-kind. Much of this was in assistance to "current" refugees and for the alleviation of grave social problems resulting from past migrations.

Particular emphasis in 1967 was laid on the plight of refugees in Vietnam and the Middle East, with personnel and funds contributed through the Vietnam Christian Service and the Near East Council of Churches Committee on Refugee Work. Another major effort of the year was in continued support of India Famine Relief and in behalf of victims of other natural disasters. Projects including refugee service were supported in several Central African states, in Algeria, in India (e.g., for Tibetans) and Pakistan, in eight Latin American countries, in the Far East, in Italy and Central Europe. Special programs were also maintained in Greece and Lebanon, continuing the interest of the Congregational Christian Service Committee.

With major emphasis on self-help projects, especially in agriculture and community development, the Board did most of its work cooperatively through other agencies.

Unitarian Universalist Service Committee, Inc.
78 Beacon Street
Boston, Massachusetts 02108

The Unitarian Universalist Service Committee has been giving assistance to Spanish Republican refugees since 1939. The headquarters of this program are at 93 rue Riquet, Toulouse, serving refugees in Toulouse and in rural areas in the South of France.

Services fall into four categories—food parcels, clothing, emergency cash and scholarships.

United Lithuanian Relief Fund of America, Inc.
105 Grand Street
Brooklyn, New York 11211

Close to 10,000 Lithuanian refugees still remain in Europe and receive aid which includes clothing, books and medicines from the Fund. The Fund also supports two secondary schools for Lithuanian refugee children. Some Lithuanian refugees in Poland and deportees to Siberia were helped.

United Hias Service
200 Park Avenue South
New York, New York 10003

United Hias Service provided rescue, resettlement and related services last year to approximately 53,500 Jewish men, women and children. More than 8,800 of these people were assisted to resettle in the United States and other Western Countries.

During the past year, the active caseload of persons registered to emigrate reached 23,608; premigration services in the United States and Latin America were rendered to 16,319 relatives and sponsors of prospective migrants: 3,030 were aided in the United States with such post-migration services as naturalization, adjustment of status, and prevention of deportation and jeopardy; 1,549 persons were located throughout the world; and in Latin America, 158 migrants who arrived in prior years received agency assistance.

United Ukrainian American Relief Committee, Inc.
5020 Old York Road
Philadelphia, Pennsylvania 19141

Through its headquarters in Munich, Germany, UUARC helped about 20,000 refugees and escapees to resettle in new homes and to find employment. The agency continues its immigration program and, for the past year, has been engaged in a program for Ukrainian students in Poland. In cooperation with other Ukrainian agencies, UUARC is planning to organize a nursing home for the aged in the United States.

Volunteers for International Technical Assistance, Inc.
College Campus
Schenectady, New York 12308

VITA provides a person-to-person technical information service to agencies working with refugees and to others working in developing areas.

VITA's method complements the methods of other organizations in international development. VITA's Inquiry Service receives technical problems by mail and sends them to skilled scientists, engineers, businessmen and educators. The more than 2,500 volunteers on the VITA roster come from every state in the United States and from fifty other countries. They are associated with 410 corporations and 150 universities, agencies and institutes. Working relationships with the Institute of Food Technologists, the American Institute of Nutrition and the American Society of Agricultural Engineers give access through VITA to their entire memberships.

World Rehabilitation Fund, Inc.
400 EAst 34th Street
New York, New York 10016

The World Rehabilitation Fund, Inc. is involved directly in services to refugees in two projects in Hong Kong and one in Vietnam.

In Hong Kong under a grant from the U.S. Department of State, the World Rehabilitation Fund, Inc., has completed and equipped the John F. Kennedy Memorial Centre for Spastic Children. The building was dedicated in March 1967. This center has facilities for 60 inpatients and 120 outpatients. It is being managed by the Hong Kong Red Cross with financial assistance for a two-year period from the World Rehabilitation Fund.

The World Rehabilitation Fund Day Centre was, at the time of writing, under construction. Completion was expected in the spring of 1968. This center is to provide vocational rehabilitation training and sheltered employment for 450 persons daily. Personnel for the center have been trained by the World Rehabilitation Fund. The equipment and assistance in the operational expenses for a two-year period will be contributed by the Fund.

In Vietnam, under a contract with the U.S. Agency for International Development, the World Rehabilitation Fund is helping to expand the National Rehabilitation Institute in Saigon, equip staff and provide supplies for four smaller satellite centers. Two of these centers were opened in the spring of 1967, openings for two more were, at the time of writing, expected by the fall of 1968.

N.A.E. World Relief Commission, Inc.
33-10 36th Avenue
Long Island City, New York 11106

The major refugee programs of this agency are in Korea, Burundi and South Vietnam. Services include distribution of food, clothing and medicines to needy refugees in those areas, along with projects for the handicapped.

World University Service
20 West 40th Street
New York, New York 10018

Needy students from all over the world benefit from WUS aid. Last year WUS continued five major programs aimed at refugee students. They were as follows:

Algeria: Financial assistance was given to 150 Algerians studying in Europe, and food, drugs, clothing, etc., were provided to refugee students in Tunisia and Morocco. WUS is helping the Algerian university community get back on its feet, which requires $30,000 in assistance.

Angola: Financial aid has been provided on a continuing basis to refugee students from Angola and other African territories ruled by Portugal.

China: More than $600,000 have been provided over the past six years to foster higher education in Hong Kong. This program continued last year, in addition to which sixteen Chinese professors were brought to the U.S. to study and five students sent to Canada.

Hungary: Since 1956, WUS has been aiding Hungarian refugee students, a six million dollar program. They have been able to continue their studies in new countries of residence.

South Africa: Grants to colleges in neighboring countries enable nonwhite refugee students from South Africa to continue their education.

World Vision, Inc.
919 West Huntington Drive
Monrovia, California 91016

In nineteen countries of the world—in South and Central America, in Southeast Asia, India, Africa, and Hong Kong—World Vision, Inc. maintains orphanages ministering to the medical, social, spiritual and vocational needs of orphans, many of them refugees.

Through its Emergency Mission of the Month program, sponsored by friends of the organization who contribute $10 monthly, funds are sent to alleviate the plight of refugees around the world. Refugees also benefit from relief goods shipped with the assistance of AID to many underdeveloped countries.

World Vision, Inc. is an interdenominational service agency, approved by the Advisory Committee on Voluntary Foreign Aid under the AID program.

Young Men's Christian Association of the United States
291 Broadway
New York, New York 10007

The YMCA acts as a complementary service to the large global programs of the United Nations, governments and churches; it is concerned with the morale of refugees. It is not a relief or migration agency. It tries to establish community centers in the camps and through them help in the organization of educational, recreational, religious programs, etc. The agency also offers language training, orientation courses for prospective migrants, camping for the children, together with handicrafts and vocational training.

Young Women's Christian Association
600 Lexington Avenue
New York, New York 10022

[472] BEHAVIOR IN NEW ENVIRONMENTS

The World YWCA, in cooperation with the National YWCAs, is providing services to refugee women and girls in Egypt, Greece, Hong Kong, Jordan, Lebanon and Pakistan. In Austria a refugee program is carried out jointly with the World Alliance of YMCAs. In Vietnam the World YWCA serves women and girls in cooperation with the Asian Christian Service. The YWCA, in its refugee service, puts great emphasis on the provision of basic education, vocational training and cottage industries in order to help the women and girls to become self-supporting.

The Authors

EUGENE B. BRODY is Professor of Psychiatry and Chairman of that department at the University of Maryland and Director of the University's Psychiatric Institute and Community Mental Health Center. He is Editor-in-Chief of the *Journal of Nervous and Mental Disease* and on the editorial boards of a number of other publications. A 1944 graduate of Harvard Medical School, trained in psychiatry at Yale, and a graduate of the New York Psychoanalytic Institute, his research interests have progressed from an early focus on psychophysiology through studies on psychotherapy and psychoanalysis to a concentration in recent years on problems of human behavior in relation to culture and society. He is the author of approximately 100 scientific papers and editor or co-editor of *Psychotherapy with Schizophrenics: A Symposium (1952), Psychiatric Epidemiology and Mental Health Planning (1967),* and *Minority Group Adolescents in the United States (1968).* He is the author of *Social Forces and Disordered Behavior in a Latin American Metropolis*, currently in press.

HENRY P. DAVID is Associate Director of the International Research Institute in the Washington Office of the American Institutes for Research. Previously he served for two years in Geneva, Switzerland, as Associate

Director of the World Federation for Mental Health (1963-65). Dr. David is the editor of *Migration, Mental Health, and Community Services* (1968) and *International Trends* in Mental Health (1966). He is currently preparing a volume on family planning in Eastern Europe.

ROBERT L. DERBYSHIRE is Associate Professor of Sociology in the Department of Psychiatry at the University of Maryland Medical School. The social aspects of health and illness, family and marriage relations, medical education, minority groups, and urban poverty are his major interests. In these areas he has published studies involving Mexican American and Negro populations. Dr. Derbyshire was Associate Director of the Social and Community Psychiatry Training Program at UCLA. Some of his research results are published in two recent books of readings: *Minority Group Adolescence in the United States*, edited by E. B. Brody, and *Interdisciplinary Perspectives on Poverty*, edited by Weaver, Falk, and Magid.

HORACIO FABREGA is Associate Professor of Psychiatry and Anthropology at Michigan State University. He was recently director of a project dealing with the adjustment problems of low-income urban residents in Houston, Texas, giving special emphasis to medical concerns. Dr. Fabrega is the author of numerous articles focusing on the relationship between ethnicity or culture and medical phenomena.

MARC FRIED is Research Professor in the Institute of Human Sciences, Boston College, and Director of the Program on Deprivation and Social Transition. He is the author of numerous articles on the working class, mental health and illness, and urban relocation. He is currently completing a volume on the urban working class and is doing a study of the impact of rural-to-urban migration on Negro occupational status.

ABOUT THE AUTHORS

ROBERT C. HANSON is Professor of Sociology and Director of the Research Program on Social Processes, Institute of Behavioral Science, University of Colorado. Currently he is Principal Investigator of the project, "Urbanization of the Migrant: Processes and Outcomes." He is a co-author (with R. Jessor, T. Graves, and S. Jessor) of *Society, Personality, and Deviant Behavior* (1968). W. N. McPHEE is Professor of Sociology and Computing Science at the University of Colorado. He is author of *Formal Theories of Mass Behavior* and co-author of *Voting* (with B. Berelson and P. Lazarsfeld) and of *Public Opinion and Congressional Elections* (with W. Glaser). ROBERT J. POTTER is Chairman and Professor of Sociology at the State University College, Brockport, New York. He was Research Associate on the project, "Urbanization of the Migrant: Processes and Outcomes." OZZIE G. SIMMONS is Program Advisor of the Ford Foundation's Office for Latin America and the Caribbean. He was Professor of Sociology and Director of the Institute of Behavioral Science, University of Colorado, from 1961 to 1968. He is author of *Work and Mental Illness* (1965), co-author (with Howard E. Freeman) of *The Mental Patient Comes Home* (1963), and author of several monographs and many research articles in the behavioral and health sciences. He has served as a member of the Health Services Research Study Section of the National Institutes of Health, as chairman of the Alcohol and Alcohol Problems Review Committee, National Institute of Mental Health, and as associate editor of the *American Sociological Review.* He is an associate editor of the *Journal of Health and Social Behavior* and of *Social Science and Medicine*. Dr. Simmons is presently working as program advisor in the Ford Foundation's program in Chile, and is writing a book reporting his research in Peru on relationships between drinking behavior and social structure. JULES J. WANDERER is Associate Professor of Sociology at the University of Colorado. He is engaged in research on urban violence, and on problems in the sociology of knowledge.

ROBERT J. KLEINER is Professor of Sociology at Temple University. He has published numerous articles concerned with the sociological and social-psychological correlates of mental disorder. His research interests and activities include the problem area of juvenile delinquency. He is currently working on a cross-cultural study of social deviance which is testing the generalizability of an anomie-alienation model of deviance.

SEYMOUR PARKER is Professor of Anthropology, University of Utah (Salt Lake City). He is the author of numerous articles on the Social Psychology of mental disorder. Many of his publications have been concerned with the cross-cultural aspects of mental health and the psychological aspects of anthropology. His current research interests are focused on the cultural aspects of poverty. Drs. Kleiner and Parker are the co-authors of *Mental Illness in the Urban Negro Community*.

ROBERT LEON is Professor and Chairman of the Department of Psychiatry at the University of Texas Medical School in San Antonio. During a period of service, with the U.S. Public Health Services he acted as a consultant to the Seattle Relocation Center.

HARRY W. MARTIN, Professor of Psychiatry (Sociology) at the University of Texas Medical School at San Antonio, received his Ph.D. at the University of North Carolina. He was formerly Professor of Psychiatry, University of Texas Southwestern Medical School at Dallas. He has served as program and staff training consultant to community agencies serving underprivileged people, mental hospitals, and the Bureau of Indian Affairs. A principal theme of this work has been the impact of the social environment on behavior.

RICHARD L. MEIER is Professor of Environmental Design at the University of California, Berkeley. His two most recent books are *Developmental Planning* and *Science and Economic Development*. He has recently published a monograph on Asian urban development, and a number of papers on organization theory, urban simulation, and the impact of science on human affairs. Current research is focused on simulations of social institutions and the design of resource-conserving urbanism.

LLOYD N. ROGLER is Professor of Sociology at Case Western Reserve University. He is co-author (with August B. Hollingshead) of *Trapped: Families and Schizophrenia* (1965) and articles on social psychiatry, family dynamics, migration and its consequences, the culture of poverty, and urbanization. He has served as a consultant in the development of research projects in Puerto Rico, Colombia, and other Latin American countries. A book on the organizational problems of Puerto Rican migrants in an American city is in process.

VICTOR D. SANUA is Associate Professor in the Department of Social and Psychological Foundations of Education at City College, City University of New York. He has written in the field of minority groups, rehabilitation and the sociocultural aspects of mental illness. His writings include "The Sociocultural Aspects of Schizophrenia: A review of the Literature" in *The Schizophrenic Syndrome*, edited by L. Bellak and L. Loeb (1969), and *The Rehabilitation Problems of Puerto Ricans*, Monograph XII, New York University Bellevue Medical Center. He formerly worked on the Midtown Manhattan Study at Cornell and taught at the Center of Social Psychiatry of the University of Paris (Sorbonne). He is consultant to the Veterans Administration in clinical psychology, and to the *Mental Health Book Review Index*, and holds the office of Vice-President of the Interamerican Society of Psychology.

HARRY K. SCHWARZWELLER is Benedum Professor of Sociology at West Virginia University. He has co-authored (with Joseph J. Mangalam and James S. Brown) a number of articles on rural-to-urban migration and a book dealing with migration from an Appalachian mountain neighborhood (The Beech Creek Study). He is currently studying rural migration in the context of a south German village and is directing a cross-cultural study of the career plans and values of rural young people in three nations.
JAMES S. BROWN is Professor of Rural Sociology and of Sociology at the University of Kentucky. During his twenty-three years as a member of the Agricultural Experiment Station staff there he has specialized on eastern

Kentucky and has written many articles and bulletins on its history, population, institutions, changes, problems and resources. His greatest interests have been the kinship structure, the process of migration and the adjustment of Kentucky Mountain migrants in southern Ohio. Much of his work (like the present paper) has been on Beech Creek, an area in eastern Kentucky where he first studied more than twenty-five years ago. The follow-up study of Beech Creekers twenty years later was made with Harry K. Schwarzweller and Joseph J. Mangalam. A summary book entitled *Beech Creek Families in Transition: A Case Study of Kentucky Mountain Migration* by these three will soon be published. M. JAY CROWE is Associate Professor of Sociology, and Associate Chairman of the Department of Sociology, at the Denver Center of the University of Colorado. He is a member of the advisory board of the Institute of Urban Affairs at the Denver Center. He has conducted research on the social aspects of air pollution and his current research includes problems of urban education and urban change.

LYLE W. SHANNON is Professor and Chairman of the Department of Sociology and Anthropology at the University of Iowa. He is the editor of *Underdeveloped Areas: A Book of Readings and Research, 1957*, and author of numerous articles on the economic absorption and cultural integration of inmigrant Mexican Americans and Negroes to northern urban industrial areas. More recently he has been author, along with John Stratton and Joy Randall, of *A Community Self-Survey System* (published in four parts), a study conducted for the Iowa Urban Community Research Center and the Division of Extension, University of Iowa, 1968. He is continuing his research on the process of absorption and integration of migrants.

ELMER L. STREUNING is Director of the New York State Department of Mental Hygiene Research Unit, Division of Epidemiology, Columbia University. Formerly he was Director of Research, Mental Health Services, Research Unit, Albert Einstein College of Medicine, Lincoln Hospital. Dr.

ABOUT THE AUTHORS [479]

JUDITH G. RABKIN and Dr. STANLEY LEHMANN were Dr. Streuning's research associates. HARRIS B. PECK is Associate Professor of Psychiatry and Director of the Lincoln Hospital Mental Health Services at the Albert Einstein College of Medicine. He is also the Associate Director of the college's Division of Social and Community Psychiatry. Dr. Peck is one of the founders of the American Group Psychotherapy Association and Editor of the *International Journal of Group Psychotherapy*. He also serves as Chairman of the American Psychiatric Association's Task Force on Poverty. Before coming to Einstein, Dr. Peck was Director of the Bureau of Mental Health Services of the New York City Court of Domestic Relations. He is co-author of the book, *The Treatment of the Delinquent Adolescent*.

FREDERICK B. WAISANEN is Professor of Sociology and Communication at Michigan State University, where he is also affiliated with the International Communication Institute and the Latin American Studies Center. He has been a consultant to UNESCO and Director of Research (1963-1966) for the American International Association's Inter-American Communication Program. His research interests and recent publications focus upon the social psychology of modernization.